Elementary
Particles

SCIENCE,
TECHNOLOGY,
AND SOCIETY

Contributors

J. P. BLEWETT

V. P. BOND

WILLIAM A. FOWLER

SHELDON KAUFMAN

TILL A. KIRSTEN

H. D. MACCABEE

M. A. RUDERMAN

OLIVER A. SCHAEFFER

R. P. SHUTT

G. M. TISLJAR-LENTULIS

C. A. TOBIAS

Elementary Particles

SCIENCE, TECHNOLOGY, AND SOCIETY

Edited by

Luke C. L. Yuan

Brookhaven National Laboratory
Upton, New York

1971
ACADEMIC PRESS
New York and London

ACADEMIC PRESS, INC.
111 Fifth Avenue, New York, New York 10003

F11

United Kingdom Edition published by
ACADEMIC PRESS, INC. (LONDON) LTD.
24/28 Oval Road, London NW1 7DD

LIBRARY OF CONGRESS CATALOG CARD NUMBER: 77-84252

PRINTED IN THE UNITED STATES OF AMERICA

Contents

CHAPTER 4.   Interaction of Elementary Particle Research with
              Chemistry

                        Sheldon Kaufman

CHAPTER 5.   Accelerated Particles in Biological Research

                   C. A. Tobias and H. D. Maccabee

CHAPTER 6.   Possible Use of Densely Ionizing Radiations in
              Radiotherapy

                   V. P. Bond and G. M. Tisljar-Lentulis

CHAPTER 7.   Interactions between Elementary Particle
              Research and Engineering

                        J. P. Blewett

List of Contributors

Numbers in parentheses refer to pages on which authors' contributions begin.

J. P. BLEWETT, Brookhaven National Laboratory, Upton, New York (290)

V. P. BOND, Medical Department, Brookhaven National Laboratory, Upton, New York (242)

WILLIAM A. FOWLER, California Institute of Technology, Pasadena, California,* and Institute of Theoretical Astronomy, University of Cambridge, Cambridge, England (50)

SHELDON KAUFMAN, Argonne National Laboratory, Argonne, Illinois (159)

TILL A. KIRSTEN, Department of Earth and Space Sciences, State University of New York at Stony Brook, Stony Brook, New York (76) †

H. D. MACCABEE, Lawrence Livermore Laboratory, University of California, Livermore, California (197)

M. A. RUDERMAN, Department of Physics, Imperial College, London, England (50) ‡

OLIVER A. SCHAEFFER, Department of Earth and Space Sciences, State University of New York at Stony Brook, Stony Brook, New York (76)

R. P. SHUTT, Physics Department, Brookhaven National Laboratory, Upton, New York (2)

G. M. TISLJAR-LENTULIS, Medical Department, Brookhaven National Laboratory, Upton, New York (242) §

C. A. TOBIAS, Donner Laboratory, University of California, Berkeley, California (197)

* Present address.

† Present address: Max-Planck-Institut für Kernphysik, Heidelberg, West Germany.

‡ Present address: Department of Physics, Columbia University, New York, New York.

§ Present address: Strahlenbiologisches Institut der Universität München, Munich, West Germany.

Preface

It is well recognized that particle physics holds the key to understanding the fundamental nature of matter. Intensive and vigorous research in the past few decades by scientists throughout the world has resulted in tremendous advance in this field. To further advance our understanding of this foundation of all the natural sciences, we shall have to dig ever deeper into the atomic, nuclear, and subnuclear domains.

In the early 1930's, particle accelerators joined natural high-energy cosmic ray particles as a source of powerful particle projectiles for penetrating into the subnuclear domain. Particle accelerating machines, such as the cyclotron, first invented by E. O. Lawrence and his co-workers, synchrocyclotron, proton synchrotron, alternate gradient synchrotron, linac, etc., were successfully designed and built through the ingenuity and resourcefulness of scientists and engineers both in this country and abroad. These advanced machines, generally huge in size, are high in cost both in terms of money and highly skilled technical manpower. Thanks to the farsightedness and understanding of the U.S. Government, generous support has been provided to construct such machines in the U.S. and to provide adequate facilities for scientists to carry out their research programs.

Since the basic nature of matter is at the root of every aspect of science and technology, a better understanding of this fundamental knowledge would certainly advance correspondingly the understanding in all the branches of science and technology.

Owing to the detailed nature and complexity of each field of modern science, most scientists are so immersed in their own branch of specialization that they often are not fully cognizant of the origin of significant advances in the other branches of science relevant to their own. Such has been the case with particle physics. Thus, it seems extremely worthwhile to gather together a comprehensive review of some of the important and interesting developments of recent years in science and technology which were brought forth either directly or indirectly by the advancement in our understanding of the basic nature of matter, which in turn was brought forward as a result of the valuable accomplishments of the particle physics research. The subject matter involved in the present review is as follows: physics and particle physics, interactions in astrophysics, interactions in chemistry, interactions

in accelerator and engineering applications, interactions in biology, interactions in medicine, and interactions in geochemistry. It is a hope of this
editor that this endeavor provides a comprehensive review of such advances
and accomplishments under one cover for the benefit of scientists and non-
scientists alike.

This volume was made possible through the generous cooperation and
contribution by a number of authors who are authorities in their own fields
and who share the objective and purpose of this book. The editor wishes to
express his gratitude to Drs. R. P. Shutt, W. A. Fowler, M. A. Ruderman,
S. Kaufman, J. P. Blewett, C. Tobias, V. P. Bond, G. M. Tisljar-Lentulis,
O. Schaefer, and Till Kirsten for their cooperation. It is understandable that
each author has his own style of presentation on the particular subject.
Naturally it would be extremely difficult to completely harmonize all these
articles. For instance, some articles are somewhat more involved with technical details than others. But it is hoped that the objective of bringing these
articles to a general level which will be understandable to most readers
with some scientific background, has to a large degree been accomplished.

There is also some unavoidable overlapping of some of the chapters.
However, this overlapping presents slightly different viewpoints and so they
tend to compliment the alternate presentation of the subject matter.

It is also hoped that this volume will serve to demonstrate how closely
all branches of science and technology are related to each other. Any advancement in the basic understanding in one branch would evidently benefit the others. The continuous pursuit toward a better understanding of the
basic nature of matter would certainly have a lasting beneficial effect on all
science and technology as are evidenced in the presentations contained in
this book. With the 200–400-BeV proton synchrotron at Batavia, Illinois in
1972, the completion of the 30-BeV intersecting storage rings at CERN,
Geneva, Switzerland, and an earth orbital space station for particle physics
research using the extremely high-energy particle source of cosmic rays, one
can certainly expect very interesting results and a number of surprises to
add to our basic understanding of nature.

1   Science, Physics, and Particle Physics

R. P. Shutt   *Physics Department*
Brookhaven National Laboratory
Upton, New York

In this chapter, the role of science, physics, and especially particle physics among human efforts is examined. The aims and methods of particle physics, or high-energy physics, are stated and the value of this field in attaining short- and long-range human goals is discussed. The need is emphasized for developing arguments, beyond those based on extrapolation from the past, for pursuing this science very vigorously.

I. INTRODUCTION

When ancient man, some tens of thousands of years ago, created tools for hunting, he had to experiment with various materials and configurations— different kinds of wood and stone for bows and arrows, useful arrangements for traps—to achieve success. Thus, it was necessary that he perform research tasks to satisfy certain needs—he did applied research. He probably had neither the time nor the inclination to meditate upon the reasons behind the performance of his tools, nor the forces that made an animal fall into his trap. Occasionally, however, he must have wondered about natural phenomena—lightning and thunder, the moon and the stars. He did not understand them, but gradually noticed that there were recurrences in the weather and even more regular recurrences in the sky. He accumulated information, at first from curiosity—here was something to see, to observe, to discuss, to remember, to believe in—later his observations of the stars enabled him to navigate on water and land. Land and water themselves offered an ever-present challenge: mountains to climb, lakes and rivers to cross, previously unknown animals and plants to identify. This all must have been exciting and refreshing to him.

Later this information became most useful, when the tribe had multiplied and needed new hunting grounds and shelter; his previous curiosity for the new and unknown was rewarded—without it they might not have survived. Perhaps the young exploring son who had found the land, now so useful, had been reproached by his elders upon return; he had consumed time and valuable food, not to perform his daily chores, but to wander off into unknown regions. Very much later other lands were discovered by explorers who either merely wished to explore or had a definite goal—to find the world's end or a new route to known lands. In our time some of those discoveries have indeed evolved into great countries.

Our ancestors, some thousands of years ago, formed a society cohesive and affluent enough to make possible time to search the unknown, to think about the world, to separate "superstition" from "truth," to contrast man with his environment. The sophisticated disciplines of religion, philosophy, the arts, and scientific endeavor evolved. Certainly, in those ages wild countries still existed where life had continued much as it had been tens of thousands of years before, but these are not the countries which are remembered for their contributions to our civilization, our present material and spiritual, or physical and mental, well-being. It seems significant that the impact of the contributions of these ancestors of some thousands of years ago has survived long-intervening interludes of devastating wars and decadence. What will be the heritage of those who will follow us some hundred or thousand years hence? Which of our efforts will have contributed to their well-being, to their civilization? What countries of our present world will be remembered for their contributions? Had there been no previous advanced civilizations, would we still be setting traps for wild game to feed ourselves? Had there been no advanced thought on abstract and concrete matters, would we still be old men and women at 30 years of age, rarely have enough to eat, usually be too cold or too warm, often be scared out of our wits by our own superstitions, murder our competitors in the tribe when we have an opportunity? Much of the practical-minded civilization of Sparta seems to be remembered as an absurdity, while the men of Athens have laid the foundation for many of the principles we live by. Did the population of Athens suffer because men there involved themselves in advanced thought? Did the Athenians suffer more than the Spartans? Should all those men of Athens and other great civilizations have been forced to turn their thoughts solely to practical contemporary, however important, problems concerning food, housing, and good health for all, and also to the most effective pursuit of wars? Should the Acropolis never have been built because it was a waste of money? Or will civilization in the future depend sufficiently on all previous performance so that leading men in any age should encourage advanced thought that benefits their contemporaries as well as later generations? Life in the future will certainly depend on present achievements. It is the human privilege to have our lives depend on achievements desired for the future.

To ask why the human race exists is similar to asking why the universe exists. To ask the aims of the human race is of more immediate importance, and coping with this question underlies, directly or indirectly, most human efforts that go beyond satisfying those bare necessities of survival shared by all living creatures. Since the human mind is able to project and plan into the future, we are concerned with that which will occur in the future: we may wish to create an environment which allows more than just bare survival, for more and better food, more comfortable shelter, better relations with contemporaries. We may wish to improve our appearance and to have a more attractive shelter, to decorate our person and shelter; we may wish to gather with contemporaries for debate, for which larger shelters are needed that must be more impressive to more persons. For debate words are needed. Languages—beyond mere grunts—are formed to express thoughts. The thoughts are to be preserved. The written word is needed. Thus, communication of thought not only between contemporaries but between generations is established. Most thoughts, information, experience, knowledge, and understanding accumulate through the ages with a huge compounded interest for the benefit of all who follow. Literature, architecture, music, sculpting, painting, religion, medicine, philosophy, technology, sociology, to name a few forms of expression of human creativity—all satisfy human needs of some sort and therefore contribute to man's well-being. Those who create usually enjoy doing so, as do those who recreate, for instance, by performing a great play, or who recreate by just admiring or by understanding a great sculpture of some past age. Many elementary needs are satisfied by technology, such as the supply of goods, transportation, and communication, and without technology the arts could not have flourished; tools and paints are necessary for the painter, instruments for the musician—an understanding of his vocal system for the singer—building materials and techniques for the architect. Human creative efforts are interdependent, but exploration precedes all effort, beginning with the initial attempts of the newborn. Pushing exploration and creativity to the limit of human ability has resulted in the greatest works of literature and the arts, the greatest advances in all other fields of endeavor, and the greatest strides toward satisfaction of the needs of mankind.

II. SCIENCE

Almost everything in our near and far environment has been worth exploring, such as the lands and seas, the sky, the earth and rocks, the atmosphere, plants, animals, the human anatomy, relations of all of these to each other, the obvious and the less obvious, and we know that practically all exploration has led to the satisfaction of human needs. Extrapolating into an unknown future, one can expect that such rewards of exploration will continue, that elementary and not so elementary needs will be satisfied, that needs for change will be satisfied. For our benefit and that of all who follow us, one cannot doubt the necessity for continuing the greatest effort toward exploration, creativity, and human advancement.

When methods and results of exploration in some direction multiply, become unique or specialized, and become well organized, they have been called a "science." In science, observations and thus accumulated knowledge are correlated in order to establish which regularities or "laws" might exist. Such regularities, for example, might be recognized as geometrical or numerical patterns, or as recurrences when procedures of investigation are varied. From regularities the existence of further phenomena may be predicted. The predictions can be tested, and the correctness of hypotheses proved or disproved. In the process of such tests, usually still other facts—often not those predicted—emerge, and so on. Eventually, those principles and laws by which nature operates are determined. Often they can be expressed symbolically in mathematical terms, sometimes only in language, every word or symbol having been defined very carefully. Therefore, science is a human creative effort of exploration involving observation, definition of terms relating to the observations, and correlation of observations, usually resulting in the need for still further observations, a feedback process reminiscent of the work of the sculptor who gradually works his way into a block of stone—he chisels and he observes, and chisels and observes, until he has found that which he had searched for, what was in his mind only as an image during his labors. A scientist having uncovered, through painstaking labor, a new principle operating in nature may feel the same satisfaction as the artist who has finished his latest work. Quite similar criteria apply. His own need to create is satisfied. His work almost surely will be of benefit, if not directly to all of his contemporaries, at least to later generations. To those working in other areas, the methods and results of a science may often seem obscure due to the necessary abstract language or mathematical terms, but the motivation, the mental processes, the rewards to all, do not seem to differ from those in all other human efforts. As individuals, scientists differ from each other almost as much as do all other human beings engaged in other endeavors. Their science is their common interest and they usually have to spend many years to learn principles, facts, and methods already developed in their field; they have to use a common "jargon," a collection of well-defined words, in order to communicate with each other about their subject in a finite time, but as in all other fields, and even as in modern "big science" where group effort predominates, advances are made by individuals in small contributions by each doing his share, and in truly large contributions of historic significance by the few men of great genius, who are born only a few times in a century.

We have stated that human needs in the past have been satisfied by exploration and creativity in all areas of effort, and, extrapolating into the future, there is no obvious reason why this process should not continue, unless we wish to return to a life in caves or to other primitive forms of existence. Science is one of the human efforts that has been successful in filling many of our needs, so that by the same extrapolation, presumably, scientific effort should also be continued, and, as with all best efforts, it should be continued with full vigor, even without guarantees for immediate practical benefits, which can hardly be given in any field. A certain amount of faith in such extrapolations must exist, just as a belief in success must exist in all efforts. Creative effort cannot blossom with partial momentum. Humans cannot put forth their best, and their enthusiasm, initiative, and drive cannot be sustained, when their movements are inhibited. Creative effort must spearhead into the unknown with full speed. Some research work is so time consuming that a sizable portion of a human life span is necessary to advance a significant step. Artificially slowing down such work could stop it altogether.

Time is an important ingredient in scientific as well as in all other work. The artist, the writer, the scientist, as well as the politician and the businessman —all must wish to create in the shortest possible time so that they can test the results of their efforts and so that they can turn to the next step in their life's work. This urgency in creative effort must not be called "impatience," a word which rather signifies a weakness of character, but this quality is an essential part of human striving, drive, and enthusiasm. Clearly, there are other motives for hurrying: the career-minded wish to advance quickly, wish to attain recognition and possibly prizes, perhaps "beat the competition," in quite a similar manner to the politician who wishes to attain high office, or the businessman who wishes to become wealthy. As stated above, human effort is exerted by individuals satisfying needs—those of their contemporaries, their successors, and certainly also their own. It is not surprising, therefore, that besides the ideal and lofty motives of the "mountain climber," there are also the immediate material or practical satisfactions which provide a portion of the necessary incentive.

III. PHYSICS, THE SCIENCE OF FORCES

Physics is the science that explores the regularities, or laws, occurring in nature which seem to govern the observable behavior of matter. Such matter might form stars or planets or rocks, or water or air, or leaves of plants or nerve strands of animals. Such matter might move or burn, break or grow, or conduct electric charge; it might be bright or dark, or hot or cold. Humans seem to perceive with five senses which they can use for making observations, but the human mind can reason far beyond what our senses perceive, and thus we can imagine, describe, and often understand that which we cannot touch or see or hear, that which is too far away or too small to perceive, or takes too long a time to test. We can even think about that which might be beyond our comprehension—the "beginning" or the "end" of time or space, the origin of matter, the infinitely small described by a mere point of no extension. One can believe that everything in existence is governed by the laws of nature discovered by physics research. These laws, when expressed in the terms used in physics, then are merely ways to describe and correlate phenomena occurring in nature, making use of a minimum of assumptions and parameters.

To what have these laws reference? It appears that all processes occurring in our universe are promoted by "forces"—physics is the science that explores these forces in the widest possible sense. Gravitational forces determine the course of planets and satellites; electromagnetic forces drive electric motors, produce electric currents, or radio waves, or light; and nuclear forces have produced the largest sources of concentrated power yet available to man. Electric forces between atoms also keep bridges from falling, or they make water, evaporated from the oceans, condense so that rain drops will fall, attracted to earth by the force of gravity. When a force drives an object through space, it produces energy, and the longer the time it acts in some direction, the higher a momentum results. Thus force, energy, and momentum are very closely related to each other. Due to the electric forces between atoms, energy is released when fuel burns in a combustion engine. The compounded molecules thus move at high speeds and exert a force on the piston of the engine when colliding with it. Similarly, rockets attain their momentum. Energy is consumed when water evaporates, and released when it condenses, a process which profoundly affects atmospheric phenomena. Energy is consumed and released in our bodies as mostly electric forces drive biological processes in the proper directions.

Physics is an analytic field, where one tries to divide and subdivide, and thus to penetrate the unknown, until an understanding of the driving forces is reached. There are laws governing the fall of a whole rock, but in order to understand completely the rock's resistance to breaking on impact, one would have to divide the rock into small particles, perhaps into atoms. But the whole rock—or metal shaft, or transistor, or plastic material—is still of great interest because the forces that exist between aggregates of atoms bind large molecules, crystals, and all the other larger structures which exhibit their own aggregate characteristics. Similarly, the laws of physics underlie the chemical processes occurring in living cells, but to subdivide the cell into atoms would kill the cell; it must be studied as an entity. Thus there are sciences of unique importance that must study given structures or must even build up such structures before studying their aggregate behavior; they must synthesize. In astronomy, it is not possible for us to affect the course of the stars for an experiment; only observation is possible—with results of sweeping interest. The physicist, whose aim is to discover and understand the forces underlying all natural processes, has had to subdivide, first into molecules, then atoms, then nuclei—then into subnuclear particles, and the end is not in sight. In physics we now consider objects extending through tiny fractions of a billionth of an inch in space, and often existing only for tiny fractions of a billionth of a second. In astronomy distances extend up to many billions of light years in space (light travels 6000 billion miles in one year) and periods of development lasting many billions of years are inferred. In one field we seem to be not far from the mentioned point without extension in space, in the other not far from what one might call infinitely large. Yet, there is no end to numbers; a billion can be multiplied over and over by other billions and a billionth can be divided forever by further billions. Exact knowledge of actual limits set by nature would be of crucial interest and importance in man's effort to understand the universe.

Physics has uncovered that matter and energy are equivalent. As a rocket moves through the atmosphere, it becomes warmer due to friction with air. The generated heat represents energy; the body of the rocket becomes more "massive," although only by a tiny amount. In fact, according to relativistic theory, the energy of motion of the rocket already made it somewhat more massive than it was at standstill. A ray of light consists of electromagnetic energy; therefore, it must have mass. Of course, it could never stand still and remain a ray of light; it must move with light velocity. Light is studied in that part of physics called "optics." One found quite early that some phenomena observed in optics could be understood as light "rays," like trajectories of particles, but to understand other phenomena, it was required to assume that light consists of waves. Now one knows that all energy, and therefore matter, must be described mathematically with the aid of a wave picture (whose deep significance is not yet understood but which might be imagined as a kind of statistical phenomenon). The wavelength associated with a bit of matter decreases as its mass increases. This is not surprising since in optics, acoustics, mechanics, or radar, the shorter waves (of some given amplitude) require more energy to produce, and contain more energy. Thus, only an elementary particle of infinite mass—containing infinite energy—can be associated with an infinitely small wavelength. Since the wavelength is interpreted as the range of its existence in space (a particle must "fit" within its wavelength), the infinitely "heavy" particle would have no extension in space; it would fit onto a mathematical point, or it would represent a "singularity." It is of interest to note that in one theory in cosmology, where one tries to understand the origin, development, and future of the universe, it is surmised that after maximum expansion the universe, or at least a part of the universe, will begin to collapse due to gravitational forces. It could continue to do so at increasing speed until its practically infinite energy is contained in the mentioned point "singularity" of practically zero wavelength. Afterwards expansion would begin again. It is indeed striking that the laws of physics, as now understood, would allow all matter in the universe to be concentrated at possibly one single point, thus containing almost infinite energy. It is also striking that this collapse could be caused by the weakest known force, namely gravitation, which is enormously weaker than the strongest. There may be limits on "smallness" in nature so that gravitational collapse would be less complete, though qualitatively unaltered. (If it should occur, one can assume that at the time of such a universal collapse, itself taking billions of years, our sun would have cooled off sufficiently so that the very narrow range of conditions for life as we know it will long have ceased to exist on earth.)

Since forces drive all processes in nature, physics, as the science studying these forces, can be considered the most basic. Without the discoveries that have been made in physics, no observations in any other field of science could be truly understood. If research in physics were slowed down or stopped altogether, all other sciences would have to suffer eventually, with immeasurable consequences for civilization now and later. Furthermore, much of the apparatus used in any scientific investigation was developed as physics research equipment—spectrometers, x-ray tubes, photographic lenses, vacuum systems, electric instruments, to name a few examples, and even the most recent high energy accelerators used in particle physics are also used for chemical, biological, and medical studies.

The source of technology is scientific investigation. Without scientific understanding, technology at best would be based on "trial and error," a procedure which, because of an almost infinite multitude of possible but not necessarily successful paths, could take an almost infinite time and be nearly infinitely expensive. Certainly, scientific method makes use of trial procedures, plots corrected courses leading to improved trials, and so on, until convergence on a new principle of nature has been achieved, but realization of this principle, where applicable, will eliminate the necessity for "trial and error" in all future cases; the investment in scientific method thus pays incalculably high dividends.

IV. THE BRANCHES OF PHYSICS

The law describing gravitational force was discovered when the motion of the planets around the sun, and also the trajectories of objects thrown by hand, were explored. The branch of physics called "dynamics" thus evolved. Force, energy, momentum, mass, velocity, acceleration—all were here related to each other and to distance in space and duration in time, as functions of each other and as functions of space and time coordinates. Motion could be described in symbolic mathematical terms and therefore also in numerical terms when not only the gravitational force was involved but also any other kind of force. Absence of motion could also be described: static equilibrium. Stresses in bridge structures could be calculated so that they would not collapse due to the forces of weight or wind pressure. By far the most important laws of physics and probably of all sciences concern the "conservation" of some of the quantities defined in physics. Energy cannot disappear, it can only be transformed into other forms of energy; chemical energy released in a gasoline engine becomes the energy of motion of a vehicle and of heat in the engine. Momentum cannot be acquired without acquiring an equal and opposite momentum; thus rockets are propelled, and a game of billiards is played. Similarly the conserved "angular momentum" of the gyroscope controls the motion of a ship. It is thought that the laws of conservation of energy and momentum are consequences of the properties of space and time, where no preferred location, instant, or direction seem to exist. Electric charge is also conserved; creation of a positive charge must be accompanied by the creation of an equal negative one. An important constituent of atomic nuclei, the proton, cannot be destroyed, except by annihilation by its antiparticle, the antiproton, together with which it must also be created. Of course, when the proton is annihilated by an antiproton, energy and momentum are still conserved because the annihilation results in radiation of other particles with exactly the balancing energies and momenta.

From dynamics, results were immediately applicable, such as an understanding of mechanical vibration, sound, motion in air or fluids, stability of ships, or reciprocating and rotating machinery. Mechanical technology still generates more applications and improvements at all times. But since physics is concerned with the study of forces, and since forces produce motion, the results of dynamics are also applied in all other branches of physics. Similar to other sciences, results from physics research thus are fed back toward further advanced studies. The physicist, as any other scientist, must work upon an ever-increasing foundation of accumulated knowledge and techniques. Not only must established principles and laws be used in attempting to discover new ones and to establish the limits of validity of old ones, but also instruments once developed purely for purposes of experimentation in the laboratory have come into general use in technology and now are used as routine equipment for research by the physicist of the next generation. Similarly, mathematical and computing methods once developed for a specific purpose have become a part of the general research technique.

In addition to clear definitions of terms used, their substance, and the goal which is to be attained, in any science certain assumptions upon which further work is based must at first be made. Such assumptions usually are "reasonable," or consistent with the everyday experience of the researcher. If there are no obvious reasons against an assumption concerning, for instance, symmetries of up and down, right and left, or before and after, or concerning maximum velocity or minimum energy, then such an assumption may be made, and the results of the work that follow may be quite correct within ranges where these assumptions apply. However, going beyond everyday experience, many of the assumptions that applied sufficiently well in special cases turned out to be wrong more generally; the theories of relativity and quantum mechanics had to be conceived, and certain "obvious" symmetries given up to describe new phenomena. Some of the greatest advances in physics were made only when experimentation proved that previous assumptions were no longer justified in the new realms under investigation. However, new descriptions of physics phenomena always revert to the older ones found to be correct where assumed conditions applied properly. For instance, the realization that energy or matter cannot be propagated faster than with light velocity led to the by now very well verified "special theory of relativity" which, whenever very high velocities occur in nature, makes predictions that are completely different from those made by a nonrelativistic treatment where velocity is not limited. Yet, at the lower velocities familiar to us—even at the velocities of our space vehicles—relativity results only in tiny, negligible corrections to older calculations. The spectra of light, due to motion of electrons which are parts of atoms or molecules, could not be understood until one ascribed to them certain (quantized) possible energy levels only, prohibiting states with intermediate energies. At very high atomic energy levels, however, quantization again leads to negligible corrections of previously applicable calculations. Esthetics requires certain symmetries in the arts, and constructions that are mirror images of other constructions are quite acceptable to humans, but they are not always acceptable to nature on a nuclear scale; observed nuclear processes are not always also possible according to their mirror images, or to space inversion: parity is not always conserved.

The branch of physics called "thermodynamics" treats a collection of phenomena concerning heat energy and the effects of temperature and pressure changes. At first the physical picture was one where heat seemed to behave like a substance with certain flow characteristics. The famous first and second laws of thermodynamics described, respectively, how energy was to be conserved and in what direction heat could flow in a system. While still very useful, these thermodynamic statements seem somewhat empirical in the context of physics at this time, since thermodynamic observations were understood only when matter was subdivided and it was realized that the observed phenomena were aggregate effects of very large numbers of atoms or molecules in motion, with many different velocities in many different directions; statistical mechanics was very successful in describing most observations. But the functioning of a steam turbine, combustion engine, or refrigerator can be described very well in terms of the original thermodynamics, even though they are driven by statistically distributed mechanical effects due to many small particles moving *almost* with random velocities in random directions. In recent years, effects observed at very low temperatures have commanded major interest. The technological field of cryogenics, dealing with materials at low temperatures, has resulted from this work. Space exploration would have been very much less successful so far without cryogenic techniques. Similarly, numerous important medical advances could not have been made without the availability of very-low-temperature equipment.

"Electromagnetism" describes electric and magnetic phenomena which actually are inseparable except for the special condition of absence of motion (in a macroscopic sense). A flow of electric charges produces a magnetic field which can drive an electric motor. A fast-changing magnetic field produces a fast-changing electric field which can drive electric charges, or produce an electric current, such as in a power transformer. A very fast-changing magnetic field produces a very fast-changing electric field which in turn produces a very fast-changing magnetic field, and so on, a transfer of energy which is characteristic of a wave motion through space. As is well known, such waves can be detected as signals in radio communication, in radar, but also in infrared and x-ray photography. The theory of "electrodynamics" describes and correlates a very wide range of electromagnetic phenomena, but it must again be modified as the length of waves is decreased below those of the shortest radio waves. The laws of quantum mechanics must apply. As a result, electromagnetic waves, under the proper conditions, behave like particles, as if all the energy of the wave distributed in space were suddenly concentrated in one spot. Conversely, as already mentioned, a wavelength is associated with any particle, and a traveling wave with the particle when it moves. A beam of particles exhibits properties quite similar to those of waves: there can be interference, dispersion, all sorts of effects only possible when waves are involved. Such "peculiar" behavior, or dualism of matter can be interpreted qualitatively in several ways, none of which are satisfactory. Presently, it seems best to accept the facts of quantum mechanics as dictated by nature and to work with the various mathematical formulations expressing them. The existence of forces and the laws of conservation of energy and momentum are more familiar to us than quantum mechanical phenomena, but really they are no less "peculiar." In fact, as pointed out above, too much familiarity can lead to taking things "for granted" and therefore possibly to erroneous assumptions in scientific procedures. After all, most incomprehensible so far has been the existence of the huge universe, pieced together of such countless infinitesimally small parts, all interacting in some way through the various kinds of forces. If we can explain our findings only in terms which are at first unfamiliar to us, we should not be surprised. If we cannot comprehend them by means of pictures or models that we can describe, this is due to our limitations. We can strive to comprehend the meaning of everything around us; we can try to understand the laws of physics in terms of more profound and unifying laws. In the process we most likely will make more discoveries, and encounter more surprises. Ultimately, we might arrive at a completely "closed" system of laws and phenomena, where the existence of each is explained by applying all the others, with no link missing. Then our search would be ended. Therefore, this should remain our aim. It could, however, also happen that ever new formulations are necessary in describing natural phenomena. Then our science would remain open ended, and we could only understand a part of nature, never all. In this case research in physics might at some time come to an end temporarily because of a lack of further methods of investigation.

A digression into "metaphysical" phenomena may be permissible. Experiences claimed by spiritual media, by extrasensory perceptionists, by palmists, by astrologers, all have in common the fact that they are very difficult to prove. Thus far none seem to have been proved by generally accepted scientific methods, meaning that the results must be repeatable under proper conditions and must be sufficiently beyond experimental error. If their existence were ever proved, however, these phenomena would not have to be considered as "beyond physics." As natural phenomena, they also would be due to certain causes or forces for which one could search until an explanation is found. If hitherto unknown forces were involved, they would merely be added to the range of physical phenomena. We are aware with only five senses; our nerves can receive and signal to our brain a small range of electromagnetic wavelengths, called "light," and another small range of infrared radiation perceptible as heat. Similarly, we can receive, within limits, waves moving through gases or fluids as sound, and we can perceive fast-moving molecules as heat. We can also perceive mechanical pressure on our skin, and we can distinguish some aggregates of molecules from others by reaction with our smell and taste organs. Thus, we can interact directly with only a very narrow range of natural phenomena. Yet, through our instruments we can observe far beyond the range of our senses, and our brain is able to comprehend far beyond what its senses transmit directly; our five senses merely transmit information and are not to be confused with the functioning or limitations of our brain. Quite possibly there may be a few additional kinds of phenomena with which we can interact but that we are not aware of, or so far unknown forces may even exist which can affect us in some not yet understood manner. It would seem that any scientifically proven phenomenon can be included in our physical picture of the world. Thus, the existence of anything now called "metaphysical," when proved, becomes a part of this picture as we define it. We may never comprehend it completely due to some limitation of our mind, but that does not exclude it from existence. Even ghosts and spirits, if they really existed, would also be a part of our world, whether or not we could actually prove their existence, and although an explanation of their existence, even if proved, may fail us. "Metaphysics" merely seems to cover what is believed by only a few, since it has not been proved to the satisfaction of all. The definition of physics should remain open to cover any physical phenomena that exist, but physics as a science must demand scientific proof beyond mere belief.

"Optics" deals with a portion of electromagnetism. It concerns phenomena involving "light" that may or may not be directly visible to the human eye. Through observation of light spectra produced by dispersion of light into lines of different colors, or wavelengths or energy contents, more was learned about atoms, and later also about nuclei, than in any other way previously known. Light is produced by stimulating atoms to emit electromagnetic energy in the process of rearranging their electronic structure. Such stimulation might be produced by mutual collisions of atoms in hot substances. The more energetic x rays, with their short wavelengths, can be produced by bombardment of atoms in metals with electrons. As one increases their velocity, or their energy and momentum, the bombarding electrons can knock electrons contained in an atom completely out of the atomic structure. Then no light or x rays are produced, but in addition to the primary bombarding electrons, free secondary electrons can be observed as a result of these collisions. As the energy of the bombarding particles is further increased, the heavy nucleus of an atom can also be affected. It may be knocked out of its surroundings, flying off at some velocity, or it may even be shattered into fragments—protons, neutrons, and subnuclear particles. This is the concern of "nuclear physics," where "high-energy physics" or "particle physics" had its beginning.

Atoms can combine into aggregates called molecules, and molecules into crystals and living cells, leading, for instance, to the vast disciplines of chemistry, metallurgy, and biology. Nuclei can also combine with each other, with very spectacular releases of energy, a fact which, for example, is applied in "atomic power" reactors producing heat which can be used to generate electric power, but which also leads to the creation of radioactive nuclei that, after a certain average lifetime, decay into other nuclei and often into electrons or energetic x rays (gamma rays). The energy released in the decay of such radioactive materials is transported by the fragments which might be stopped, for instance, in cell tissue, destroying or altering it, or which might merely signal the location of the radioactive material in the tissue. These are powerful tools of biological and medical science, and of many other fields.

V. PARTICLE PHYSICS, FRONTIER WITH THE UNKNOWN

During his search for the forces underlying the processes of nature, the physicist has discovered many phenomena which are caused by these forces and which are therefore a part of physics. But the central goal is still to find and understand all of the driving forces.

The existence of four kinds of forces has been proved thus far. By far the weakest is the gravitational force which, as mentioned above, is strong enough that it could some day produce the rebirth of the universe after completely collapsing it. Although gravity was discovered first, it is not well understood in detail. Next in strength is the weak nuclear interaction force which seems to act universally between all particles including the neutrino, a particle that has only a mass, or exists only, when it moves with the velocity of light, like the light quantum particle (or photon). This force was discovered most recently; its existence had been postulated to explain details observed in some modes of radioactive decay, and it is perhaps better, though far from completely, understood than the previously discovered strong nuclear interaction. This may be so because the mathematical approximations that must be used to solve the equations of quantum mechanics apply much better if the interaction force is weak. Next in strength is the electromagnetic force which, since it is also much weaker than the strong nuclear interaction, can also be very well described. It is perhaps the best understood because it is a "long-range force" like the gravitational force and is sufficiently stronger than the latter to permit effective experimentation in the laboratory. Least understood is the strongest force, the "strong nuclear interaction" or "strong short-range force" which most strongly affects interactions between nuclei and between subnuclear particles, and may well be the composite of several different sorts of forces, for instance, mathematically depending in different ways on relative distance and orientation of particles in space. It is extremely strong near the "edge" of a nucleus but further away does not exist, quite in contrast with forces due to gravity or due to electric charges which, although much weaker at a short distance, decrease with the square of the distance and therefore exist even far away. The strongest of the four mentioned forces is 10^{39} (a thousand trillion trillion trillion) times as strong as the weakest. Thus, a huge range of interaction strengths is covered. Therefore, the gravitational force seems to play a negligible role in nuclear interactions. On the other hand, as just stated, the strength of nuclear interaction forces decreases so fast with distance that "outside" a nucleus, or less than a trillionth of an inch away from it, there is none, and it is the electromagnetic force that holds atoms and molecules together. At a large distance from an atom or molecule only the gravitational force is left to act, since there the effect of the positive electric charge of the nucleus is neutralized by the negative charges of atomic electrons. Since nuclear interaction forces are strong and their range is short, they produce the greatest complexity of observed phenomena. Anything seems to occur in nature that is energetically and kinematically possible and is not forbidden by special restrictions due to existing laws of conservation of electric charge and of protons and of quantum numbers called "spin," "parity," "isotopic spin," and "strangeness." Thus, with the large amounts of energy available in nuclear interactions, a multitude of new phenomena is found.

The four above-mentioned kinds of forces can act simultaneously between particles and therefore their effects can interfere with each other. For instance, weak nuclear interactions participate in predominantly strong nuclear interactions. The electromagnetic interaction usually participates because many interacting particles are electrically charged or may exhibit properties of magnets in addition to their other properties. A universal theory would describe all forces from a unified point of view, perhaps deriving all from further still unknown principles, and it could predict correctly all observations that might be made under given experimental conditions. No attempts at such a unified theory have succeeded thus far, but it remains a goal.

The aim in high-energy physics is to find the laws and forces governing nuclear interactions and their relation to other laws and forces. In quantum mechanics, forces are thought to be transmitted by the exchange of particles. For instance, somewhere during the very short time while a bombarding electron is close to another electron, a light quantum travels from one electron to the other, transferring the necessary energy and momentum to change the course of both electrons in accordance with the laws of conservation of energy and momentum. Similarly, in strong interactions particles such as pi mesons, K mesons, and others are thought to be exchanged. Then, to understand and to describe a force would require knowing enough about the mechanism of exchange and about all particles that may be exchanged. As mentioned before, in all such interactions certain quantum numbers must be conserved; for instance, the sum total of all spins and angular momenta existing before a collision must not be changed afterwards, just as total energy and momentum must be conserved, or the ("transverse") electromagnetic wave constituting the light quantum exchanged between electrons must be properly oriented. Thus, conservation laws, symmetries in space and time, and families of particles are all very closely related.

Because particles are so readily produced when one studies nuclear interactions at high energies, this field of investigation has been called "elementary particle physics." Of the more than 100 "elementary" nuclear and subnuclear particles found so far, however, only a few are stable, the others decay radioactively more or less rapidly. The proton, electron, and neutrino, and their antiparticles, the antiproton, positron (or positive electron), and antineutrino are stable. Even these stable particles can be transformed into each other and into high-energy x rays (or gamma rays) when a particle collides with its antiparticle and annihilation takes place, but their total number in the universe is thought to be conserved, considering that for every particle created or annihilated, its antiparticle must also be created or annihilated; every particle "neutralizes" the existence of one of its antiparticles but, of course, their equivalent energy is conserved. K mesons are not stable, but they must also be created in pairs so that the quantum number called "strangeness" is conserved, which had to be invoked to explain their "strangely" long decay times. In contrast, pi mesons, rho, omega, phi, eta, f, and other mesons, do not have to be created in pairs; their number in the universe is not conserved, but ultimately they all decay into stable particles. The meaning of "elementary particle" is debatable. Perhaps all particles can be changed into each other under proper conditions so that none is more "elementary" than others, or perhaps all particles observed thus far are composites of a few truly basic building blocks not yet discovered, although their existence is suspected. Therefore, merely "particle physics" seems a proper name for the field. Its other name, "high-energy physics," must not be misleading. The total energy released in an atomic power reactor is extremely high but it is not a result of high-energy physics phenomena. Based on studies of individual low-energy nuclear reactions carried out several decades ago, with the reactor use is made of tremendously large numbers of relatively small nuclear energy releases combining to produce a very large total energy release. In high-energy physics the energy involved in a single collision between two particles is very large, many thousand times as large as the energy released in a low-energy nuclear reaction, but it is of course still far smaller than the total produced in the reactor. Under investigation in high-energy physics are individual reactions between colliding particles with high energies, and of primary interest is the production of new particles and related manifestations of the nuclear interaction forces. Because of the short-range character of nuclear forces, their detailed study is possible only in high-energy physics. As stated above, high energy is associated with short wavelength in quantum mechanics, and a wave of short length only can transmit a signal within a small confined space; a short antenna can effectively receive only a short radio wave. Therefore, only particles of high energy can be effective in probing the phenomena locked inside nuclei, or in probing for nuclear forces.

It seems proper to say that particle physics is the main frontier of physics, and therefore of all sciences which are based on physics. If the answers could be found to the many questions that have been raised here, a complete picture concerning the processes of nature may well present itself. Human knowledge and experience would then have advanced to a point where man might be master of his existence to a much greater extent than he is now. What has become possible due to that which has already been learned in physics would have been very hard to imagine some hundred years ago. What may well become possible in the future due to further knowledge to be gained is now very hard to imagine.

Some of the results of physics, as mentioned above, could be applied immediately in technology. For others it took longer until either the technical ability or the demand existed. For instance, new applications of the results of electrodynamics are still being found, even a century after the laws of electrodynamics were formulated in complete form. From any science there is "technological fallout," almost immediately applicable technological advances made during scientific investigations. The scientist must often be an inventor in order to obtain the apparatus necessary for his work. He may become interested in technical problems related to his research and thus practice "applied science," though not exactly large-scale technology. There is a continuous flow of ideas from a "pure" to an "applied" science, to technology, to other sciences, and back to the "pure" science, an obvious advantage of widespread dissemination of research results by scientific journals, conferences, and modern means of easy day-by-day communication over long distances.

VI. PARTICLE PHYSICS, THEORY AND EXPERIMENTATION

Experimentation in particle physics requires a large amount of apparatus. First, a source is needed for the high-energy particles which are to interact with available nuclei contained in targets. The frequency with which a particular kind of interaction occurs determines the interaction "cross section." The properties of particles originating from such interactions are determined by means of particle detectors, where data are gathered on the numbers and kinds of particles, their electric charges, directions of flight, momenta, lifetimes, secondary interactions, and so forth, data which then are analyzed, often by means of additional equipment especially designed for data processing purposes. Finally, computations are carried out by means of large digital computers. New results are very carefully tested for validity, compared with previous results which were perhaps obtained by a different method, and are interpreted in terms of the most advanced theoretical thinking. If agreement with a theory is found, the latter may contain elements of truth, although it may not yet be generally valid. If no agreement is found, the experiment can be rechecked for validity of the data. If the data are found to be valid, one must conclude that the part of the theory that was to be tested, in fact does not correspond to the processes of nature. The assumptions underlying the theory may not be applicable to this case, the approximations used for the calculations may be insufficient, or the formulation of the theory did not take into account some effect that could have been overlooked or was not known, or the formulation did not proceed in the proper direction.

Successful theoretical research in particle physics requires of the theorist outstanding ability and experience in the fields of physics and mathematics, and an unusual amount of imagination, insight, and abstract thinking. Essentially equivalent wave-mechanical, matrix, or group theoretical formulations of quantum mechanics are possible. "Field theory" represents probably the most complete mathematical formulation of quantum mechanics in its most advanced form, but attempts at mathematical solutions for specific cases have led to grave difficulties which have not been overcome thus far. Therefore, many partial formulations have been attempted, all far from complete, and many only "phenomenological" in character; they merely correlate data, do not relate observations to the forces that caused them but directly use observed results to predict other results that can be tested, or apply only to a particular set of results but not generally. Such phenomenological approaches are by no means "simple minded." In fact, since no complete, logically produced solutions exist, they require considerable ingenuity and subtlety to conceive and have been extremely fruitful.

In particle physics, over one hundred particles have been discovered so far, all differing from each other in one or another of the quantities that characterize a particle, such as its mass, charge, spin, parity, and strangeness. As with the optical line spectra under intense study at the beginning of this century, which led to the initial formulation of quantum mechanics, one must first try to "catalog" these particles; one must find a way to place them within some numerical pattern. For instance, as the mass of particles in a "family" increases by a certain amount, their spin might also increase by one unit, with a simultaneous reversal of parity, but with the strangeness remaining constant. This simple example is incorrect, and a much more complicated pattern, based on group theory, has evolved during recent years into which indeed many of the so far observed particles can be fitted. It must be realized, however, that such patterns, although of great interest and of primary importance, are still only the *result* of the forces occurring in nature and can only show the way to a complete theoretical formulation of the laws of physics. A complete theory must involve the basic forces that lead to the production, interactions, and properties of all of the observed particles. Field theory attempts this, but it cannot be used generally because of the mentioned difficulties and so it is not known whether it is generally correct or complete. Its application, when limited to electrodynamic phenomena alone, has been quite successful.

The velocities of particles dealt with in high-energy physics are usually high enough that the special theory of relativity applies, and this has represented a further major complication in developing a proper theory of quantum mechanics. The particles can only be produced in collisions between particles at high relative velocities, and their masses are high enough that, for instance, even in the decay of a particle that is at rest in the laboratory fast particles can result. To summarize, while experimental observations made in particle physics are beginning to fall into certain patterns, a proper relativistic formulation of a theory involving the forces leading to the observations must still be found as well as its mathematical solutions. To overcome the possibly conceptual and probably mathematical complexities may require a very major scientific breakthrough. To be completely correct, quite likely such a theory will have to take into account all possible forces and quantum numbers at once. At the present time it appears that such a theory is still a long way off and that very many additional experimental data will be needed to lead to the theory and to test its validity once it has been proposed.

To obtain the experimental results that may finally lead to a complete understanding of the universe, apparatus is needed to make valid observations possible. In particle physics, observations become possible when enough energy is available to accelerate particles to very high velocities, very near the velocity of light. Only such particles with very high momenta are associated with the short wavelengths that enable them to interact at very close range with other particles; they are the probes that can enter the possibly quite complex structure of a proton, and they can produce the quite heavy particles that may be exchanged in the mentioned weak nuclear interactions, or they can make it possible to observe very heavy particles that might constitute the building blocks from which all the other particles so far seen are made, held together by some basic force. The latter concept, to humans, seems especially satisfying; one likes to believe that all of nature is made of a few basic ingredients combined in different ways, a belief which, however, could be quite false and may be due only to our own limitations. Similarly to a perhaps infinitely large universe existing for an infinitely long time span, nature might have preferred an infinite complexity in "furnishing" the space that is a part of the universe. On the other hand, according to our experience Nature has satisfied our craving quite well for what we call "beauty"; her household seems uncluttered by unnecessary items, her laws are as simple and symmetrical as conceivable, and her processes as economical as possible. Therefore, our belief in ultimate simplicity could be justified. Such simplicity, however, could still lie far beyond our comprehension and therefore may continue to seem complex to us.

How could a proton be made of building blocks each of which is much heavier than the proton itself? We must assume that the building blocks exert very strong forces on each other. Therefore, as they approach each other to form a proton, their relative ("potential") energy must decrease as the forces move them toward each other. The excess energy is emitted as radiation of some kind. In the end, the building blocks' original masses, each equivalent to a certain amount of energy, reduced by the "binding" energy emitted during the proton formation process, are left. Thus, the equivalent mass of the resulting proton is less than that for the sum of the building blocks, and the missing, possibly very large energy must be supplied by the particle bombarding a proton if one wishes to observe one of the building blocks alone, and not just the aggregate of several bound together. This is reminiscent of an iceberg, nine-tenths of which is under water and not directly visible; it is bound to the earth by gravity. To see the whole iceberg and to explore its properties, unencumbered by the presence of ocean water, one must lift it out of the water, requiring a large amount of energy.

The heavier such possible building blocks are, the more energy is necessary to make them observable, the shorter is the wavelength associated with them, and the shorter, therefore, is the wavelength of the particle needed to produce them; as a consequence, the higher must be the energy to which this particle has to be accelerated first—and therefore the larger the apparatus needed. To produce an infinitely heavy particle would require infinite energy associated with infinitely short wavelength, but perhaps nature does not really allow infinitely short lengths, as perhaps the universe is not infinitely large but is limited in some multidimensional geometry. If a cutoff existed at some minimum length, one, of course, would not have to (nor be able to) probe any further. The end would then have been reached for the need to increase the size of high-energy physics apparatus. At the present time, there is no evidence that one is near such a limit, it is rather that one is so far from it that one can only hope that physical phenomena will be understood without first having to reach any limiting short distance.

It is obvious then that particle physics research is possible only if sources of particles of high energy are available and that the size of the necessary apparatus must increase as the wavelength required for probing the structure of particles decreases. Of course, one can expect that technological breakthroughs will reduce the scale and cost of such equipment, just as a breakthrough in theoretical understanding may finally limit the energy where tests are still needed. Several such precipitous advances in experimentation and theory may very well bring convergence to this field where presently no end is in sight, however well defined the aims may appear.

The results obtained in particle physics, as in any other science, become valid and useful only when they are free from systematic errors and are sufficiently accurate. To avoid systematic errors which will shift the average of a number of experimental observations away from its proper value, is one of the most important tasks for the experimentalist and requires great care and a responsible attitude. To obtain sufficient accuracy, so that different phenomena can be resolved from each other—like a set of closely spaced fine lines placed sufficiently close to a human eye—requires that all or as much as possible of the information that is pertinent to an observation becomes available in an experiment. For instance, one would prefer that all of the particles produced or involved in an interaction are observed if one wishes to understand this interaction completely. Any particle that is not observed, but was present, reduces the probability that an interpretation is correct. Furthermore, the measurements (of relative locations and angles in space, of time intervals, or of energies and momenta) must be sufficiently accurate, for example, to permit distinction between two particles of somewhat different masses. However, measurements without some kind of error are impossible; the measuring or observing procedure will always perturb the system to be measured to some degree since physical procedures are used which must involve physical forces which by their interactions, however weak, produce the measurement. In addition, measuring equipment cannot be made perfect, and, even if it were, the operator of the equipment cannot operate it with perfection. In order to eliminate the effect of experimental errors as much as possible, one averages at least several and often a large number of observations in an experiment. In order to obtain a result that is most likely to correspond to the true process under investigation, one performs high-statistics experiments. For these various reasons, the source must not only accelerate particles to high energies, but it must also accelerate many of them—it must have high intensity. Also, the targets for the high-energy particles must contain many nuclei to be bombarded—they must be sufficiently large. Finally the particle detectors must permit good measurements—they must not introduce distortions of angles or distances and they must contain a sufficient number of units of equipment to provide all the necessary information. Therefore, detectors also become large and complex, and their construction and operation, *in toto,* demand an effort quite comparable to that required for the particle accelerator.

VII. PARTICLE PHYSICS APPARATUS

Investigations in high-energy physics originated in cosmic ray research where particles of sufficient energies were readily available. Primary cosmic radiation consists mostly of protons which have escaped from various bodies of particles in space—for example, stars—and which gradually have been accelerated, by a combination of extremely weak but extensive electric and magnetic fields existing in space, to energies that range much higher than those thus far produced in the laboratory. When these protons enter the earth's atmosphere, secondary particles are produced in collisions with the nitrogen and oxygen nuclei in air. All of the particles observed in cosmic ray studies, and many others, have now been produced at accelerator laboratories. The cosmic radiation has low intensity and cannot be controlled with respect to the kind, number, direction, and momentum of the particles incident on detecting apparatus. Although very significant discoveries were made, and still may be made because of the available high energies, cosmic ray studies therefore could not be pursued in depth, which, as stated above, requires sufficient numbers of interaction events (good statistics) and as complete as possible data on the colliding and produced particles. Therefore, particle accelerators were built, which permit excellent selection of experimental conditions. The development went from the cyclotron and betatron to the synchrocyclotron, to the synchrotron, and finally to the alternating gradient synchrotron, a *relatively* inexpensive machine due to its underlying principle of "strong focusing" which produces a very fine, well-collimated beam of particles requiring relatively small units of equipment. In such machines protons, with positive electric charge, are produced by the stripping of negatively charged electrons from hydrogen atoms colliding at low energy in an ion source. The protons are accelerated to an energy of several hundred thousand electron volts by passing through an electric field of several hundred thousand volts, produced by rectification of the alternating voltage appearing at the terminals of a high-voltage transformer. Next, the protons are accelerated to not far from 100 MeV by means of an electromagnetic wave traveling in a "linear accelerator." Like a surfer riding an ocean wave, the protons ride on the electromagnetic wave whose velocity is regulated so that the proton remains on the leading side of the wave. This beam of protons is injected into a ring-shaped magnet, the radius of which is equal to that of the circular proton trajectories produced by interaction of the magnetic field with the moving charged particles. The magnet not only produces the circular trajectories but it is so constructed that its units periodically and strongly refocus the beam of particles so that it remains small in cross section and will not diverge as it would if left alone. The accelerator ring contains stations where again electric fields are set up so that a proton is accelerated somewhat every time it passes through the station while traveling along its circular path. As the proton's momentum increases, the accelerator's magnetic field is also increased in synchronization so that the proton can remain in the magnet until it has reached the energy for which the accelerator is designed. On these principles is based the Brookhaven National Laboratory (New York) 33-billion electron volt (abbreviated BeV or, recently, GeV) alternating-gradient synchrotron (AGS), the European (CERN) 28-GeV machine located in Switzerland, the 70-GeV Russian machine which has begun operation at Serpukhov, and the 200-GeV machine at Batavia, Illinois, projected for completion in 1971. Also, a machine is approved for 10 GeV in Japan, and a machine for 300 GeV at CERN. The sizes and costs of future machines may be much reduced from extrapolated estimates by making use of electrically superconducting materials capable of producing high magnetic fields very efficiently.

Beams of high-energy electrons can be produced best by means of a long linear accelerator, such as is used also for injection of protons into a synchrotron. A 20-GeV electron accelerator (SLAC) is in operation at Palo Alto, California.

Unfortunately, only a fraction of the energy of the particles produced in an accelerator is available for experimentation; when a ball of wax collides with another ball of wax that is at rest in the laboratory, both balls may stick together and continue to fly together at some velocity which is required by the law of conservation of momentum. The energy corresponding to this velocity is not available to produce deformation and heat—other forms of energy—in this collision of wax balls. At high energies, the energy available, for instance, for producing new particles is proportional to the square root of the laboratory energy of a bombarding particle. About 14 GeV are available in a collision of a proton with a proton at rest when the energy of the incident proton is 100 GeV. Therefore, very heavy particles, to be produced in very close collisions, require extremely high laboratory energies. If two balls of wax of equal mass and with *equal* velocities collided head on, they could stick together, be deformed, become heated, but would remain at rest in the laboratory; *all* the energy of the colliding particles is now available to produce other forms of energy. Machines producing colliding beams of protons or electrons, therefore, are also of great interest. They permit studies of very energetic collisions between protons or between electrons. The structure of a proton appears to be so complex that interactions between two such complex structures might need special interpretation. Results may be obtained more easily in collisions of beams of mesons with protons or neutrons. Nevertheless, because of the large amount of energy available, very interesting and important results may be obtained by CERN's ISR (intersecting storage rings), under construction for producing high-energy colliding beams of protons from their 28-GeV accelerator. In the future, colliding beam machines may be the *only* economically feasible means of making available much higher energies for further fruitful studies.

In some experiments the whole or a fraction of the beam of high-energy protons (or electrons) produced by the accelerator is allowed to strike a target, and then the particles produced in the target are analyzed directly. In many other experiments, however, a beam of secondary particles is first selected from those emerging from the target. This makes it possible to allow particles other than protons to strike a second target where interactions can then be studied. At the first target, all sorts of particles are produced, such as π and K mesons, positively or negatively charged or with no charge at all, and also antiprotons, neutrons, protons, and x rays. These particles emerge from the target in all directions with all possible velocities. The charged π and K mesons, for instance, are unstable but live long enough so that one can pass them through a magnet which changes their direction of flight, the resulting deflection depending on their electric charge and being largest for the smallest momenta. A slit through a metal slab will then select particles of needed momentum and charge. The particles can also pass through a strongly focusing magnet, similar to an accelerator magnet, which will keep the beam from diverging. Because of differences in their masses, one can finally separate one kind of meson from the other kinds of particles so far contained in the beam. This is achieved by letting the beam pass through strong electric fields, which produce further deflections now depending on the masses. Thus, it is possible to separate a beam of only the particles wanted for a particular investigation with as much purification as necessary. Obviously a large amount of equipment is also necessary for beam selection.

The beam of particles that has been selected according to charge, momentum, and mass at last strikes a second target which might consist of a thin-walled container filled with liquid hydrogen or deuterium, or of a slab of solid material. Hydrogen and deuterium are preferred targets since the hydrogen atom contains the simplest nucleus, the proton, and deuterium, besides a proton, contains only a neutron which is also a target of interest. The interactions observed in high-energy physics are so complicated even in the simplest nuclei that in heavier nuclei, consisting of many protons and neutrons, most effects of interest become hopelessly confused by the secondary interactions that are possible before a produced particle can leave the nucleus.

Most detectors make use of the fact that, while traveling through matter, electrically charged high-energy particles pass close enough by atoms so that electrons may be torn away from many of the atoms. The energy received by these electrons produces a small amount of light in some materials called "scintillators." This light can be collected and its effect vastly amplified into an electric signal in a photomultiplier. A number of small scintillating "counters" can be arranged so that the path of a charged particle can be determined accurately. When a particle passes through two sets of scintillators spaced some distance apart, two signals are emitted at different times, allowing determination of the velocity of the particle. Limits of velocity can also be determined by a Čerenkov counter where light produced by a different effect is also collected.

The path taken by a high-energy particle through a material is lined by atoms that have been torn apart into electrons and "ions." If the material is a gas, one can apply a high electric voltage to two plates or two sets of wires which will accelerate the "loose" electrons until a spark is formed between the plates. The spark is photographed or its location recorded in some other way, for instance, electromagnetically in the wire spark chamber. If many sets of closely spaced plates are present a track of sparks results, only interrupted by the plates, or if the plates are far apart, a track of predischarge "streamers" is produced under the proper conditions. Counters and spark chambers are two of the three most useful detecting methods in particle physics. The third again makes use of free electrons or ions which can initiate the growth of droplets in gases saturated with vapor, as in the now rather obsolete cloud chambers, or, if the electrons carry enough energy, the growth of bubbles in superheated liquids, as in bubble chambers. A liquid, for instance, liquid hydrogen, is superheated momentarily by lowering its pressure sufficiently without changing its temperature. The tracks of bubbles thus produced by high-energy particles, passing through the chamber at this time, are illuminated and recorded on photographic film. The liquid in a bubble chamber can act simultaneously as the target for the incident particles and as the detector. A magnetic field applied to the chamber volume produces circular curvatures in the paths of the charged particles which can be used to measure momenta. Each of the three mentioned methods has its own specific advantages and therefore its range of successful applications; as methods, they usually complement each other and rarely compete. Technically, bubble chambers are the most complex and costly, but one chamber can provide the research data for large numbers of research groups. Thus, they are operated as facilities, as accelerators are. Counters and spark chambers, so far, usually have been arranged in particular geometrical configurations by individual research groups for specific experiments.

The data from some electronic detectors (scintillators, Čerenkov counters, and wire spark chambers) can be analyzed very quickly and completely if a digital computer of sufficient capacity can be connected directly ("on-line") to the detecting array when it is exposed to the particle beam. From the signals sent into the computer, the data necessary for further interpretation of the observed quantities are calculated. Where photographs of tracks are obtained, such as in spark and bubble chambers, the film first is carefully scanned for interaction events of interest, then very accurate measurements of locations of track points on the film are performed and recorded on magnetic tape which again provides the input to a large digital computer. Analysis of photographs still is a time-consuming procedure, and many efforts are under way to automate recognition of events (pattern recognition) and their measurement.

To summarize, detection, analysis, and computing equipment must keep pace with the energies available at new accelerators and also must become more powerful, of higher resolution, as energies are increased, if a new accelerator is to be used effectively. With increasing energy, both the accelerator and the detection equipment increase in size and complexity. In spite of many advances and "breakthroughs" in concepts, design, construction, and efficiency of use, the effort in building and operating apparatus for particle physics research has had to increase. Since high-energy particles have to be produced for bombardment, it has been necessary to build accelerators of increasingly greater size requiring much land, large construction and operating staffs, complex controls, and difficult logistics for most efficient use. Similarly, much equipment and effort is required to provide the proper experimental conditions in the form of useful beams of particles incident on the detecting apparatus. Again, with increased energy, such beam-handling equipment must become larger in size and more complex. Finally, the detectors must determine many more parameters as the energy is increased. For instance, more particles whose energies, momenta, and other physical parameters must all be measured, are produced in the average event at higher energy. To obtain results within acceptable experimental errors, the detectors must also provide more accurate measurements and must be more efficient with respect to the number of useful events detected. Therefore, they, too, have had to increase in size and complexity.

VIII. ARGUMENTS FOR PARTICLE PHYSICS

An end for the need for large apparatus in particle physics is not in sight. Ideally, particle physics research will end when natural phenomena are understood so completely that a mathematical theory can be formulated which unifies all forces that occur and can predict all possible phenomena without further assumptions. If this aim cannot be reached, then particle physics will end when no further advance is possible because of human limitations. Such an end, however, will probably be just temporary, because it is not conceivable that man will rest as long as unexplored territory remains.

Because of the multitude of large elaborate apparatus needed and because of the large number of personnel needed to develop and operate the equipment, particle physics research has become expensive. This has created a problem which is rather new for a field of pure science. Up to this point most of our contemporaries probably have agreed that pure science is a necessary effort to produce knowledge of and experience in the ways of nature, most of which has become applicable in fulfilling human needs of many kinds. As long as the investment was not large, it was quite clear that human progress needed the effort invested in scientific research. The urge to explore, learn, and understand is one of the most important and productive human characteristics, and scientific research had resulted in many short- and long-range benefits in fields such as technology, medicine, and many others that are considered vital to modern everyday life. Everyone was willing to extrapolate into the future and to accept that great strides will continue if scientific research can progress. But the necessary effort has had to increase in order to produce further results, so that at some point the funds required for a few pure sciences, including high-energy physics, began to be comparable with the funds demanded for other large and perhaps politically or socially important projects. Reexamination of the value of science relative to other efforts thus has become a necessity.

In past and still in present times, funds for scientific research have been obtained from many sources. At first, probably, an individual financed his own research, which was particularly easy if he was somewhat well-to-do. He thus satisfied his own curiosity completely by his own or by his family's efforts. Then there were times when a king or other potentate employed a scientist, as he would employ a court jester. Perhaps the king hoped that the scientist would discover how to make gold for him, or he might indeed have been truly interested in education, the furtherance of knowledge, and general human advance. The latter goals, of course, became the concern of institutions of learning, and thus much pure research was sponsored by universities and by philanthropical institutions. Eventually manufacturing concerns recognized the advantage of research efforts, and then a certain amount of pure research was also sponsored by industry. But the costs of some kinds of research increased beyond the capabilities of individuals and of private institutions, so that whole communities, whole countries, and even groups of countries, had to take over sponsorship. One cannot expect, however, that all members of a population will understand and agree to accept the need for continuing scientific research on a large scale. Many of the members have more immediate needs for food, shelter, roads, improved health, and the like. The leaders in the population should and often do understand the longer-range benefits of scientific effort and they may be in favor of it. But politics is a complicated and not necessarily logical process, and therefore it seems indicated for scientists, and those favoring science, also to think about the arguments for continuation of their work at the rate and scale required for reaching ultimate success. The first such argument, again, is to point to previous developments resulting from scientific efforts and to point out that there is no reason why such rewards should not continue into the future. While sufficient so far, this argument involves extrapolation and appears weaker when contrasted with immediate problems confronting society and perhaps thought to compete for available funds. It is most certainly urgent that no one remains hungry, homeless, without education, or without medical attention when needed. Upon purely superficial comparison with such social problems, continuation of large-scale scientific effort with full momentum appears to be of little importance at any particular time. It is vigorous scientific effort, however, which has made it possible, in the relatively short time of a few decades, for so many of our contemporaries not to suffer any longer from the mentioned conditions that we must strive to eliminate for all. It is scientific effort that has produced peace of mind and prosperity for a great many and that can be expected to produce these conditions for ever increasing numbers. In countries where scientific effort is minimal or nonexistent, living conditions usually are at the lowest level. Conversely, of course, a poor country cannot afford much scientific research, but once research can be started it feeds back enormous dividends into all areas of our existence. Ideological and political developments play an all-important role in the fate of countries. Besides other rewards, scientific thinking also promotes objectivity, and many political and social problems would benefit from much-increased objectivity.

Science has been and will be vital in creating the conditions necessary for many improvements, and in promoting objective thought about the many problems that remain, but many of these problems cannot be solved by purely scientific methods at the present time. There is no truly scientific nor "other" technique available to eliminate selfishness, hatred, discrimination, laziness, lust for power, intolerance, and all the other undesirable features of the human character. Perhaps a "pill" will be developed at some time which could cure all "sickness" of human character, but how many desirable characteristics might it also eliminate, and are not even some of the "undesirable" characteristics partially necessary—such as enough "selfishness" for survival, enough "lust for power" for healthy ambition, enough "laziness" for relaxation? What can cure the worst catastrophe between humans —seemingly forever recurring wars? It does not seem possible that material benefits from scientific effort alone will be able to resolve these problems. Neither can fast solutions be forced by means of direct application of large amounts of money which on first sight seems most easily converted to material benefits. Money in however large amounts cannot directly affect any of the mentioned undesirable human characteristics—in fact, it may reinforce them. Money can give *temporary* material relief to some, but it cannot cure our "sickness." This has been shown on numerous occasions in history. Money, however, can be used to promote education which seems to be by far the most promising method for solving many or all of our problems. Again, short, quick courses to learn some craft will only give temporary relief and will not cure many problems. *Education* is meant to promote the profound change in attitudes that is needed to improve our world. This is hardly possible even in one generation, and certainly not in a few years. Such education will require continuing effort through generations, it will require much money, and, of course, it will "generate" much money due to vastly improved productivity of everyone.

Education in science is necessary to promote scientific—and objective— thinking and to pass our huge heritage of knowledge on to later generations. Assume that progress in a field of science were now slowed down considerably by a government that does not appreciate its value. The consequences would be, first and immediately, generally less research in the afflicted field, perhaps in proportion to the withheld funds; second, due to the necessity to be more selective, some promising research projects stopped completely—note that in science it is extremely difficult to predict what efforts will "pay off" in important results—third, readjustments and disruptions of work in the field, much larger than in proportion to the withheld funds, due to the smaller number of scientific research and supporting positions available; fourth, fewer students of the afflicted science due to decreased availability of research facilities and due to great frustrations to be expected for those entering the field; and fifth, as a result, fewer qualified teachers and researchers, which completes the first phase of a general downturn of the field. If the best potential experimenters should not find it worthwhile any longer to enter the field of particle physics, the results produced by mediocre ones are unlikely to provide the data necessary to test new theories. Therefore, the best theorists will find it useless to enter the field—pure theorizing, without experimental verification, belongs in the realm of philosophy and certainly cannot go far in particle physics. A substantial slowdown of our efforts in particle physics, similarly to a slowdown in other sciences, could start a vicious circle which eventually may stop the field altogether. In fact, to mature and to produce the hoped-for results, a scientific field must continue to grow with its needs. Without particle physics, physics would have no true frontier, and in a larger sense, science would have no frontier. Since no truly new principles or laws of nature could be discovered any longer, all science would soon have to become "applied science" and ultimately new developments in technology and education in science might cease, and thereby education in scientific method, in objective discipline, and in abstract thinking, all of which are so badly needed to help in bringing about the required profound changes in human attitudes.

We conclude that in truth there cannot exist any competition for funds between science and social or political problems; science itself contributes very heavily toward solution of these very same problems. A statement that "until no one remains hungry, no further scientific apparatus can be bought" seems most fallacious. Funds are certainly needed to provide whatever short-term relief is possible for social problems, but a permanent cure can be found only through a long-term effort to change attitudes through education—not mere short term "indoctrination"—in the various fields that constitute our heritage, our culture, and civilization. For education in a science, existence of vigorous research efforts in this science is a prerequisite. Funds therefore are certainly needed to continue these efforts, as an important part of the total required effort. Meanwhile, scientific effort will also provide the material environment in which our eventually changed attitudes will be able to persist.

Those who practice scientific reasoning acquire a certain amount of objectivity, although, since scientists are members of their contemporary society, they can also acquire useless characteristics, such as intolerance of deviation or speculation, professional jealousy, bigotry, or one-sided over-specialization. Objectivity in one subject does not immediately produce humans free from faults, but objective reasoning does seem to promote improved, sounder relations between people and must remain a part of all education; in spite of big differences in political ideology, it has already been possible for professionals from many countries to meet to exchange views in their fields and thus to begin to understand each other better.

Is it possible to make promises based on arguments more definite than extrapolation from the past into the future, that many kinds of immediately useful "technological fallout" will always result from future research, that the ultimate answers, beside their educational values, will hold the greatest rewards for all of mankind? Could such arguments aid a population and its leaders to decide that the promised advances are indeed worthwhile and are needed as soon as possible, just as they decide that a social or political effort, including a war, may be necessary? Or should merely a kind of faith in science be promoted, similar to the religious faiths, that created the impressively beautiful and, without a doubt, extremely expensive cathedrals of medieval times?

Many scientists are "purists" and do not wish to speculate on the practical aspects of their science. They are exploring nature because nature is here to be explored and they seemed quite justified in their attitude since successful research effort often does not tolerate any distractions, but if a research effort were slowed down because no one is inclined to develop the arguments necessary to obtain funds for vigorous pursuit of a science, then a change in approach would be needed. Science is too valuable an effort to be retarded by lack of convincing arguments which so obviously exist but need proper formulation. If individuals with the talents of some of the existing scientists would discuss, however speculatively, the possible course that their science might take, where it might lead scientifically and practically, such discussions might not only be of great value to their own efforts but also might evolve additional arguments for sponsoring their field to proceed at full speed. Of course, all arguments, old and new, should be made widely known and explained, so that the members of the population, who ultimately sponsor the research with the taxes they pay, are properly informed and can show their agreement or disagreement. Thus far, the arguments for continuing particle physics at full speed, raised by some individuals and by important committees, have been entirely based on extrapolation from the past into the future. Although there is no reason not to believe in them, such arguments are not necessarily compelling to everyone. The question asked here is whether more effort in developing arguments might not be possible. Would speculative discussions help? Are there any scientists who might be willing to spend a continuing portion of their time to speculate on the values and applicability of their science? How might discussions on early applications of their field be set up between workers in different sciences and those concerned with technology? Should scientists, including particle physicists, pay more attention to immediately practical fields by, in fact, spending a portion of their time on practical problems? Among the scientists within a field, there are the just-mentioned "purists" who wish to pursue their studies unencumbered by "extraneous" problems, although in particle physics, for example, such "extraneous" matters might concern the development and construction of accelerators, and of beam and detection equipment, the very apparatus that the "purist" needs for his work. Much of the development and construction work thus has had to be carried out by the "engineer type" scientist supported and often guided by the experience of great numbers of engineers and technicians. Of course, not all scientists would be able to think in practical terms, even if exposed more to practical problems during their educational period, but many have successfully participated in solving practical problems and even have acquired patents on inventions. The advantages of increased concern with practical applications of their science generally and of their special field in particular would consist of (1) closer contact between individuals involved in different fields of human effort, (2) more immediate and wider recognition of the importance and value of a special field to all humanity, (3) earlier exploitation of such values due to practical advances achieved, and (4) an increased "horizon" and there-fore possibly new ideas in his own field for the individual scientist; such improved relations between the specialist and his contemporaries, and breaking down of the, often fictitious, "ivory tower" atmosphere, could result in enormous mutual benefits, including much-improved general will-ingness to support the specialist's research. Formation of exclusive societies of specialists, which has been an only too common occurrence, can hardly be expected to contribute to their own or to anyone else's well-being. Ex-clusive specialization should therefore be counteracted, within reasonable bounds, as part of an education.

Besides the discussed contributions of extreme general significance to education and to civilization, what specific consequences might a continued strong effort in particle physics research and a final understanding of the various forces observed to act in nature have on our lives and on the lives of those existing in some future generation? In order to reply to this ques-tion one should first define what might be worthwhile objectives for humanity to attain. What might be a near-ideal world? Might it be one where all who are born will be guaranteed long, happy lives? "Long life" and "happy life" are of course not two independent goals, but for this discussion it is sufficient to say that "long life" is largely related to physical well-being, where science already has made important strides and where much more improvement can be expected, assuming further research, more advanced tools, and a saner approach to many existing problems. "Happy life" is more difficult to define. Would it suffice to consider it a great improvement if everyone could be educated and could work and create according to his abilities and interests for his own and his contemporaries' and successors' benefit? Hopefully, due to automation no routine tasks would be left and no one left to perform them, since through education and scientific methods all humans might develop very great intelligence and sophistication. Could a world then be created where political upheavals and wars would no longer occur, where there is no fertile soil left for misguided ideologies, where all people live and work for each other as well as for themselves? Happiness, however, cannot be equated with stagnancy. Continuing change seems always necessary. Thus, development must continue and still further improvements must occur. It seems that activity and more creativity will always remain necessary. What seems nearly ideal today may need changes tomorrow. This continuing flow of needs is the most difficult to predict or satisfy. Perhaps the ultimate "frustration" of death will always be hard to accept for humans—and stagnant eternal life without continuing growth seems equally unacceptable—but a substantial reduction or elimination of all the other frustrations occurring in a lifetime might produce a sufficiently stable situation and environment for generally available happiness. How might such aims be helped by research and an understanding of the forces acting in our universe?

Physics has produced the "atomic bomb," a very large, almost instantaneous energy release due to nuclear forces. However destructive, perhaps this development has already averted another very big war. Presumably this result is beneficial but it seems a perversion of the human character if it is caused by fear—the fear of retaliation and the fear of making our world uninhabitable. This would not appear to promote the stable equilibrium necessary for the kind of world spoken of above. Most knowledge can be used destructively—germ, chemical, atomic, and space warfare, the electric chair, misuse of the spoken or printed word, of karate, of narcotics, to name a few applications.

Science will provide an understanding of our world and will provide a great multitude of tools. If these tools can be used to improve our material world, will this help to create a happier world? It should, if the tools will be used intelligently. It is up to us humans to apply the results of science properly once they are available. Obviously, there is a continuous feedback between all human activities. Thus, a gradual development of the human mind made science and all other fields of effort possible, and science helps to improve the environment for further development of the human mind and character, which in turn will produce more results from our efforts. The interactions between human endeavors, between all humans, are hugely complicated. It seems that advances must come in many areas. Thus, if science provides understanding and tools, scientific and all other education must show us the way to intelligent application, or generally must teach us to think properly, logically, and creatively.

An immediate result of money spent on any government-sponsored effort is, of course, that through such funds, a very large number of people can be employed. The funds are not wasted. Directly needed are the designers, engineers, technicians, administrators, and physicists to conceive the ideas, design, construct, assemble, operate, and maintain particle physics apparatus. Many large components are further designed and manufactured by industrial concerns who usually consider such projects worthwhile for experience gained, although to them relatively little direct material profit may be involved because major particle physics equipment rarely is ordered in large quantities. There are also the countless available standard items needed for assembly of the equipment and for control instrumentation, all purchased from industry, and similarly there are all of the expendable materials such as gases, liquids, photographic film, electronic parts, and electric power needed to drive the equipment, all of which certainly contributes to employment of a great many people in private industry. A gain of somewhat longer range consists of the experience gathered by all of the technical personnel involved in particle physics research. The needed apparatus is complex and of high precision and requires great care in operation, an excellent training ground in many techniques that, once acquired, should remain useful throughout a lifetime even in other more technical fields. Furthermore, exposure to this field has stimulated many to broaden their education further—technicians to become competent engineers, data processing aides to become computer programmers. Thus, the development of talent is stimulated. Next in range are the results of "technological fallout," direct or indirect industrial applications of apparatus originally developed for scientific investigations, and stimulation of industrial development by research needs for not yet available materials and equipment. Eventually, the basic understanding gained by means of scientific investigations can be applied, possibly in many fields where a correlation was not even expected. Again, probably the most important result concerns the contribution of scientific effort to education in its profound and lasting sense, and to present and future civilization and culture.

The amount of direct "technological fallout" from high-energy physics is indicated by the large number of patents in numerous technical areas that have been granted to laboratories. These concern new measuring and control instruments and novel designs and methods, all usable in many fields. Of even greater importance, however, seems to be the stimulation that particle physics provides for industrial development. For instance, the great need for high magnetic fields, in order to produce high-energy particles economically, select the proper ones for study, and to measure their interactions, has provided industry with one of its incentives to develop superconducting materials which can conduct very large currents at very small electric power. As a result, already electric power transmission lines with minimal losses are under serious discussion. Further developments will not only result in economically operable magnets for use in particle physics research, but also in superconducting radio-frequency cavities capable of transmitting very powerful signals, superconducting elements for digital computers, and countless other applications. The processing of particle physics data requires very precise measuring equipment and very complex digital computer programming for the most advanced computers. The great quantity of data that must be analyzed further requires automation guided by computers. Such developments, originally carried out for the purposes of particle physics research, are obviously applicable in other fields of science and in industry. Operation at the low temperatures required for liquid hydrogen targets, bubble chambers, or superconducting magnets requires large but economic cryogenic refrigerators whose development thus is further stimulated. Nonmetallic but extremely tough materials, molded or otherwise composed, are needed for use at these low temperatures that also occur in space outside our atmosphere. High-voltage and radio-frequency equipment had to be developed for the above-mentioned linear accelerators and beam separators. Sophisticated circuitry is needed for electronic counter work. Wide-angle photographic lenses, with minimum distortions of images on film, are being developed for bubble chamber photography, and their existence should stimulate many applications. Including, finally, the needs for ever more accurate and reliable electric, hydraulic, and pneumatic measuring and control instruments, one is justified in expecting continuously increasing mutual stimulation for further development between particle physics and industry.

IX. SOME SPECULATIONS

Development of nondestructive applications of "atomic energy," or of the nuclear chain reactions first used for production of an atomic bomb, is still in an early stage. To be sure, useful (as well as useless) radioactive materials have been produced in great quantities, reactors are used as furnaces to produce heat to drive electric power plants, medical and biological research have profited considerably, and so has further research in nuclear physics, but basically, what has been achieved is to release a large amount of energy and radioactivity, well controlled in time. The energy from a reactor is used very much like the energy in a well-controlled coal or oil furnace, which serves to produce heat to drive engines on thermodynamic principles, to forge and alloy metals, or to produce electric light; the amount of energy produced per unit of time can be controlled, but it has no preferred direction in space. So far it has been difficult to produce from nuclear energy well-directed forces such as those produced by the propulsion engine of a rocket where the hot combustion products of well-contained chemical reactions are directed "backwards" to produce the necessary "forward" momentum of the rocket. Fuel for reactors is expensive and therefore must be efficiently used. This seems to be possible in reactors such as applied for "furnaces." Therefore, reactor-powered ships are cruising the oceans without having to refuel for very long periods. But such efficiency seems not to have been achieved in other applications of nuclear chain reactions. The importance of inducing energy to flow in preferred directions can be illustrated by means of the quite well understood electromagnetic forces. Electromagnetic energy, such as visible light, infrared heat radiation, radio waves, can be focused by means of lenses or reflectors and can be collimated. A fair amount of directionality, limited only by the physical principle of diffraction, can therefore be achieved, and thus high resolution photography, radar, astronomical telescopes, auto headlights, and television via satellite are possible. Flying saucers, if they existed, might be propelled by very intense electromagnetic radiation. One of the most impressive developments during the last few years, however, has been the light laser (or maser for short radio waves), which allows one to produce extremely precisely directed electromagnetic waves. While usually light is emitted by atoms individually in all directions at random, at various uncontrolled times —or incoherently—in the laser one succeeds in stimulating atoms to emit their light waves so that they are correlated in space and time—or, they are "coherent." Such coherent light can then be made extremely intense and well directed. Besides its great scientific interest, during the few years since its inception the laser has become a very important tool. One of its most impressive uses is for repair work on the human retina, but there are many other very important applications in medical, technological, and other research and development work. The speed with which a tool providing coherent electromagnetic radiation was applied in so many areas seems truly significant. Directionality and extreme resolution in space have provided a tool much finer than a surgeon's scalpel, compared to the mere "footprint" left by a release of undirected energy. Availability of coherent x-rays would provide another enormous advance. Here, however, a difficulty requiring a new idea is encountered in the short wavelengths of x rays which would be much harder to stimulate coherently than the long waves due to transitions of electrons in the outer shells of atoms.

Might a further understanding of nuclear forces lead to tools producing coherent radiation of nuclear particles? Indeed, as mentioned here repeatedly, particles also obey a wave picture, their interactions can be incoherent or coherent (with many particles at once). In nuclear interactions, there occur interferences and diffraction of the waves representing particles, just as with light or sound and all other means for transmission of energy involving waves. But the wavelengths and ranges of interaction of nuclear particles are very short. Perhaps very dense nuclear matter must first be available, such as seems to exist on dwarf stars, which could possibly be manufactured, even without gravitational collapse, if nuclear forces could be understood.

In electron microscopes, incoherent beams of electrons are focused by magnetic fields, and due to the relatively short wavelength of the electrons in the beam very large magnifications can be obtained, so that structures as small as complex molecules can be probed. Similarly, intense incoherent, but directed, nuclear radiation could become an extremely powerful tool, which to contain would require quite new materials that must also first be produced. To illustrate, if 10 lb of matter *per day* were very efficiently converted into directed radiation traveling with a velocity near that of light, then a thrust would be developed which would move a one-ton weight 30 million miles from a starting point during the first day, and, since acceleration—here about one *g* which is equal to the gravitational acceleration on the earth's surface—would continue, much greater distances during the following days, truly immense distances compared to those possible with rockets driven by chemical combustion. On the 20th day a speed of about 10,000 miles/sec would be reached, and almost a billion miles would be covered during this day by which a total of only 200 lb of matter has been converted. Then it might even become possible to reach some other star systems in a humanly reasonable time! It would also become possible to reach any point on earth in a matter of minutes. If the nuclei in a material to be converted were polarized (oriented in space), a certain amount of directed radiation could already be obtained.

Elimination of unwanted and potentially harmful radioactivity from nuclear interactions has been and will remain a difficult task, similar to the formation of soot by furnaces. Complete choice and control of nuclear reactions would facilitate selection of only those types of high-energy radiation which are completely useful for advances in other fields, such as biology, medicine, agriculture, forestry, and animal breeding. An example of a very efficient and relatively "clean" nuclear reaction consists of annihilation of nucleons (protons or neutrons) by antinucleons. Here, within about one millionth of a second after an annihilation, only weakly interacting neutrinos and electromagnetically interacting electrons and x rays remain. It is not necessary (nor possible) to contain the neutrinos because they hardly interact at all, and electrons and x rays are much more easily contained than many other kinds of radiation. An "annihilation" furnace would "burn" much cleaner and more efficiently than present-day reactors.

The very-high-velocity "combustion" products represent an excellent propellant if they could be directed. However, the necessary antimatter must be produced and transported first, which, on a large scale, may well remain a very expensive process. Therefore, antimatter may be usable only in special cases. Nevertheless, one can imagine that antimatter and high-energy reactions in general might be used to trigger other nuclear reactions which need very high energy to be initiated. This would be similar to the use of a match to start a fire or to the use of the plutonium atomic bomb reaction to start the even more powerful hydrogen fusion bomb reaction. The "power" produced in a proton accelerator accelerating 10 trillion (10^{13}) electrically charged particles once a second to 30 GeV is 50 kW, which presently is small compared to the power, amounting to millions of watts, needed to drive the accelerator's magnet; an accelerator can never be an efficient power source, but the 50 kW of produced power are of a very special nature: the 10^{13} particles are available as short "bursts" lasting much less than a thousandth of a second; they travel nearly with light velocity, as if they had been accelerated by an electric potential difference of 30 billion volts. Lightning is produced by potentials of billions of volts, but it never produces particles near light velocity that are capable of initiating nuclear reactions since gas discharges consist of very large numbers of small energy transfers in collisions between atoms. Therefore, high-energy particle beams and high-energy nuclear reactions have unique properties that might be usable for unique purposes.

It must be emphasized again that at this time a discussion of such applications is purely speculative. If any were now possible, they would surely be already under investigation, but speculation of some kind is needed at the beginning of any new effort. While it appears absurd to assume that the force exerted by an individual nuclear particle could ever be turned on or off by an experimenter, one nevertheless can hope to gain control over aggregate forces and reactions and one can put these forces to work for our various purposes, in the form of tools vastly improved over any that we now encounter in all areas of life. Understanding of these forces and of all interactions between particles should lead to their control in space, time, and direction, from all of which hitherto undreamed-of applications will become possible.

X. SUMMARY

In detail, the needs and aims of modern man are extremely complex, but perhaps they can be expressed quite generally by "longevity" and "happiness" for everyone. Longevity is made possible by the proper environment and by physical health. Happiness could be due to satisfaction with the circumstances of one's existence, concerning needs for work, rest, stress and the relief of stress, stimulation of interest, and interactions with others living now, before us or after us. Certainly, the latter is of particular importance because we learn from those present and before us in time and work for those present and after us, and our aims can only be reached by common effort.

To strive for fulfillment of our needs is an effort as old as humanity itself. At first, there were mostly the basic instincts, born out of the urge for survival among at once creative and destructive processes of nature. Later, there developed an awareness of our near and far environment, of the past and the future, and of threats to our existence, such as floods and storms, heat and cold, other beasts, famine, sickness—and man himself as murderer and warmaker. Gradually, man began to explore his environment and means for protection against destructive processes and found that many of nature's phenomena—even potentially destructive ones, such as fire, wind, and flood —could be used to create improved living conditions. Once, by exploration, man had lifted himself out of the circle of other creatures, he found that he could not stand still. More activities were created, forever to be channeled into the stream of development and improvement.

Exploration is the basis for everything that has been and will be achieved in all fields of human effort. Certain kinds of exploration are called "science," when they deal with well-defined terms and contain an accumulation of well-organized results which are reproducible and are well substantiated by observations that are believed by many. All science explores nature of which everything is a part. All of nature's processes are driven by forces. Knowledge and understanding of these forces and their effects is therefore the basis for all scientific effort. The science called "physics" deals with exploration of all forces and with many of the phenomena due to them. In pursuing its objectives, research in physics has been led into a submicroscopic world of great complexity. At its frontier with the unknown it has had to deal with ever smaller objects, shorter distances, and shorter times, in contrast with the frontier of astrophysics which deals with the largest objects, largest distances, and longest times. Yet, the two frontiers are related in many ways. Other sciences use the basic concept of force but deal with aggregate effects, such as formation and properties of molecules, crystals, living cells, and even groups of people. All sciences, just like the arts, languages, and all other human activities, satisfy needs of individuals and of groups of individuals, now and later, and therefore help us to find our way to common goals.

Physics interacts with other sciences and they with it; there is a continuous exchange in all human efforts. Physics is divided into areas according to the strengths and effects of forces. Established thus far have been the gravitational, electromagnetic, and the weak and strong short-range nuclear forces. There may be more, especially the strong nuclear force may consist of several components that can be distinguished by their strengths, orientations, and dependences on distances to interacting objects. The aim of physics is to determine the existence and properties of all forces, correlate them, and, if possible, produce a mathematical description, or theory, which needs no further assumptions, and from which any conceivable natural phenomenon can in principle be explained or predicted by calculation, however complex the latter may have to be. Understanding the source of these forces may mean to understand the origin and the destiny of our universe. It may permit man to understand himself and to control his own destiny to a very great and perhaps sufficient degree.

During the process of exploration, a huge number of phenomena that are due to interactions by forces have come to light; even if a final understanding can never be reached, many of these phenomena have been and will be useful materially, or useful in other areas of effort, and have led to applied science and technology. This has been a reward for the effort of exploring the unknown, having advanced humanity enormously since the era of the cave man. One can expect that these rewards of scientific effort will continue in the future, but the wisdom to use these rewards must also be developed. Thus, changes in human attitudes must be accomplished by education in many fields. To this end science contributes objectivity of thought, an essential part of any kind of advancement.

Exploration of nuclear forces has led to "high-energy" or "particle" physics. The theory of quantum mechanics had to be conceived to explain atomic phenomena. A picture involving waves in association with particles had to be invoked. A dualism, that waves and particles are equivalent, was established, as energy is equivalent to mass. This is the way nature operates. It seems mysterious to us according to familiar experience with larger objects. Our known senses perceive directly only a very small portion of natural phenomena, but the human mind can reason very far beyond direct perception. Yet it can also be misled by what its five senses transmit. Here objective thinking is needed and must be developed and trained. The universe still seems so complex and beyond our grasp that a true understanding of all acting forces and their effects may be beyond the capacity of the human brain, but such possible limitations have not kept us from making huge strides which once must have been beyond imagination. Perhaps such limitations merely affect our speed of advance and not its scope. Perhaps we merely lack sufficient information which we must attempt to obtain by further experimentation.

The apparatus needed for experimentation in particle physics is large, complex, and expensive; it is a fact of nature that the "smaller" a particle is to be explored, the greater is the energy necessary to probe for it, produce it, and study it. Since forces and particles are very closely related, the study of particles should lead to an understanding of forces. Therefore, further progress in the field of physics, first, requires equipment to produce particles of high energy—and short wavelength—to act as probes, and, second, equipment to detect and measure their interactions and to detect and measure the properties of new, hitherto unknown particles produced in these interactions. Research in particle physics requires particle accelerators and different kinds of detectors, such as electronic counters, spark chambers, bubble chambers, and much ancillary apparatus.

The expenses for this equipment and for its operation are large enough that only entire nations, through their revenues, seem to be able to sponsor research in particle physics. Practically no other sources of funds, which in the past had sponsored most scientific research, seem to be able to afford the expenses required by large-scale research, although in an affluent country private sponsorship would not seem impossible, if, say, a number of institutions contributed toward a common fund. Whoever sponsors research on a large scale must be convinced that it is needed, that it is useful, that it advances humanity. Generally, the arguments for scientific effort, listed in order of increasing range of importance and time, are that (1), far from being wasted, the funds granted for research enable people of many professions to be employed and trained in advanced thought and skill; (2) immediate applications, or technological fallout, result since many of the techniques and instruments developed for research are also useful for other purposes; (3) scientific knowledge gained in one field is applicable in many other fields, particularly knowledge about physical forces that are ultimately responsible for any phenomenon of nature has very wide applications; (4) knowledge is a part of our heritage and civilization which must be advanced and passed on to future generations, and scientifically objective procedure is an important part of education and should promote the improvement in human attitudes that is very much needed if substantial gains are to be accomplished by mankind in the future; (5) the greatest achievement may lie ahead when, with the help of science, man might begin to understand himself, his purpose, and his destiny.

The effect of artificially retarding a scientific field can be to stop it altogether in not too long a time. Research is done by people who must have initiative, drive, enthusiasm, and patience in addition to their special talents and education. As humans, they can also be frustrated. To be successful, every field must include among its ranks some of the most talented people.

Once these and others feel that because of lacking support their efforts in a field cannot lead to success, they can change their efforts to other fields that may even be of less interest to them but that might promise more assurance for success. In a field that offers less opportunity for research there will also be fewer students of high caliber, and, since there exists an interdependence of sciences, the consequence of a slowdown in a major field will soon cause disruptions in other fields and generally in education and progress.

Particle physics and large social projects should not even be considered to be in competition for funds. Science itself is a part of our efforts for advancement. Science in very many ways contributes toward the solution of short-range and long-range social problems. Money expended on social projects without accompanying effort can at best produce stop-gap measures of no durability. Of course, acute difficulties must be relieved, but it seems clear that few can be cured at the same time. For a cure, effort is needed.

Science is not for the curiosity and amusement of a few, but it is a part of our culture and civilization and of our striving for the advancement of humanity. The scientist does have a deep curiosity about the processes of nature but he no less must feel that his work serves a purpose other than his own. This may not always be obvious; in particle physics there is technical jargon, intricate mathematical and experimental procedures, and specialization. There has rarely been an inclination to speculate on the eventual usefulness of the field, there is merely the belief, well justified by past experience that the field is important and may lead to great rewards for all of mankind; it is there to be explored. Study and research in particle physics require very much of of a physicist's time, but the need for convincing arguments to acquire sufficient funds for support of this work must not be ignored. Concern by more particle physicists with practical problems in their own as well as in related fields would establish much improved contact with society as a whole, which, after all, is called upon to support this research. Such a range of interests in the practical as well as in the ephemeral should be stimulated during the educational period of the physicist. Efforts would seem well spent on explanation of particle physics and its aims in terms accessible to many, and on speculation about any human advances to be gained from the field. This does not at all mean that particle physicists should enter fields such as technology, philosophy, or popular writing—it is the *concern* with such matters that is needed and that might produce additional arguments for continuation of high-energy physics at an accelerated rate. Such arguments may gain the proper support, but their development may also lead to new approaches and techniques in this and in other fields.

Physics is a link in a chain of human efforts. If a link is broken, a chain is worthless. The frontier field of high-energy or particle physics, therefore, must be continued until man's search for understanding his environment has been successful. The means seem to exist for advancement toward our human aims. We must continue to search for these means.

XI. CONCLUSION

The deep significance of the forces observed to act in nature can perhaps best be recognized by imagining what might happen, if, one by one, these forces were turned off—if they did not exist.

If the strength of the gravitational force were smaller than it is, the size of the orbit of the earth around the sun would have to be smaller or the earth would have to rotate more slowly around the sun. Such a change of the *weakest* known force alone could result in a change of conditions (temperature, air pressure) on the earth's surface so that man could not exist here. If gravity were reduced gradually, the earth's orbit would become unstable, gases will escape into space, liquids would evaporate faster and their remainders cool off and freeze. Suns would partially disintegrate, their surfaces would tend to freeze, but their internal pressures would cause huge explosions. The earth's surface similarly would be broken up by bottled-up pressures. At zero gravity only solid fragments would remain (or such liquids or gases that are still trapped within warm solids). These solid fragments would gradually sublime or be broken up by collisions with other fragments so that eventually just gas or dust would be left in the universe. Gravity is needed for our existence. If the strength of gravity were different than we know it to be, life probably would not exist on earth but it might exist on some planet of some other sun where now life cannot exist, where the proper equilibrium of conditions for life cannot now be established.

Turning off the electric or electromagnetic force makes the formation of atoms and molecules impossible. Only certain neutral particles would exist, no electrons, no protons, and probably no neutrons, which, when free, decay into a proton, electron, and neutrino, the latter being perhaps the only "truly neutral" stable particle. Light would not exist because it is electromagnetic radiation. Whatever neutral matter could exist, may be compressed by gravity to extreme density, somewhat like dwarf stars, if gravity could still be present without existence of electric forces (a relation between gravity, electric and other forces *might* exist). Electric forces at the same time "brace" matter against gravity and "bind" matter by forming chemical compounds; electric forces can be repulsive or attractive because of existence of both negative and positive electric charges. If electric forces were considerably stronger than they are observed to be, compound nuclei could not be formed with existing short-range nuclear forces, since the repulsive electric force between the positively charged protons that in addition to neutrons constitute nuclei may then be too large. Thus, there had to exist some kind of balance between existing forces if processes as we know them were to be possible.

The nuclear short-range force forms nuclei out of nucleons—protons and neutrons—and perhaps it forms protons out of still other particles. Particles such as π mesons or K mesons are thought to be exchanged in strong interactions. When such a meson, or also a neutron, escapes from a nucleus, it decays by a weak nuclear or an electromagnetic interaction force; it does not get very far after its escape. If it had not decayed, it might have traveled to another part of the universe and interacted there; besides their other functions, perhaps these forces contribute to the stability of the universe by "cleaning up" strongly interacting "debris." Without the strong interaction force no nucleons would exist, nor could the constituents that might form nucleons be expected to exist. If any particles at all were to exist, but no forces of any kind, one must ask the question "What are these forceless particles themselves made of, what force holds them together?" It appears that matter cannot exist without forces. Therefore, explorations of particles and of forces are equivalent.

Having turned off all forces, we may thus have arrived at "Nothing." In "Nothing" there is no matter, no interaction, no light, and it cannot have any boundaries which would have to consist of "Something." "Nothing" would be infinite in size or without bounds in other ways, like moving around on the surface of a sphere. Perhaps it is then not at all mysterious that the universe, containing almost "Nothing," seems boundless. Perhaps the only boundaries of the universe consist of matter and force fields—they are "internal boundaries," somewhat like surfaces of liquid drops bounding that which is outside the drops. Perhaps, without containing matter, space has no meaning.

How did forces, or matter, start in the nothingness of an unbounded universe? Did the world start with many different kinds of forces, but only a few of them had the proper relative strengths to lead to our existence? Was it a random phenomenon or was there some profound purpose? Whose purpose? How did He start out of "Nothing"? Obviously we have arrived at questions underlying religion and philosophy. Might not an understanding of all forces occurring in nature reveal how they began? Reveal their purpose? Reveal *our* purpose? And might not an understanding of our purpose at last provide us with the proper attitudes in leading our lives as they were intended, thus resolving humanity's struggles? Does not "knowing" go very far beyond "believing"?

Religious books were written by men of their times, in terms of the world they knew. The world we know is different. A superhuman Being, having created the forces of nature, might well be "expected" to make use of these very forces to continue His creations. Random processes, gradual chance combinations of matter that remained stable, may very slowly have created man, but such random processes are observed every day in nature, they are a part or a tool of nature's way of creating; they are not at all in contradiction with religious beliefs if one simply assumes that "nature's processes are God's tools." If man evolved very slowly, that seems of no importance, since only man is and must be in a hurry in his performance—only his time is limited. To accumulate knowledge about our world, to learn to understand it, must be a part of our evolution, it cannot be against proper religious concepts. To understand how we and everything else were made should bring us *closer* to the God of any religious belief. By understanding His world we may also begin to understand His purpose, to know rather than to believe. Many equate God with that which remains mysterious. If some day no mysteries remained, such gods will have been merely an invention of the human mind. And why should we not be eager to know that, too, whether or not we now believe in a god?

To continue these speculations, in empty space, in "Nothing," the concept of "time" has no meaning. Only when forces and matter are introduced, can one define "time" by, for instance, the number of heart beats it takes for a person to walk some distance, or the number of vibrations a pendulum makes between sunrise and sunset, or the number of oscillations an electron makes in an atom while a rocket travels from Earth to Venus. The concept of time has meaning as long as forces exist and can produce motion. Time started when matter started; it will continue as long as matter exists. If "time," however, is an invention of the human mind, which is useful for describing motion in relation to other motion, then to talk about the "beginning" or "end" of the universe may not be meaningful; time would have no absolute duration, just as space presumably has no absolute distance. Then time, as well as space, would be without bounds. In space, distance between particles of matter is measured only in terms of distance between other particles. In time, relative motion between particles is measured in terms of relative motion between other particles. Forces form particles and produce motion. Particles and motion are forms of energy. Motion is best represented by momentum. Energy and momentum seem to be conserved anywhere in the universe, independently of location, direction, or motion in space. These symmetries may *require* the absence of bounds for space and time. Due to the existing forces, matter may occasionally move very violently in space, leading to complete destruction of all structures as we know them, but new eras of evolution would seem *always* to be guaranteed if indeed certain particles are also conserved, such as the proton, electron, or neutrino, which then would be responsible for the long-range stability of the universe.

Whether or not matter, or a set of properly balanced interaction forces, was introduced into space by chance or by purpose, or merely exists in space in relative motion, all that follows might be explained by random phenomena. There were strongly interacting nucleons, perhaps having been formed before by their constituents, colliding sometimes and forming compound nuclei. The nuclei sometimes met electrons, forming atoms. Gravitational forces pulled atoms closer together, promoting their interaction and formation of chemical compounds due to the electrical forces. Some compounds could be liquids that suspended many other chemicals which could form more complicated compounds. Some grew to form crystals. Others grew in more complicated ways until they fell apart under the influence of destructive trends that were also a part of random processes. At some time some such structure happened to meet another different structure or just a complex molecule which enabled the first structure to grow until it divided into two similar structures. Reproduction had started, no doubt only after very many other random meetings between existing structures that resulted in nothing new. Plants started. Life started. Still other molecules suspended in the liquid gradually attached themselves to that which already was alive. Cell structures could begin to move, begin to help themselves to find food and to survive destructive trends. But motion needs control. Brain functions, at first purely involuntary or automatic, later developed to the higher voluntary forms. Man is probably the most advanced product of evolution known to us in our vicinity in space. Evolution by billions upon billions of random processes has taken many billions of years, a long time only by human standards. Possibly, if information on all forces and particles existing in our world provided the input to a very powerful digital computer, a calculation involving all probabilities of possible random interactions might predict our evolution, or its probability, and also all that is to follow us. Thus far we do not have the necessary information nor the necessary computer.

An understanding of particles or forces enables man to speed up evolution in his own vicinity; favorable conditions can be promoted for nature's random processes to occur at a very much higher rate. The results have been and are expected to be beneficial to man, but of much more significance may be the insight that will be gained by understanding more about that which made this world. To gain such insight, we must first know of all existing particles or forces, we must know of their relations to each other, their relative strengths—we must know what we are dealing with. We can imagine their purpose, we can understand the evolution of our world, we could be able to unravel some basic mysteries: why do forces exist, is there a beginning and an end to their existence, what is the structure of the universe, is evolution of matter and of life a random phenomenon? By knowing the truth, rather than believing what appeals to us, we may at last understand our responsibilities so that we can all move coherently toward an existence which, we know, should be possible but that we do not seem to be able to find yet. The convergence and culmination of all human effort could thus be reached. Such is the promise of our science. Does the effort seem worthwhile?

2   Elementary Particle Interactions in Astrophysics

M. A. Ruderman* †   *Department of Physics*
Imperial College
London, England

William A. Fowler‡   *California Institute of Technology*
Pasadena, California
and
Institute of Theoretical Astronomy
University of Cambridge
Cambridge, England

* Permanent address: Department of Physics, Columbia University, New York, New York.
    † Research supported in part by the National Science Foundation.
    ‡ Research supported in part by the Office of Naval Research Nonr-220(47).

I. INTRODUCTION

In the world around us we observe and participate in continual change: ice melts, wood burns, and children mature. These phenomena are a reflection of continual rearrangements among atoms and molecules, chemical reactions which characteristically involve energies of less than an electron volt (eV $\cong 10^{-12}$ erg) for each atomic interaction. To describe and understand the immediate natural world it is not necessary to use the information acquired in present high-energy physics laboratories where elementary particle interactions are studied at many billions of electron volts, and where particles which generally survive for 10^{-6} to 10^{-22} sec are produced and studied.

However, our local natural world is a rather rare place in the universe, which, at least temporarily, possesses the very special kind of environment which can support life. To know the nature of the sun, the stars, and the rest of the universe, to discover their history and future, to reveal the synthesis of the elements and even the kinds of matter which may exist it becomes essential to exploit the knowledge and the tools of elementary particle physics. It is this branch of physics, which includes classical nuclear physics, and describes the nature and interaction of individual electrons, photons, protons, neutrons, and other particles that has given the necessary basis for extrapolating physical theory into so many regimes of astrophysical interest. In the past few years observations of radiogalaxies, quasi-stellar objects, and cosmic x rays suggest that phenomena involving enormously high energies per particle prevail in many areas of the universe:* the contributions of high-energy elementary particle physics to astrophysics will surely grow even more significant in the future.

Elementary particle physics consists of the study and knowledge of the most basic known constituents of the universe. Only fifty years ago the "elementary" particles were electrons, nuclei, and photons. Classical nuclear physics became a principle domain of elementary particle physics until, with the construction of very high-energy accelerators beginning around 1950, emphasis moved on to investigations of the structure of individual nucleons, mesons, and a host of newly discovered elementary particles. The very detailed knowledge of the interactions among protons and neutrons and their weak coupling to electrons and neutrinos, which is evolving from contemporary elementary particle physics, underlies an accurate description of

---

* This essay was written in October 1967, before the discovery of pulsars. The authors beg the reader's forgiveness for adding several footnotes in proof.

the behavior of nuclei in laboratory experiments. However, such knowledge is even more relevant in applications of nuclear physics to stellar interiors, where it is necessary to extrapolate into regimes where direct measurements have not been possible.

Yet there remain deep questions within elementary particle physics itself which may not be answered or even intelligently asked until a more complete description of the present and past universe becomes available. How, if at all, does the nature of the universe affect the elementary particle physics of the laboratory? Would experiments give exactly the same data if the outside universe were denser, composed of antimatter, younger, or expanding more rapidly? In helping to explore such questions future astrophysics may more than repay its debt to elementary particle physics.

In the following sections we shall explore some of the ways in which elementary particle physics has contributed to astrophysics in the past and speculate upon some of those areas in which developments may couple them in the future.

II. NUCLEAR ASTROPHYSICS

A. The Sun

The understanding of the structure of stars, their evolution and future, is among the most remarkable intellectual achievement of this age of science. Stellar observations, together with the development of nuclear physics, have altered our view of stars from faint points of light to enormous collections of nuclei whose abundances, densities, velocities, and interactions are known with great confidence and whose changing combinations control stellar evolution and finally stellar death. The necessary basis for this remarkable insight has been the experimental and theoretical investigation of nuclear reactions presumed to take place deep in stellar interiors. The conditions under which these reactions are studied in the laboratory are necessarily quite different from those which occur in stars.

In the central parts of the sun the average kinetic energy of a proton is about 10^3 eV; it has a sufficiently intimate and energetic (about 6×10^3 eV) collision with another proton to initiate a nuclear reaction which results in the formation of deuterium (one of the key steps in the nuclear burning which supplies the sun's energy) only once in every 5×10^9 years. A beam of 10^3-eV protons of 1 A intensity can, of course, be produced in the laboratory and introduced into a liquid hydrogen target, but this beam would produce less than one such reaction every trillion years. Thus, the proton fusion reaction cannot be measured in the laboratory when the energies of the inter-acting protons are comparable to those in the sun. The exceedingly small laboratory reaction rate arises because the probability for the occurrence of the interaction is intrinsically small and also because a beam of 10^3-eV protons in cold condensed matter has its kinetic energy greatly reduced after about 10^{-13} sec and is no longer effectively able to initiate the reaction. Not only do higher-energy beams exist longer in cold matter but in most cases nuclear reactions relevant to the stellar interior are much more abundant at higher energies. Therefore, the related laboratory experimental work is con-ducted at hundreds and thousands of times higher energies than those found within a star. The extrapolation to astrophysically significant domains in-vokes nuclear theory which has been developed from the entire body of data acquired in nuclear physics experiments; the field of nuclear astrophysics could hardly have evolved into a fundamental branch of science before sophisticated experiments and theories had become the rule in most branches of nuclear physics.

Except for parts of the earth's surface, more is known about the interior of the sun than any other part of the universe. The radiated solar energy is generated at the sun's center by a chain of nuclear reactions in which four hydrogen atoms are fused into a helium atom plus two neutrinos, which immediately escape from the sun, and some radiation (initially γ rays) which slowly diffuses to the sun's surface. The rate of energy generation, approxi-mately one erg/g-sec, is only about $1/10,000$ of the metabolic rate of the astrophysicist who studies it.

The central temperature of the sun necessary to sustain the nuclear fusion reactions is 1.5×10^7 °K. The central density is about 150 g/cm³, and the significant nuclear reactions are

$$
\begin{align*}
p + p &\rightarrow D + e^+ + \nu \quad \text{(neutrino)}, \\
p + D &\rightarrow {}^3He + \gamma \quad \text{(γ ray)}, \tag{1} \\
{}^3He + {}^3He &\rightarrow {}^4He + p + p.
\end{align*}
$$

Since the final reaction takes place in the presence of $10–10^2$ g/cm³ of ^{4}He, the last step competes with

$$
{}^3He + {}^4He \rightarrow {}^7Be + \gamma,
$$

followed by either

$$
\begin{align*}
{}^7Be + e^- &\rightarrow {}^7Li + \nu, \\
{}^7Li + p &\rightarrow {}^4He + {}^4He, \tag{2}
\end{align*}
$$

or

$$^7\text{Be} + \text{p} \rightarrow {}^8\text{B} + \gamma,$$
$$^8\text{B} \rightarrow {}^8\text{Be} + \text{e}^+ + \nu, \tag{3}$$
$$^8\text{Be} \rightarrow {}^4\text{He} + {}^4\text{He}.$$

Heavier, brighter main sequence (i.e., hydrogen-burning) stars have smaller central densities and slightly higher central temperatures. Hydrogen burning then takes place through the cycle catalyzed by C, N, and O.

$$^{12}\text{C} + \text{p} \rightarrow {}^{13}\text{N} + \gamma,$$
$$^{13}\text{N} \rightarrow {}^{13}\text{C} + \text{e}^+ + \nu,$$
$$^{13}\text{C} + \text{p} \rightarrow {}^{14}\text{N} + \gamma,$$
$$^{14}\text{N} + \text{p} \rightarrow {}^{15}\text{O} + \gamma, \tag{4}$$
$$^{15}\text{O} \rightarrow {}^{15}\text{N} + \text{e}^+ + \nu,$$
$$^{15}\text{N} + \text{p} \rightarrow {}^{12}\text{C} + {}^4\text{He}.$$

The net result is, of course, equivalent to that obtained through chains 1, 2, and 3. The reaction rates are sufficiently well known that the energy production inside of the sun can be predicted with considerable confidence. Chains 1 (mainly) and 2 are expected to account for about 95% of the solar energy generation. Chain 3 and cycle 4 generate $\sim 10^{-2}\%$ and 5%, respectively.

Close to 3% of the sun's energy is radiated as neutrinos. Unlike the radiation which diffuses out slowly, neutrinos produced at the solar center escape without further interaction. About 10^{11} solar neutrinos per second irradiate and pass, day and night, through each square centimeter of the earth. Because they escape from the sun with negligible probability of interaction, solar neutrinos are extremely difficult to detect. An attempted observation by Davis of Brookhaven exploits the reaction

$$\nu + {}^{37}\text{Cl} \rightarrow {}^{37}\text{Ar} + \text{e}^-$$

to detect neutrino fluxes. The ^{37}Ar can be measured by its radioactive decay even when only a few such nuclei are present. Absorption by ^{37}Cl is significant only for relatively high-energy neutrinos. For the expected neutrino spectrum from the sun, 90% of the ^{37}Cl transformations would be induced by that small fraction ($\sim 10^{-5}$) of neutrinos of high energy (~ 7 MeV) which come from the decay of boron in chain 3. Davis[*] is attempting to detect this solar neutrino flux in a tank of 4×10^5 gal of a cleaning fluid (C_2Cl_4) within a deep gold mine to minimize cosmic ray (nonneutrino) backgrounds. A measurement of the solar neutrino flux, even to within a factor of 2, is an extremely sensitive test of stellar models.

[*] At the Washington meeting of the American Physical Society in April 1971, Davis announced the detection of a neutrino flux corresponding to approximately 10^6 neutrinos from solar ^8B per cm²-sec at the earth's surface.

B. Stellar Evolution

In the absence of nuclear reactions the energy for stellar radiation comes from gravitational attraction; about half of the change in gravitational potential energy is radiated away and half contributes to raising the internal temperature of the star. The onset of nuclear reactions temporarily halts this contraction (most of a star's life is spent in the hydrogen-burning phase). When most of the hydrogen has been burnt to helium in the central regions of a star, the central contraction and internal heating resume until the temperature reaches about 10^8 °K, which is hot enough to burn He to carbon and some heavier elements. Subsequent contraction and rise in the central temperature then ignites these elements until, finally, at a temperature approaching 5×10^9 °K, the core of the star becomes mainly iron (with small amounts of Ti, V, Cr, Mn, Co, and Ni). No more heat can be generated by further burning since these iron-group elements have the most tightly bound nuclei. Further contraction and heating generates such an intense flux of photons in equilibrium with the iron-group nuclei that these nuclei are broken down by the photons into their constituent He nuclei plus a few odd neutrons; this reaction which is the opposite of nuclear fusion, uses up energy. Simultaneously, an enormous neutrino–antineutrino flux is expected from the recombination of electrons and positrons produced at these high temperatures, and the maintenance of this neutrino flux also uses up energy very rapidly. The only source for this energy required for iron breakup and neutrino emission is gravitational contraction, but such energy cannot be supplied in a time less than the free-fall time of stellar matter from the outer to the inner regions of the star, and this is not fast enough to give energy to the core at the rate in which it is expended. At this point, the star is no longer in quasi-static equilibrium as the contracting core cannot supply the pressure needed to support the outer layers, and an implosion is expected to result. It is the onset of this instability which is presumed in many cases to give rise to a spectacular supernova explosion. Some of the heavier elements that have been made deep within the star's interior during the course of stellar evolution then are forcefully injected back into interstellar space to be incorporated into newer and yet uncondensed stars. This violent end is not the expected fate of all stars. Stellar contraction can be halted temporarily by nuclear reactions and permanently by electron quantum mechanical "degeneracy pressure," which is caused by the necessarily rapid motion of otherwise identical electrons in any region of very high density. However, such electron pressure can halt the contraction only for stars less massive than about one and one-half suns. The more massive stars, which burn brighter and faster than those of that size, are the nuclear cauldrons that become unstable.

In the evolution of stars classical low- and intermediate-energy nuclear physics play a decisive role in determining not only the time scale but also many details of the behavior of the luminosity and surface temperature with time (in most cases these are essentially the only observables). In theoretical descriptions of stellar evolution, the nuclear physics aspects are generally much better founded than those of hydrodynamics, especially when it is necessary to include convection and the related chemical mixing in and near the core, mass loss, or rotation.

In addition to the various reactions for the burning of hydrogen into helium, stellar evolution theory must draw on that of nuclear physics for detailed cross sections of reactions such as

$$3 \,^{4}\mathrm{He} \rightarrow \,^{12}\mathrm{C} + \gamma,$$
$$^{12}\mathrm{C} + \,^{4}\mathrm{He} \rightarrow \,^{16}\mathrm{O} + \gamma,$$
$$^{16}\mathrm{O} + \,^{4}\mathrm{He} \rightarrow \,^{20}\mathrm{Ne} + \gamma,$$
$$^{20}\mathrm{Ne} + \,^{4}\mathrm{He} \rightarrow \,^{24}\mathrm{Mg} + \gamma,$$

all of which are involved in the helium-burning (red giant) phase, which follows hydrogen burning (main sequence). At temperatures above 10^{8} °K $^{12}\mathrm{C}$ and $^{16}\mathrm{O}$ begin to burn to the intermediate mass nuclei $^{20}\mathrm{Ne}$, $^{24}\mathrm{Mg}$, $^{28}\mathrm{Si}$, and $^{32}\mathrm{S}$. At temperatures above 10^{9} °K, photodisintegration and radiative capture (e.g., $\gamma + \,^{28}\mathrm{Si} \rightleftarrows \,^{24}\mathrm{Mg} + \,^{4}\mathrm{He}$) rates for $^{24}\mathrm{Mg}$, $^{28}\mathrm{Si}$, and $^{32}\mathrm{S}$ play key roles in energy generation by the stellar core. The net result is most probably the formation of $^{56}\mathrm{Ni}$, which decays in two steps to stable $^{56}\mathrm{Fe}$.

In summary it can be stated that the nuclear aspects of stellar evolution are quantitatively understood during the stages involving hydrogen and helium burning in stars. These stages result mainly in the production of $^{12}\mathrm{C}$ and $^{16}\mathrm{O}$ in stellar interiors. Beyond this point very few empirical reaction rates are known with the precision necessary to chart the further course of stellar evolution with sufficient certainty for meaningful comparison with astronomical observations. Many more precise measurements in intermediate-energy nuclear physics are needed to specify the exact conditions of temperature and density which hold during the final quasi-stable stages of stellar evolution. Knowledge of these stages is required to determine the boundary conditions at the beginning of the collapse of the stellar interior which seems to precede the explosive ejection of the external layers. There is clear evidence for an ultimate, violent stage in stellar evolution in the observations on supernovas, and perhaps also in quasi-stellar objects, and the nuclei of certain types of galaxies. This violent stage certainly must be described by ultrahigh-energy physics and the preceding stable stages clearly involve intermediate-energy nuclear physics. Low-energy nuclear physics has played a significant role in astrophysics in the past; in stellar evolution the forthcoming emphasis will be dominated by intermediate and high-energy nuclear physics.

C. Origin of the Elements

If the nuclear products of stellar evolution were always to remain buried within the stellar interior, only their contribution to the star's energy budget would be of significance to the astronomer. However, there is considerable evidence from both the isotopic abundances of the matter in our galaxy and visible supernova explosions that nuclear ashes from stellar evolution are a significant source and very possibly, the only source of all of the elements in our universe heavier than helium. In the expanding universe the fusion of helium marks the end of nucleosynthesis unless the matter is ultimately recompressed to high densities and temperatures, which is a natural consequence of star formation. Since there are no nuclei of mass 5 (or 8), the most suggestive path for the nuclear fusion of helium proceeds by three-body combination into ^{12}C which requires both high density so that there is an appreciable probability for finding three helium nuclei very close together, and high temperature so that their Coulomb repulsion does not prevent this association.

The amounts of elements heavier than helium which must be accounted for in a description of the history of our universe are not large. In the sun, for example, all elements heavier than iron appear to constitute only about one part per million (by mass) of the total; the elements C, N, O, and Ne comprise about 2×10^{-2}, and all other elements constitute much less. Therefore, the stellar processes which may account for them need not be the significant ones for energy generation but may be relatively rare and ancillary to the main events in stellar evolution.

It is likely that various groups of elements were produced in quite different ways and at very different epochs in the history of a star. Thus, the observed abundances of the elements around iron (Ti, V, Cr, Mn, Fe, Co, and Ni) are remarkably characteristic of what would be expected for matter which had been kept at a temperature $T \sim 4 \times 10^9$ °K and high density $\rho \simeq 10^6$ g/cm^3 for about 10 hr. This is ample time for the reclustering of protons and neutrons into nuclei according to thermal equilibrium but it is only long enough for electron capture to convert 2 protons out of every 28 into neutrons (e.g., $^{56}_{28}Ni + 2e^- \rightarrow {}^{56}_{26}Fe + 2\nu$). This is typical of the conditions and time scale expected for the central parts of a star just before it becomes a supernova.

The detailed prediction of the abundances of the other elements and the mechanism and environment involved in their production depends upon laboratory measurements of a great variety of nuclear interactions and decays. For example, a newer star (such as our sun) which has condensed from interstellar gas already has some ^{12}C from earlier nucleosynthesis in stars. In the CNO-catalyzed reaction for hydrogen burning, ^{13}C is one of the transient nuclei in the cycle; usually the ^{13}C is destroyed by protons in the reaction which leads to ^{14}N. However, some will survive into the hotter helium-burning phase, where ^{13}C will be attacked by helium nuclei, leading to the production of free neutrons:

$$^{13}C + {}^{4}He \rightarrow n + {}^{16}O.$$

The neutrons will be captured mainly by whatever heavier elements are present and add to their atomic masses. If the new nucleus is stable, it can absorb still another neutron. If it is unstable against the emission of an electron–neutrino pair (β decay), which converts the neutron within the nucleus to a proton, there is sufficient time at this stage of stellar evolution for the decay to take place before the next neutron is absorbed. Calculated abundances of the isotopes are in reasonable agreement with such a model.

The most significant direct observational evidence that weak neutron fluxes actually exist in evolving stars consists in the observation of spectral lines of technetium in the atmospheres of some stars. Technetium is not found on earth because it has no long-lived isotopes. The technetium isotope expected to be produced by successive captures from a weak neutron flux is ^{99}Tc whose half-life is only 2×10^{5} years, much less than the age of the stars in which it is seen. Therefore, its observation strongly suggests that ^{99}Tc is presently being produced by neutron fluxes within the star and is brought to the surface by convection which causes some mixing of core and photospheric nuclei.

Not all the observed heavier elements can be built up in this way. Some natural nuclear isotopes do not have the appropriate stable neighbor less massive by one neutron or proton to serve as the needed neutron absorber. Further, the building of heavier elements by neutron capture at a slow rate terminates at ^{209}Bi. For heavier elements α-particle decay (emissions of an He nucleus) as well as β decay can occur between neutron captures, and the mass of the unstable nucleus decreases by about that of the emitted helium nucleus. The elements between mass 210 and 238 found on earth are presumably not produced by weak neutron fluxes. However, if the neutron flux is so intense that successive neutron captures are sufficiently rapid as to preclude (usually) time for the β decay and α decay of unstable isotopes, then the course of nucleus building is altered in such a way as to predict the generation of all of the observed neutron-rich isotopes which cannot be done with slow absorption of neutrons. Elements can be built of ever-increasing mass until slow neutron-induced fission dominates. The required intense neutron fluxes are expected in the dense, hot presupernova core. At temperatures in excess of 10^9 °K, neutrons are copiously ejected from nuclei by photons of a few million electron volts, the neutron flux can exceed an astonishing 10^{34}/cm²-sec. Remarkably, there is evidence from the decay time of the light emission of supernovas that two unstable fission isotopes of californium, $^{252}Cf_{2.2\ yr}$ and $^{254}Cf_{55\ days}$, are present in great abundance. These isotopes have been produced on earth with significant abundance only in hydrogen bomb explosions, where ^{238}U has been subject to a brief but enormous nuclear flux!

Classical nuclear physics not only supplies a basis for the story of the origin of the elements and the evolution of stars but also contributes significant clues to the time scale for the development of our solar system and the elements from which it is made. Both natural isotopes of uranium, ^{235}U (mean lifetime = 1.0 billion years) and ^{238}U (mean lifetime = 6.5 billion years), are unstable against α-particle decay. They both give birth to unstable offspring, whose daughters in turn inherit this instability, and no stable member of the family tree is reached until ^{207}Pb terminates the chain begun by ^{235}U and ^{206}Pb terminates that initiated by ^{238}U. A similar decay chain initiated by ^{232}Th (mean lifetime, 20 billion years) ends at ^{208}Pb. From measurements of present meteorite abundances of these uranium, lead, and thorium isotopes, in combination with laboratory data on how these ratios vary with time, the age of formation of meteorites can be specified as being close to 4.6 billion years.* A similar age comes from measurements of the helium content if it is assumed that all of it is due to α-particle decay from the uranium and thorium chains. A variety of other isotope ratios also contribute support to this estimate. It may be compared comfortably with quite independent analyses of the age of the sun, which estimate this to be approximately 5 billion years.

Present isotope abundances can give both the age of the meteorites and the isotope abundances when the meteorites become sufficiently solid that different chemical elements could not diffuse away from each other. These primeval abundances of unstable isotopes can in turn be used to estimate the age of the elements for different hypotheses for their production. Because various unstable isotopes of uranium, thorium, and other elements have different half-lives, their ratios vary; but if these ratios are known at any epoch, they can be extrapolated back in time until they hopefully correspond to those of some model for their production. If the synthesis of these isotopes is assumed to be the building of nuclei by rapid absorption of neutrons from intense neutron fluxes, the measured abundances imply that

* Lunar soil brought back by the Apollo astronauts also has a model age equal to 4.6 billion years. There is now little doubt that the solar system formed in the interval 4.7–4.6 billion years ago.

the elements in the solar system were made in a series of such events extending over the period from 5 to 12 billion years ago, somewhat before the sun was born and after the universe had begun its own evolution. An exciting inference from the measurement of their ^{129}Xe content is that meteorites may have been formed about 100–200 million years after the end of element production. These are such satisfactory results that they give some solid support to the whole picture of nucleosynthesis.

There has already been an astonishing richness of applications of classical nuclear physics to the theory of stellar evolution and the history of the universe. There still remain new areas in which laboratory nuclear physics will contribute to the understanding of astronomical observations. For example, reasons for the abundance of deuterium relative to hydrogen on the earth will continue to be explored. There seems to be evidence here that regions of the surfaces of some stars may behave like giant particle accelerators, a phenomenon already evident on a lesser scale in the sun: some stars (peculiar A stars) which possess large magnetic fields have in their atmospheres a great abundance of ^{3}He which far exceeds that of ^{4}He. The observation of violent galactic explosions suggests other possible sources for nucleosynthesis. Some quasi-stellar objects, if cosmologically distant, are very ancient and yet seem to have elemental abundances typical of newer stars. Whatever its future, the past and present of nuclear astrophysics has seen triumphs seldom equaled in science.

D. Astrophysical Neutrinos

The neutrino has no charge; like the photon it has no rest mass and its interaction with matter is so weak that a 10^6-eV neutrino can pass through a billion earths as easily as light can pass through a pane of glass; 10^{15} neutrinos pass through our body each second with no effect. Yet neutrinos certainly do exist. They have been detected and they play a role in stellar evolution which at times is of much greater importance than the emission of light. It is the spectacular weakness of the interaction of neutrinos that may enable them to carry away more energy than that radiated in photons, which are always created much more copiously deep within the star. The photons are quickly reabsorbed and reemitted until the electromagnetic radiation from a star finally becomes thermal emission from its relatively cool surface. A neutrino, however, once emitted in the hot stellar core, will almost always escape without further interaction.

The significant property of neutrinos in stellar evolution is the conjectured existence of a weak direct electron–neutrino pair coupling which would give rise to such emission. Such an interaction would imply that an accelerated electron could radiate neutrino–antineutrino pairs in a manner quite analogous to that in which it radiates photons, except with greatly reduced probability. The evidence for such an interaction is not yet compelling; it is suggested by the observed symmetries among pairs of weakly interacting particles and it is contained in almost all models which attempt to describe weak interactions. Classical low-energy nuclear physics is concerned only with neutrinos emitted when a nucleus emits an electron or positron, or captures one of its own atomic electrons. Such data are insufficient to relate to the presumed existence of the electron–neutrino pair coupling. Most of our knowledge about neutrinos is derived from the use of very high-energy accelerators, with which neutrino beams exceeding 10^9 eV in energy have been produced. With these accelerators μ mesons were created that emitted neutrino pairs when they decayed into electrons, and unstable particles were produced that included neutrinos among their decay products. Also, sufficient fluxes of negatively charged μ mesons were supplied so that their neutrino-producing capture by nuclei could be studied in detail. Such essentially high-energy experiments have suggested but have not yet proved a universal kind of neutrino coupling which would include the weak, direct electron—neutrino pair interaction. Because neutrino interactions become stronger with increasing energy while confusing background events become rarer, high-energy experiments involving the scattering of neutrino beams by electrons will probably be the definitive test of such a coupling.

A variety of mechanisms for neutrino-pair emission by electrons has been proposed which would have relevance in different regimes of stellar interior density and temperature. Hydrogen-burning stars would be negligibly affected. However, after helium burning, when the central core temperature exceeds 10^8 °K, neutrino-pair emission would exceed photon radiation and stellar evolution would be greatly accelerated. Finally in the last hours of the noncataclysmic life of a star in which oxygen burns to silicon and silicon to nickel, the neutrino-pair emission would grow so intense that it would be not only hundreds of millions of times stronger than the light emission from the star but also a thousand times stronger than the light emission from our entire galaxy.

Indirect evidence for a significant role in stellar evolution for neutrino-pair emission comes from a consideration of the genesis of white dwarf stars—small, extraordinarily dense, rather hot stars—which, having finished nuclear burning, are slowly cooling. They are sufficiently light (usually somewhat less than a solar mass) that the degeneracy pressure of their electrons is sufficient to balance the contracting pull of gravity. Although white dwarfs with central temperatures of 1 or 2×10^7 °K are common, the much hotter stars which should have been their immediate ancestors (ultraviolet dwarfs) are not. The implication is that ultraviolet dwarfs are cooling so rapidly that it is rare to see one until it has become a canonical white dwarf—this cooling rate is too fast to be explained by light emission but quite compatible with neutrino-pair emission. The predicted energy for such pairs is about 5000 eV. It is characteristic that the definitive experiment to establish the existence of such neutrinos will probably use artificial beams of almost a million times more energy per neutrino.

E. Very-High-Energy Interactions

In recent years it has become abundantly clear that natural high-energy accelerators exist in a variety of astronomical objects. Analyses of the radio waves and visible light from supernova remnants point to the existence of electrons in such objects whose energies exceed 10^{12} eV; the shape of their recently discovered x-ray spectra indicates electron energies greater than 10^{14} eV. The observed electromagnetic radiation seem to originate in the acceleration of such electrons in magnetic fields. Radiation from radio-galaxies and quasi-stellar objects also suggests the presence of similar magnetic fields and very high-energy electrons. The energies of protons and nuclei in cosmic rays can exceed 10^{19} eV; they may originate mainly in supernova explosions or, perhaps less likely, in galactic explosions or quasi-stellar objects. Even the relatively quiet sun occasionally produces a burst of protons whose energy approaches 10^9 eV.

The combination of magnetic fields and ultrarelativistic particles seems to be very generally associated with large-scale violent events in the universe. They may well be a natural consequence of any very turbulent long-lived plasma. Certainly theory suggests that there are very natural mechanisms in such regions for the amplification of magnetic fields and for the continued acceleration of some particles by electric fields until they escape. If this is the way in which nature almost indifferently generates ultrahigh-energy particles, then the role of elementary particle physics consists only of examining their subsequent history. There already appears to be a rather rich field of applications of data and theory acquired from accelerator experiments to astrophysics.

The ultrahigh-energy electrons in cosmic sources may be accelerated directly or they may be the ultimate particles of very rapid decay chains in which collisions of high-energy protons produce π mesons or K mesons which decay to μ mesons, which in turn decay to electrons. The high-energy γ rays from the decay of neutral π mesons which would also be expected from the secondary production process still generally fall below experimental upper limits for extragalactic sources so that observations do not yet suggest either mechanism. However, the observational upper bound for high-energy (greater than 5×10^{12} eV) γ rays from the Crab Nebula, a supernova remnant, is a strong argument against proton collisions being the ultimate source of the Crab Nebula's high-energy electrons. Measurements of the positron-to-electron ratio (about 25%) in cosmic rays and the shape of the electron energy spectrum are also not compatible with a model in which the cosmic electrons are produced by collisions of the nuclear component. A similar conclusion follows from the observed spectrum of radio emission by high-energy galactic electrons.

High-energy electrons constitute a small percentage of the cosmic ray flux incident upon the earth. However, at least for this component, the flux within the galaxy seems to be much greater than that in intergalactic space. The most compelling interpretation of the isotropic millimeter radiation flux incident on the earth is that about 10^3 photons of 10^{-4} to 10^{-3} eV uniformly fill each cubic centimeter of the present universe—probably thermal equilibrium radiation left as a remnant of the initial hot dense universe. A high-energy electron can scatter inelastically from one of these "soft" photons and increase the energy of the scattered photon. An electron whose energy exceeds 10^9 eV can, in this way, make 10^3-eV (x-ray) photons. The measurement from satellites and balloons of the isotropic x-ray background puts a limit to the density of high-energy electrons which can exist in the universal microwave environment—and this limit is ten to a hundred times less than the terrestrially measured cosmic ray electron flux. In these ways quantum electrodynamics, the model theory of elementary particle physics, enters decisively into the analysis which suggests that the electron component of cosmic rays is not universal and is probably directly produced and largely contained within our own galaxy.

Much more is known about the nuclear component of cosmic rays but perhaps less can be definitively concluded. Most of the studies of cosmic-ray phenomena are concerned with the interaction with galactic and solar system magnetic fields, but in some cases a knowledge of the details of the interaction of cosmic-ray nuclei with other particles is very relevant. After their acceleration to high energies, cosmic-ray nucleons may collide with interstellar gas, starlight, or even the universal microwave radiation. Such interactions can affect the chemical composition and the energy spectrum of cosmic rays seen on earth. The combination of cosmic-ray measurements and laboratory high-energy physics contributes to knowledge of the interstellar medium and the history of cosmic rays between their creation and detection.

The relative abundances of the elements in cosmic rays differ in a rather striking way from those observed in various objects in the universe. The heavier nuclei are relatively more abundant in cosmic rays, which may reflect a more efficient acceleration mechanism than for protons or possibly a difference in their preacceleration abundances. However, the most striking feature of the composition of cosmic rays is the fraction of the light elements Li, Be, and B which compose about 10^{-3} of incident cosmic rays as compared with about 10^{-8} of the rest of the universe. The origin of the relatively enormous abundance of light elements in the incident cosmic ray flux is presumably occasional fragmentation (spallation) of the much more plentiful heavier elements when they collide with interstellar matter or with some denser gas associated with their origin. From measured spallation cross sections and abundances, it appears that the nuclear component of cosmic rays has passed through around 3–6 g/cm² of hydrogen before reaching the earth. This corresponds to a lifetime of about 300 million years in an interstellar medium consisting of 10^{-2} atoms/cm³, typical of the average density in the combined galactic disk and surrounding halo. A detailed history of cosmic rays awaits more definitive data on how the relative abundances vary with energy and on the amounts of deuterium and ³He in the incident flux.

The flux of cosmic rays in relatively empty intergalactic space (99.9% of the universe) is not revealed as easily as the electron component. Protons above 10^{17} eV can collide with starlight photons to produce π^0 mesons, which decay into γ rays whose energy can exceed 10^{16} eV. Unfortunately these high-energy photons can in turn have collisions with the plentiful low-energy universal microwave radiation which absorbs the high-energy photons and creates electron–positron pairs. Thus, the γ rays which could give proof for very high-energy cosmic rays between the galaxies are probably not detectable from the earth. A possibility of definitely identifying some universal rather than galactic component of the cosmic rays consists of observing the angular distribution, the energy dependence, and hopefully even the composition of cosmic rays whose energy approaches 10^{20} eV. Such particles are not confined by the galactic magnetic field, but they will lose much of their energy to the microwave background after going only 10 million light years. Sadly, such exciting particles are about the rarest in the universe.

III. MATTER IN THE UNIVERSE

A. Dense Matter

Among the rarest forms of matter in the universe are those kinds with which we are most familiar in our environment, solids and liquids whose density is typically a few grams per cubic centimeter and cool gases about a thousandth as dense. Perhaps 10^{-10} of all matter is in such a form—the rest comprises the interior of stars, the exceedingly thin gases between the stars, and whatever sorts of particles and radiation pervade the enormous spaces between the galaxies. Even in the laboratory the accessible regimes of density and temperature for macroscopic amounts of matter are much more limited than those which may be found in other parts of the universe.

Experimentally realizable temperatures generally fall into the domain

$$10^{-6} \ ^\circ K \gtrsim T \gtrsim 10^5 \ ^\circ K,$$

corresponding to energies of 10^{-10}–10 eV per particle. (In nuclear explosions temperatures can reach 10^8 °K but only for tens of microseconds.)

The sun, a typical hydrogen-burning (main-sequence) star, has a central temperature of 10^7 °K (10^3 eV per particle). Stars whose evolution has proceeded somewhat further are much hotter. At the end of nuclear burning, a star about to become a supernova is thought to have a central temperature of about 6×10^9 °K (10^6 eV). (The equilibrium density of photons, actually γ rays, is then 300 tons/ft³; electron–positron pairs are about twice as dense. In some proposals for the history of a supernova, temperatures are predicted to rise to 10^{12} °K and the photon density rises to 10^{11} tons/ft³.) Clearly an understanding of the processes which take place in typical stellar interiors depends upon the study of the interactions of nuclei, x rays, and electrons whose energies vary from near 10^3 to a few times 10^6 eV. Since such energies per particle are not even approached in laboratory experiments with hot bulk matter, it has been necessary to use particle beam accelerators to bring small numbers of nuclear particles to relevant energies and to explore the reactions which take place when they have collisions.

Densities achievable in laboratory experiments are severely restricted to the range from a good laboratory vacuum to the limit allowed by the mechanical strength of high-pressure equipment or shock tubes:

$$10^{-18} \ \text{g/cm}^3 \gtrsim \rho(\text{lab}) \gtrsim 50 \ \text{g/cm}^3.$$

Astrophysically relevant densities range from a presumed 10^{-29} g/cm³ in intergalactic space to the very high densities of stellar interiors: the center of the sun has a density 10^2 g/cm³, for a more highly evolved red giant it it may increase to 10^5 g/cm³, white dwarfs have central densities up to 10^7 g/cm³. Neutron stars which lie somewhere between an attractive theoretical speculation (ah, so!) and a major constituent of the universe certainly *can* exist as a stable permanent form of matter; if they are formed at all, their density considerably exceeds 10^{14} g/cm³.

Associated with high densities there is always a high average energy per particle. Forty years ago Pauli showed that many kinds of elementary particles (namely, electrons, protons, or neutrons) could not be found close together unless they had a high relative momentum (exclusion principle). Therefore, a very dense group of electrons or nucleons, of necessity, would have a great deal of kinetic energy (Fermi energy) even if there were no interactions among them and the temperature approached absolute zero. At 10^8 g/cm^3 the Fermi energy of the electrons is near 1 MeV so that even at $T = 0$ such densely packed electrons have relativistic energies. The average Fermi energy (E_F) of neutrons in a possible neutron star is at least as great as that in nuclear matter; $E_F \sim 45 \times 10^6$ eV. Possibly the neutron Fermi energy would be sufficiently high that the major constituents of a neutron star should be mesons and strange particles, as well as ordinary nucleons; a description of the state of matter in a neutron star involves a knowledge of elementary particle interactions up to some hundreds of millions of electron volts. Thus, neutron star matter is high-energy matter even in a cool star. Such matter may have remarkably exciting properties: Below a temperature of a few billion degrees it may be a superfluid with properties analogous to those of He II (which exhibits rather marvelous superfluid properties only at temperatures below 3 °K). Unfortunately not enough is yet known experimentally and theoretically concerning the interactions of groups of high-energy elementary particles to offer a reliable description of neutron star matter. Nor is it yet clear to what extent the nature of such matter is an academic question. There is no evidence that even a single neutron star really exists although neutron stars have been suggested as the endpoint in the evolution of some stars up to 50% heavier than the sun, as the stellar remnant in supernova explosions, and even as the main condensed constituent in quasi-stellar objects. However, the possibility of recognizing a neutron star depends intimately upon a satisfactory knowledge of interactions among elementary particles at high energies and the observable consequences which might result. For many stars the final stages of stellar evolution certainly do involve such enormous densities and temperatures that individual particle energies enter the domain of contemporary elementary particle research.

[*Tempus fugit!* In the short time since the completion of this chapter (October 1967) pulsars have been observed as a comparatively common star in the galaxy. An extremely strong circumstantial case can be made for the pulsar—as a rapidly rotating neutron star. Further, it appears that the pulsar is a very efficient accelerator of cosmic rays. All of the 10^{38} ergs sec of cosmic ray particles continually injected into the Crab Nebula probably are accelerated by the slowing pulsar within it. The number of pulsars and very plausible scenarios for their early history support a speculation that they may be the source of almost all cosmic rays in our galaxy. More knowledge of nuclear interactions at high energies (>300 MeV) is urgently needed to understand the structure of neutron stars.]

B. Exotic Matter

Although there are over a hundred "elementary" particles, very few are stable. Protons, neutrons in nuclei, electrons, and their (isolated) anti-particles are not unstable. All other known particles (except photons and neutrinos) decay in 1 μsec or less. However, symmetries among the observed groupings of elementary particles have encouraged the speculation that all of the hundreds of known particles are describable as combinations of perhaps three fundamental particles. In certain attractive forms of such theories these particles (quarks) have electric charges $e/3$, $e/3$, and $-2e/3$, where e is the charge on an electron. Although quarks are hypothesized to be much heavier than the known elementary particles (to account for their being undetected thus far in accelerator experiments), they cannot decay into the familiar integrally charged particles if electric charge is conserved. They, of course, may be able to decay into whichever quark is lightest since they differ by zero or unit charge among themselves, and there are known particles which could be emitted in the transition. Yet at least one quark, the lightest one, would be stable. For those theories in which fundamental triplets have integral charges it is theoretically an open question whether (super)selection rules forbid their decay. Thus, if the attractive speculation about the existence of fundamental triplets is ever confirmed, the possibility of entirely new stable forms of matter in the universe immediately takes on real significance. Such matter, being composed of positively and negatively charged quarks or perhaps quarks and electrons, would actually be only metastable since in appropriate states three quarks could combine to make a single, much lighter nucleon plus an enormous amount of emitted energy. This combination may be negligibly slow in a "normal" form of quark matter (just as all matter below aluminium in the periodic table is unstable against nuclear fusion, but at normal densities and temperatures the fusion reactions essentially never occur). Unlike the case of antimatter, isolated quark matter could give many observable signs of its existence: unusual optical and ultraviolet spectra or perhaps violent energy release associated with its intrinsic instability. (Quark matter has already been suggested as a conceivable energy source of quasars!)

One of the most attractive speculations of theoretical physics has been the hypothesis that free magnetic poles exist (and thereby automatically constrain all charges to be multiples of one electron charge). If free magnetic poles exist, conservation of magnetic charge would imply stability, and the prospect of new kinds of matter involving very strongly interacting magnetic poles would be an engaging possibility. If even a single "stable" exotic particle, a quark, a magnetic pole, or any nondecaying particle, were to be discovered in nature or in an accelerator experiment, the astrophysically urgent question would become "Where are all the others?"

Although our local environment is obviously not symmetric between matter and antimatter, there remains at least an idealistic and esthetically satisfying conjecture that exactly half of the entire universe may be antimatter. The detection of isolated antimatter in stars or gas would be enormously difficult, as its visible radiation would be identical to that from matter. It is only from weak nuclear decays which result in neutrino radiation that there is a theoretically observable difference between stars and antistars. Hydrogen-burning stars, like the sun, are effectively supplied with energy by turning hydrogen atoms into helium and neutrinos (ν):

$$2e + 4p \rightarrow {}^4He + 2\nu.$$

There is no appreciable antineutrino ($\bar{\nu}$) emission from a main sequence (hydrogen-burning) star. Contrary to popular wisdom, "there is nothing but ν under the sun."

Main sequence antistars would indeed emit $\bar{\nu}$ but the flux could be only 10^{-8}, the ν flux from the sun, even if half of all other stars were antistars. The most attractive possibility for detecting antimatter lies in observing the characteristic (~ 70 MeV) γ radiation from neutral meson (π^0) secondaries produced in nucleon–antinucleon annihilation when matter and antimatter come into contact. The upper limit for the γ-ray flux obtained from satellite-borne counters is about $3 \times 10^{-4}/\text{cm}^2\text{-sec-sterrad}$, already sufficient to rule out appreciable numbers of small (~ 0.1 light years) gas clouds of antihydrogen remaining in our galaxy, and less than the amount expected from matter–antimatter recombination in a version of steady-state theory with uniform and equal proton–antiproton creation. The expected background of energetic γ rays from the interaction of cosmic rays with known interstellar gas is about $10^{-5}/\text{cm}^2\text{-sec-sterrad}$; γ-ray telescopes with good angular resolution could detect weak sources in this background. Data from satellite-borne γ-ray detectors with sufficient angular resolution should ultimately support a more detailed exploration of regions of the universe where matter may be adjacent to antimatter.

C. Cosmology

In the canonical models of the universe which assume homogeneity and isotropy, and in most other cosmologies the universe itself has expanded from a singular region of initially infinite density. Thus, in the beginning, the primeval particles had enormous energies. The details of the way in which these particles interacted could have observable effects in our present universe. One can indeed trace the development of such matter as its great density and kinetic energy decreased.

Little has been said about the first stage when the density exceeded that of nuclear matter, but it would appear that if we apply present physical theory to such a singular state of the universe, this very high-density region has a composition dictated by thermal equilibrium considerations alone. However, as the density decreased, those neutrons present either decayed into protons, electrons, and (anti)neutrinos or were captured by protons to form deuterium, which in turn continued to interact until it was trapped in tightly bound and very stable helium nuclei, which remains today as the souvenir of those times. Here the measured cross sections for proton capture and for interactions of deuterium, tritium, and ^{3}He are needed to relate the unknown, but in principle still measurable, primeval helium abundance to the initial conditions in the universe.

When, in the very dense, energetic primeval universe the averaged density of matter exceeded its present value by more than about 10^{27}, neutrino interactions may have played a particularly significant role. Their very weak interaction implies that they were the only particles in those epochs which could travel long distances between interactions. Therefore, they were the possible means of contact between large regions of inhomogeneity or anisotropy in the early universe, and would probably play a crucial role in smoothing out any irregularities. However, the significant neutrino interaction in the early genesis of our universe is that very same weak electron–neutrino pair coupling which has not yet been confirmed in the laboratory.

Unfortunately a vital ingredient of any cosmology still remains almost unknown—the amount and nature of the matter which comprises the present universe. Except for luminous matter in and near stars and clouds of hydrogen within galaxies, it has proved very difficult to detect the constituents of the universe—perhaps there are no others. The techniques of high-energy physics are beginning to be used to explore the cosmic densities of high-energy neutrinos and of x rays and γ rays. Their densities are clearly not present in amounts sufficient to affect the dynamics of the expansion of the universe.

If the existence of quarks were to be demonstrated, a new aspect of observational cosmology would become significant. In the dense, singular stage of an expanding universe, the temperature at one time must have been sufficiently high to maintain quarks and antiquarks in thermodynamic equilibrium with other components of the universe. As the universe expanded and cooled, most of these free quarks would have disappeared either through quark–antiquark recombination or the union of quark triplets to form nucleons. However, as the density of the universe decreased, it would have become increasingly rare for a quark or an antiquark to find appropriate partners for its demise, and some of the quarks would have survived. Since the lightest members of the quark and antiquark triplets are expected to be stable, a remnant of the primeval quarks should remain today. If there had been no further quark interactions after the initial "big bang," the ratio of quarks to nucleons at present would be expected to be of the order 10^{-10} (but with great uncertainties, partly due to unknown quark properties but mainly because of unknown initial anisotropy). This is enormously greater than the upper-bound ratio already claimed for the quark content of a variety of materials including meteorites (10^{-17}) and seawater (10^{-26}). Even if quarks exist, the association of these bounds with primeval conditions is obscured at present both by the complicated chemistry of quark-bearing molecules (they always have a net charge), which may lead to their depletion in certain materials, and also by our lack of knowledge of the detailed history of local matter which may have passed through phases of stellar evolution and galactic explosions that would reduce its quark content.

The possible implications of quarks for astrophysics outlined in these last two sections, while not to be taken very seriously (at this time anyway), are at least typical of the way in which certain developments in elementary particle physics could have remarkable astrophysical interest.

IV. PROSPECTS

Because of the enormous growth in character and sensitivity of astronomical observations, the astrophysical role of elementary particle physics and especially interactions at very high energies is likely to continue to grow.

Classical astronomy is concerned with visible stars and nebulae, with objects which radiate photons of a few electron volts. Almost all of this radiation originates on stellar surfaces from the acceleration of electrons whose energies are characteristically also a few electron volts—the domain of elementary particle research over forty years ago. The past decade has seen an enormous expansion in the spectrum of radiation with which the universe is observed. The "new look" of the universe was first seen by radio telescopes where the typical photon has only 10^{-5} the energy of a visible one, but the processes which give rise to these radiophotons involve electrons many billions of times more energetic than those usually encountered in stellar surfaces. Satellite and balloon-borne detectors which rise above the screening atmosphere of the earth have extended our view of the universe to include photon energies of thousands of electron volts (x rays). Continued attempts are being made to detect sources which radiate photons between 10^8 and 10^{16} eV, sources which could only have high particle energy sources. The origin of the observed x radiation seems to be of two sorts: some sources are 10^5-eV electrons colliding with ions while others are ~10^{13}-eV electrons accelerating in magnetic fields. Thus, optical (and infrared) photons almost alone seem to be low-energy phenomena. The expansion of the detectable photon spectrum to lower as well as to higher energies has involved astrophysics decisively in the regime of very high-energy particles—the domain of contemporary elementary particle physics. Finally many remarkable optical objects, the Crab Nebula supernova remnant, quasi-stellar sources, and some radiogalaxies, are seen to emit their visible light by means of enormously energetic electrons.

Observations beyond the visible will become increasingly sophisticated; improved γ ray telescopes are to be flown, and x ray telescopes of greater sensitivity will discover more sources and some information on x ray polarization may soon become available. Radiotelescope interferometry has already exceeded the angular resolution of the best optical telescope observations despite a wavelength ratio of almost 1 million. Because they are not particularly degraded by the atmospheric irregularities which limit optical telescopes, radio telescopes can probably give a picture of the universe on a much finer scale than any other device, and this picture will be one of the universe's high-energy activity. The expected proliferation of these sorts of astronomical data will certainly continue to require detailed and available knowledge of varieties of elementary particle interactions for their interpretation. It is much less clear in what way, if at all, astrophysics may be involved in future developments in elementary particle physics.

Some kinds of information still to be verified through particle physics have been touched upon in previous sections. The conjectured weak electron–neutrino pair interaction must be tested in the laboratory. Accurate fragmentation cross sections for very high-energy collisions of medium and heavy nuclei will be needed to accompany measurements of the energy dependence of the abundance ratios in cosmic rays. It will be necessary to know nuclear reactions more accurately in order to improve stellar evolution models. Expected very high-energy neutrino production by cosmic ray secondaries in the earth's atmosphere will have to be determined accurately enough so that underground experiments which are presently detecting such neutrinos can discover if there is an extraterrestrial component. The possible existence of exotic particles and matter, such as quarks or magnetic poles, will serve to stimulate much theory and some experiment until such phenomena are found in the laboratory, or developments necessitate change in existing speculations.

Enough is already known about elementary particle interactions for most evident astrophysical applications. The search for regions of the universe in which matter and antimatter may mix by γ-ray astronomy is not noticeably limited by uncertainties in the nucleon–antinucleon interaction. In nature particle energies can greatly exceed those available in the laboratory, but extrapolations to very high energies are probably qualitatively reliable, and not much more is presently needed. There is no compelling reason to suspect that the production of ultrahigh-energy particles involves any significant elementary particle interactions beyond those which have already been discovered. However, to expect no surprises may be vanity on the part of the astrophysicist. Certainly there are high-energy astrophysical phenomena which are not yet understood: exploding galaxies, the fate of heavy dense imploding stars, and the nature of quasi-stellar objects. If, for example, quasi-stellar objects are indeed billions of light years away, the rapidly fluctuating light from them is so intense that ultrahigh-energy electrons can survive only for a matter of seconds before losing their energy. This puts an enormous and perhaps unbearable burden on models which try to explain the coherent production of these electrons over fairly large volumes. Even in the Crab Nebula it appears that ultrahigh-energy electrons continue to be produced almost 1000 years after the supernova explosion. It may well be that phenomena involved in the acceleration of very high-energy elementary particles are richer than suspected (*sic!*).

Even more provocative is the contingency or the illusion that a deeper understanding of elementary particles and cosmology will finally reveal some connections between the two, namely, that what is revealed in the laboratory depends in some way on the state of the rest of the universe or how it has developed. There are already some tantalizing experimental results and real and suggested breakdowns in the symmetry of fundamental laws between past and future and between particles and antiparticles which yet may be related conceivably to this same lack of symmetry in our part of the universe. "Particle–antiparticle symmetry" means that the measureable interactions among any group of particles is identical to that which would obtain if all the interacting particles were replaced by their antiparticles. It is an attractive hypothesis but it is known to fail (very slightly!) for the decay of a neutral K meson into a pair of π mesons. "Time-reversal invariance" means that the laws we have derived to describe the interactions among fundamental particles would have been the same if, instead of looking at the universe, we had looked at a film of it run backwards in time. Remarkably this invariance has been found to fail only for the decay of the neutral K meson (and here only slightly). The measurement of K meson decay is the most sensitive elementary particle experiment which would detect a breakdown in the two invariances and they may well fail to a very small extent in all interactions. Our environment does not exhibit these symmetries: it is composed of matter, not antimatter, and the universe expands rather than contracts. Attempts to link the invariance breakdown to the lack of symmetry of the local outside universe have not been successful, but there remains the enormously exciting possibility that some elementary particle phenomenon or theory will suggest a link between the two. Further experiments in the physics of elementary particles at very high energies will test microcausality more stringently (the requirement that signal propagation never exceed the speed of light), Lorentz invariance (relativity), some conventional concepts of space–time, and a host of conservation laws. Many of these enter into the cosmology of the universe, and a fundamental lesson we have learned from nature is to expect to be surprised.

There is perhaps an overly optimistic view of the future of physics which embodies the idea that all experimentally determinable pure numbers will somehow be derivable. The ultimate theory should then give values for the famous fine structure constant $e^2/\hbar c$, the ratio of the weak coupling constant to $h^3 M_p{}^{-2} c^{-1}$ (M_p is the proton mass) etc., and significantly also the ratio of the universal gravitational constant G to $e^2 M_p{}^{-2}$. It is suggestive that any theory which would determine this last ratio should show some connection between elementary particle physics, in which e and M_p enter so vitally, and cosmology, which at present does not comprehend elementary particles at all but is an outgrowth of gravitational theory. That there is an approximate coincidence between $e^2/GM_p{}^2$ and the square root of the number of observable nucleons in the universe has encouraged this sort of speculation: either the state of the present universe in some way determines the ratio $e^2/GM_p{}^2$ or it was somehow fixed in its present value during an early stage of evolution. In either case the extrapolation of present physical theory to very early epochs of the universe may be misleading. There may not be any deep relationship between elementary particles and the structure and evolution of the universe, but if there is, its discovery will be necessary for the ultimate understanding of both.

The marriage between elementary particle physics and astrophysics is still fairly new. What will be born from this continued intimacy, while not foreseeable, is likely to be lively, entertaining, and perhaps even beautiful.

GENERAL REFERENCES

In this rather general nontechnical review we have omitted detailed references which, to avoid injustices, would have had to be quasi-infinite. Some surveys of topics mentioned which contain abundant references are:

Nuclear Astrophysics:
    Burbridge, E. M., Burbridge, G. R., Fowler, W. A., and Hoyle, F. (1957). *Rev. Mod. Phys.* **29**, 547.

Fowler, W. A. (1967). "Nuclear Astrophysics." Am. Phil. Soc., New York (a more popular survey).

γ-Ray Astronomy:
Fazio, G. G. (1967). *Annu. Rev. Astron. Astrophys.* **5**, 481.

X-Ray Astronomy:
Gould, R. (1967). *Ann. Phys. (New York)* **35.**
Morrison, P. (1967). *Annu. Rev. Astron. Astrophys.* **5**, 325.

Cosmic Rays:
Ginzburg, V. L., and Syrovat-skii, (1966). *Sov. Phys. Usp.* **9,** 223.

Quarks in Astrophysics:
Okun, L. B., Pikelner, S. B., and Zeldovitch, Ya. B. (1965). *Sov. Phys. Usp.* **8,** 702.

Astrophysical Neutrinos:
Chiu, H. Y. (1966). *Ann. Rev. Nucl. Sci.* **16, 591.**
Ruderman, M. (1965). *Rep. Progr. Phys.* **28,** 411.

3   High-Energy Interactions in Space

Till A. Kirsten* †   *Department of Earth and Space Sciences*
Oliver A. Schaeffer†   *State University of New York at Stony Brook*
                        Stony Brook, New York

* Present address: Max-Planck-Institut für Kernphysik, Heidelberg, Germany.
 † Work supported by the U.S. Atomic Energy Commission under contract AT-(30)-3629.

The availability of artificially accelerated protons has contributed in a major way to the understanding of cosmochemical problems. In particular, the interaction of cosmic rays, which are over 80% protons, with meteorites can only be understood if carefully measured radiation cross sections are available.

The following chapter is a summary of the thin-target and thick-target cross-section measurements. The information obtained from these cross sections is related to: cosmic ray variations in time and space, cosmic ray energy spectrum in the past, composition of cosmic radiation, terrestrial age of meteorites, space erosion of meteorites, exposure ages of meteorites, cosmic dust, the lunar surface, and tektites.

I. INTRODUCTION

The planetary system is a gigantic laboratory where nature has been performing an extensive high-energy physics experiment for billions of years. Just as in one of our large accelerators, targets are bombarded by high-energy particles. However, there are some complications. We do not know very much about the targets' shape and chemical composition, except that they are very complex. In addition, the targets are distributed over a large volume of space and do not remain in one place. Only fragments of the targets become available to us. We receive these fragments by chance and they are altered in their passage through the atmosphere. Furthermore, the radiation they receive in space is not monoenergetic but has a broad spectrum. In addition, it is composed of different kinds of particles and is not focused. Finally, we do not know how long the beam has been on and how its intensity has changed during time.

However, in some cases it is possible to differentiate the parameters. Then, the disadvantage of complexity becomes an advantage. Since one gets the opportunity to study all these different parameters separately, one may learn much about the interplanetary matter and the cosmic radiation if one investigates the interaction of both.

In particular, the cosmic radiation interacts with the surface regions of the planets and their satellites, with asteroids and planetoids, meteoroids, some comets and the interplanetary gas and dust. Interactions on the earth (mainly in its atmosphere) are not considered here.

The only natural extraterrestrial objects available for laboratory studies are meteorites and cosmic dust. They will be the main theme of our discussion. This article was completed before the lunar landing. The study of high-energy products in lunar material has already led to important facts about the moon (*Science,* 1970). Other materials for labroatory investigations are recovered parts of artificial space probes.

Since the average energy of the primary cosmic radiation is in the gigavolt region and since meteorites do not contain significant amounts of fissionable heavy elements, the most important interaction mechanism is the spallation process. As a result of this process a wide scale of stable and radioactive isotopes comes into existence and causes changes in the isotopic composition of the elements. Radioactive isotopes with half-lives below 10^8 years are all of cosmogenic origin and are therefore easy to distinguish from the general composition of the meteorite. On the other hand stable isotopes yield measurable isotopic changes only if the absolute concentrations of the particular elements in the meteorite are very small. Consequently, all of the discovered *stable* cosmogenic isotopes are rare gases or isotopes of other meteoritic *trace* elements.

In order to draw quantitative conclusions from the cosmic ray-produced nuclides, a detailed knowledge of nuclear data as obtained from accelerator experiments is necessary. Theoretically, one would need the complete excitation functions for particle-induced nuclear reactions covering a wide range of energies. Because of the large scale and chemical complexity of reacting materials, one would also like to know these data for a large number of target elements and reaction products and also for different bombarding particles. Therefore, all available thin-target spallation cross sections and excitation functions are useful and necessary contributions to cosmochemical studies. Practically, a few selected reactions have more importance than others. These are the reactions which either contribute in a major way in naturally occurring interactions or which are easiest to detect in meteorites. Proton-induced reactions in the energy range between 300 and 3000 MeV are most important. The most profitably studied target elements are iron, nickel, magnesium, silicon, aluminum, calcium, and oxygen which yield reaction products such as tritium, ^{10}Be, ^{14}C, ^{22}Na, ^{26}Al, ^{36}Cl, ^{40}K, ^{41}K, ^{44}Ti, rare gases, and others. Because of the secondary reactions involved, thick-target cross sections are also valuable. In addition to experimental data, the results of theoretical calculations about the spallation mechanism and general nuclear systematics contribute to our understanding of cosmochemical processes.

One of the greatest advantages of "meteorite targets" is that they were exposed to the radiation for a very long time, inimitable by laboratory experiments. By studying the concentrations of radionuclides with very different half-lives (from hours to stable), methods have been developed to obtain information about the history of cosmic rays and meteorites. The origin and interplanetary history of meteorites, their fragmentation, erosion, and ablation as well as the variation of the cosmic ray intensity in time and space have been studied. One cannot obtain this kind of information by direct measurements of the recent cosmic ray intensity.

In the following, we will discuss these subjects and the recent progress in this field.

II. RADIATION

Three different radiation components form high-energy reaction products in space: primary galactic cosmic radiation, solar cosmic radiation, and secondary particles which are formed within the irradiated bodies as a result of primary reactions whenever the dimensions of these bodies exceed ~30 cm. Since the secondary radiation is already a part of the interaction itself, it shall be discussed in Section IV.

A. Galactic Radiation

The galactic cosmic radiation is isotropic. About 85% of the particles are protons, about 15% are α particles, and less than 2% are heavier nuclei. Near the earth, the ratio of protons to α particles to C, N, O nuclei to heavier nuclei is 1000:150:13.3:3.9 (Singer, 1958; McDonald, 1959; Arnold et al., 1961). Direct flux measurements as carried out by Webber (1961) are influenced by solar modulations. The average flux of galactic particles is about 5 nucleons/cm²-sec-sterrad during times of solar minimum and 1 nucleon/cm²-sec-sterrad during times of solar maximum (Webber, 1961; More and Tiffany, 1962). The high-energy part of the spectrum is affected by the solar modulation to a much smaller extent. The flux for particles with energies above 3 GeV is ~1 nucleon/cm²-sec-sterrad during solar minimum and ~0.5 nucleon/cm²-sec-sterrad during solar maximum (Webber, 1961; More and Tiffany, 1962).

Arnold et al. (1961) used the measurements of McDonald (1959), Vernov and Chudakov (1959), and Van Allen and Frank (1959) to calculate an average energy of 4 GeV and an omnidirectional energy flux of 20 GeV/cm²-sec-sterrad for the primary radiation. From the same data, they found that the differential energy spectrum for primary protons and α particles above 1 GeV decreases toward higher energies according to the relation

$$f(E)\, dE = f_0(1 + E)^{-2.5} \qquad (E \text{ in GeV}). \tag{1}$$

Below 1 GeV, the galactic primary energy spectrum changes considerably with time and is only poorly known. Geiss *et al.* (1962) introduced a parameter β which can reflect these changes without change of the general form of the energy spectrum given in Eq. (1). The spectrum is then written as

$$f(E)\, dE = f_0 \exp\left(-\beta E^{-2/3}\right) E^{-2.5}\, dE, \tag{2}$$

which is valid for all energies above 200 MeV if β is chosen properly. For energies above 300 MeV, spectra Eqs. (1) and (2) agree for $\beta = 1.7$. A primary energy spectrum from new data obtained by Webber (1964) is in general agreement with a β value of 1.7.

The question of the existence of long-term time variations of the primary influx rate, due to variations of the radiation sources or of the galactic magnetic fields, might be resolved by means of meteorite studies and will be discussed in Section VII,A,1.

In the inner part of the solar system, two kinds of short time variations occur:

1. In anticoincidence with the solar sunspot activity, 11-year cyclic modulation of the primary galactic radiation takes place. It mainly affects the energy region below 1 GeV.

2. The low-energy part of the galactic radiation becomes modulated during sporadic magnetic storms on the sun (Forbush modulation).

The influence of these variations decreases with increasing distance from the sun and should be negligible at a distance of ~3 a.u.

B. Solar Radiation

The solar component of cosmic radiation is emitted discontinuously during solar flares. The flux of solar protons with energies above 20 MeV averaged over a long time is about 330 p/cm²-sec-sterrad (More and Tiffany, 1962) in the vicinity of the earth. At larger distances from the sun, the effective flux becomes lower.

If compared to the galactic component, the energy of solar radiation is rather low, e.g., the average flux of protons with energies above 500 MeV is only 0.03 p/cm²-sec-sterrad. The decrease of the differential spectrum at higher energies is somewhat different from flare to flare. For the energy range from 50 to 500 MeV, a typical differential spectrum decreases proportional to $E^{-4.5}$ (Webber, 1961; Biswas and Fichtel, 1965). The chemical composition of the solar radiation also changes considerably from flare to flare. An average composition is approximately 70% protons, 30% α particles, and 1% heavier nuclei. In more detail, for energies below 400 MeV, the measured proton-to-α-particle ratios at a given energy lie mostly between 1 and 10. If particles of equal rigidity or equal total energy are compared, the ratio rarely approaches values as high as 50. For energies above 400 MeV, the p/α ratio increases.

It should be noted that very high p/α ratios often create the impression that α particles are rather unimportant. This is due to the fact that one compares protons and α particles of equal energy per nucleon rather than the total energy. Since the energy spectrum of the solar radiation decreases with $\sim E^{-4.5}$, 400-MeV α particles are indeed much less abundant than 100-MeV protons, which is not true, however, for 100-MeV α particles.

III. TARGETS

For radiochemical and mass spectrometric studies of cosmic ray interactions in space, we are restricted to laboratory experiments. Thus, we describe only meteorites, cosmic dust, space probes and the lunar surface.

There is a continuous influx of interplanetary matter on the earth, somewhere between $\sim 10^5$ tons/year (Schaeffer et al., 1964) and $\sim 10^7$ tons/year (Brocas and Picciotto, 1967). The objects entering the outer atmosphere range from submicroscopic to asteroidal size. The penetrating depth of cosmic rays in solid material is of the order of 1 m. Therefore, irradiated bodies have been exposed to radiation in space either as a whole if small or only in surface layers if large.

Unfortunately, the objects which we collect on the earth, called meteorites, do not reflect the original situation during exposure because the capture mechanism selects and alters the initial shape. Objects exceeding a weight of about 100 tons vaporize completely when impacting on the earth and form an explosion crater. Thus, objects of this size are not available for radiochemical analysis. Objects below this size are decelerated by the passage through the atmosphere without undergoing complete vaporization. However, these objects get more or less ablated, the outer surface region melts, and the characteristic melted crust of meteorites is formed. Explosions can also occur which break up the objects either in the atmosphere or on impact.

If the object entering the atmosphere is smaller than a few centimeters, it will vaporize completely and result in a meteor. Again, no material of this original size can be found on the earth's surface. Thus, the size range of recovered individual meteorites is between 1 mm and 3 m in diameter, corresponding to an original mass of a few milligrams up to ~ 100 tons.

Particles with an initial diameter below about 300 μ can enter the earth's atmosphere in a diffusionlike process and arrive on the surface of the earth (perhaps after a long time) without essential change. These particles are defined as cosmic dust.

It is generally accepted today that meteorites are members of the solar system and they may have surrounded the sun in elliptical orbits. Thus, a long time exposure to the galactic cosmic radiation may be anticipated. To get an idea which high-energy interaction products one can expect, we have to take a look at the chemical composition of these natural targets.

A. Meteorites

There are three general types: iron meteorites, stone meteorites and stony-iron meteorites. Irons average to 90% iron and 10% nickel content, with individual variations in the Fe/Ni ratio. Stones are formed mainly from silicates. They are overwhelmingly olivine [(Mg, Fe)$_2$SiO$_4$] and orthopyroxene [(Mg, Fe)SiO$_3$]. Stony irons appear as a nonhomogeneous conglomerate of stones and irons. When discussing high-energy reactions in meteorites we deal mainly with Fe, Ni, Si, Mg, and O targets. However, there is a large number of minor constituents in all three classes of meteorites which also have to be considered. This is especially true when the reaction product has a mass number not far away from the mass number of one of these minor elements. Occasionally elements like Al, Ca, S, or C can also be found as major constituents.

In addition to chemical differences there are very evident differences in the mineralogical composition of meteorites. One of the most stimulating ideas in meteoritics in the last five years was the investigation of correlations between the radiation effects in meteorites and their mineralogical classification since different origins and histories of various meteorite classes could be reflected by different radiation effects. Indeed, many such correlations were discovered. Therefore it is necessary at this point to enumerate the most important meteorite subclasses.

Iron meteorites are classified according to their crystallographic structure, as to hexahedrites (4–6% Ni content), octahedrites (6–14% Ni), and nickel-rich ataxites (>14% Ni). The average composition of all iron meteorites is Fe 90.8%, Ni 8.6%, Co 0.6% (Mason, 1962), although the nickel content of individual meteorites can be as much as 62% (meteorite from Oktibbeha County). In addition to Fe, Ni, and Co there are traces of Cr, P, S, and C in most of the iron meteorites (Mason, 1962), which sometimes also contain silicate inclusions.

The minor group of *stony irons* is divided into the subclasses pallasites and mesosiderites, depending whether the silicate consists mainly of either olivine or pyroxene.

Most of the *stone meteorites* contain chondrules, more or less spherical silicate aggregates of millimeter size. These meteorites are called *chondrites*. Stones without chondrules form the subgroup of *achondrites*.

In the first approximation, chondrites show no large variations in their chemical composition, and are classified according to mineralogical composition into the classes enstatites, bronzites, hypersthenes, pigeonites, and carbonaceous chondrites (Mason, 1962). Usually, the average elemental composition of chondrites is given either in weight percent or in atomic percent without oxygen, the most abundant element in chondrites. In considering a chondrite as the target of cosmic ray bombardment the atomic percents must be recalculated to include oxygen. On this basis, the average composition of chondrites is as follows: oxygen 55.3%, silicon 15.4%, magnesium 14.4%, iron 10.6%, sulfur 1.7%, aluminum 1.3%, calcium 0.8%, nickel 0.6% (Kirsten *et al.,* 1963; using data from Urey and Craig, 1963). The minor components (<0.5%) are in this order of abundance: sodium, chromium, manganese, potassium, phosphorus, cobalt, titanium. In the special case of the carbonaceous chondrites, the hydrogen and carbon contents become significant.

The *achondrites* are composed of the same major elements as the chondrites, but they are not as uniform in composition as the chondrites, neither from subclass to subclass nor from meteorite to meteorite within one subclass. Average numbers, therefore, have little significance. Compared with the chondritic composition, the silicon content is enriched up to 40% at the expense of iron, which is depleted to a similar extent. Magnesium content can be similar or depleted up to 60%, the aluminum content can be either lower or five times higher than that found in chondrites, and the same is true for calcium. The latter fact serves as the characteristic for the distinction between calcium-poor achondrites (aubrites, diogenites) and calcium-rich achondrites (eucrites, howardites).

Because of their heterogeneity, achondrites are especially suitable for the study of the correlations between interaction products and chemical composition of the target.

B. Cosmic Dust

Despite the fact that meteorites are the most interesting extraterrestrial objects found on the earth, they constitute only a negligible part of the total mass of interplanetary material arriving on the earth.

In terms of the total mass influx, extraterrestrial dust particles with radii below 300 μ are much more important than meteorites. Nevertheless, because of the very nature of these small particles, our knowledge about cosmic dust is much poorer than about meteorites. It is extremely difficult to distinguish cosmic dust from terrestrial material such as volcanic or industrial exhausts. Radionuclides present only in extraterrestrial material and not in terrestrial material are a powerful tool for this differentiation.

Dust samples are collected at ground level in unindustrialized areas by means of air filters, in the atmosphere by means of airplanes and balloons, and in the upper atmosphere by satellites. The total mass of dust obtainable in this way is rather small. This can be overcome if one studies the dust particles accumulated over long time periods either in ocean sediments or in arctic snow and glaciers.

According to Öpik (1956), the radius r of primary dust particles varies between ~ 1 and ~ 300 μ. The number of particles follows a size distribution which is proportional to $r^{-2.8}$. The shape is either spherical or irregular (fluffy particles).

The chemical composition of cosmic dust is not very well known. By microprobe analysis, iron, nickel, silicon, calcium, and aluminum have often been detected in individual grains presumed to be extraterrestrial. Magnesium, sulfur, potassium, titanium, chromium, and manganese are found only rarely (Grjebine et al., 1964; Park and Reid, 1964; Wright and Hodge, 1964). The extraterrestrial origin of all the particles analyzed is still disputed (Gerloff et al., 1967).

The sources of cosmic dust are believed to be primary particles (directly condensed from the interplanetary gas), asteroidal and cometary dust, meteor residues, and meteorite ablation products. This leads to the guess that most probably the chemical composition resembles either chondritic material or nickel iron. Indeed, magnetic as well as unmagnetic particles have been observed.

C. Lunar Surface and Space Probes

The first quantitative analysis of lunar surface material is the results of the Surveyor V and Surveyor VI α-scattering experiments, which are given in Table I (Turkevich et al., 1967, 1968).

The data indicate a similarity to terrestrial basalts. γ-Ray experiments from the USSR Luna 10 mission suggested also a basaltic composition (Vinogradov et al., 1966). Compared to chondritic material, basalt is enriched in silicon, aluminum, calcium, and oxygen but strongly depleted in iron and magnesium. The more detailed analyses of the Apollo returns were not available at the time this manuscript was written.

The last target considered here are recovered space craft parts. Satellite materials used for radiochemical observations were usually functional parts of the spacecrafts themselves and not especially designed for this kind of experiments. Stainless steel battery cases, lead ballast, silver in AgBr emulsions, and aluminum parts from Discoverers 17, 18, 25, 26, 29, and 36 and iron from Sputnik IV have been investigated (see Section VII,B,9).

TABLE I

CHEMICAL COMPOSITION OF THE LUNAR SURFACE
AT THE SURVEYOR V AND SURVEYOR VI
LANDING SITES[a]

Element	Surveyor V (atomic %)	Surveyor VI (atomic %)
O	58 ± 5	57 ± 5
Si	18.5 ± 3	22 ± 4
Ca ⎱	13 ± 3	6 ± 2
Fe ⎰		5 ± 2
Al	6.5 ± 2	6.5 ± 2
Mg	3 ± 3	3 ± 3
Na	<2	<2
C	<3	<2

[a] Turkevich et al. (1967, 1968).

IV. INTERACTION

A. Reaction Mechanism

Since we deal with inelastic interactions, only incident particle energies above 30 MeV will be considered. Radionuclides can be formed either by fission, spallation, fragmentation, or nuclear evaporation. Particle-induced fission plays a role only for interactions with heavy elements. The targets considered here contain heavy elements like gold, lead, mercury, or uranium usually as traces (at most a few parts per million). Therefore we can usually restrict ourselves to the spallation and fragmentaion mechanism.

In this process, the first collision of the primary particle with a target nucleus leads to the ejection of some highly excited particles, created in the so-called "knock on" phase by direct interactions. These particles undergo further reactions with other nuclei (nuclear cascade). The mechanism is essentially the same for primary protons or α particles. When all cascade particles have left the nucleus, the latter remains in an excited state. The next phase is the comparatively slow nuclear evaporation process. In this step, the residual nucleus loses its excitation energy by emission of a single or clustered nucleons and γ rays. Complex nuclear fragments can also leave the excited nucleus under special conditions (fragmentation). The remaining deexcited nuclei are spread over a wide mass range below the mass number of the target element.

Nuclear cascade and evaporation lead, in thick targets, to a complex spectrum of secondary particles, mainly protons, neutrons and π mesons. The flux of secondary particles is strongly depth dependent. The flux increases over the first few centimeters distance from the target surface and then slowly decreases with increasing depth. The spectral shape of the secondary radiation varies also with the depth. The mean absorption thickness for primary protons in the gigavolt-region is about 100 g/cm². One also has to distinguish between small and large irradiated bodies. For objects with a diameter below 30 cm, such as cosmic dust or small meteorites, secondary radiation is unimportant and fairly accurate production rates can be derived from the primary spectrum alone. In large objects the secondary spectrum becomes very important.

The development and the energy spectrum of secondary particles at different absorption thicknesses were measured in the atmosphere by comparison of the star size distributions in nuclear emulsions at different altitudes (Birnbaum *et al.*, 1952; Powell *et al.*, 1959). Arnold *et al.* (1961) gave reasons to believe that these results are transferable to massive iron bodies and should also be true for stone meteorites. In an extensive study, these authors calculated the flux and the energy spectrum of the secondaries at various depths within a large iron meteorite. The shape of the secondary spectrum at 10- and 100-g iron/cm² depth is represented by the exponents given in Table II (Arnold *et al.*, 1961). Most of the secondary radiation consists of neutrons with energies below 500 MeV. At higher energies the total flux decreases rapidly. The energy spectrum does not change very much for depths greater than 100 g/cm²; however, the total flux rapidly decreases.

TABLE II

ENERGY SPECTRA INCLUDING SECONDARY PARTICLES AT TWO DIFFERENT DEPTHS OF A THICK IRON TARGET[a]

Energy range	Spectrum at 10 g/cm² depth	Spectrum at 100 g/cm² depth
2–100 MeV	$3.7(E^{-1} + 0.01E^{-2} + 1.1 \times 10^{-5}E^{-3})$	$6.0(E^{-1} + 0.01E^{-2} + 1.1 \times 10^{-5}E^{-3})$
0.1–3.0 GeV	$7.2(0.4 + E)^{-2.5}$	$3.3(0.2 + E)^{-2.5}$
>3 GeV	$11(1 + E)^{-2.5}$	$5.8(1 + E)^{-2.5}$

[a] Arnold *et al.* (1961).

B. Reaction Products

At a certain energy, the production rate P_i of a nuclear reaction product in atoms/gram/second is given by

$$P_i = 4\pi I_0 \sigma_i N, \tag{3}$$

where σ_i is the production cross section (mb), $4\pi I_0$ is the omnidirectional flux (particles/cm²-sec-sterrad), and N is the number of target atoms in a 1-g target.

If the target is composed of different elements or isotopes j ($j = 1, \ldots, r$) with the relative atomic abundances

$$N_j = \frac{\text{atoms of the element } j \text{ in 1-g target}}{\text{total number of atoms in 1-g target}} = \frac{\text{abundance in atomic } \%}{100},$$

we get

$$P_i = 4\pi N I_0 \sum_{j=1}^{r} N_j \sigma_{ij}, \tag{4}$$

where σ_{ij} is the cross section for the production of the nuclide i from the component j.

For an energy spectrum f, we obtain

$$P_i = 4\pi N I_0 \frac{\int_0^\infty \left(\sum_{j=1}^{r} N_j \sigma_{ij}(E) \right) f(E)\, dE}{\int_0^\infty f(E)\, dE}. \tag{5}$$

For large bodies the energy spectrum becomes a function of the depth. Within a thin hollow sphere located d cm below the surface of a spherical body, we obtain

$$P_i(d) = 4\pi N I_0(d) \frac{\int_0^\infty \left(\sum_{j=1}^{r} N_j \sigma_{ij}(E) \right) f(E, d)\, dE}{\int_0^\infty f(E, d)\, dE}. \tag{6}$$

Taking into account different incident particles k ($k = 1, \ldots, s$), protons, secondary neutrons, and π mesons, we get

$$P_i(d) = 4\pi N \sum_{k=1}^{s} I_{0k}(d) \frac{\int_0^\infty \sum_{j=1}^{r} N_j \sigma_{ijk}(E) f_k(E, d)\, dE}{\int_0^\infty f_k(E, d)\, dE}. \tag{7}$$

Over long periods, both flux and energy spectrum may be time dependent. The total number of atoms Q_i of a stable radionuclide produced during the radiation time T in one gram of the target is therefore

$$Q_i(d) = 4\pi N \sum_{k=1}^{s} \int_0^T I_{0k}(d, t) \frac{\int_0^\infty \sum_{j=1}^r N_j \sigma_{ijk}(E) f_k(E, d, t) \, dE}{\int_0^\infty f_k(E, d, t) \, dE} . \tag{8}$$

If the produced radionuclide is radioactive with a half-life short compared to the radiation time T, one has the concentration P_i/λ_i at the end of the irradiation. The decay rate is therefore equal to the production rate:

$$D_i(d) = P_i(d), \tag{9}$$

where $D_i(d)$ is the decay rate of the nuclide i with the decay constant λ_i in decays per gram of target and second $(1/\lambda_i << T)$. After a time Θ following the end of the irradiation, the decay rate is

$$D_i(d, \Theta) = D_i(d, 0) \exp(-\lambda_i \Theta). \tag{10}$$

For half-lives comparable to the radiation time T, one obtains in the case of a radiation constant in time

$$D_i(d) = P_i(d)[1 - \exp(-\lambda_i T)]. \tag{9a}$$

Equation (8) can be integrated numerically if all the cross sections and energy spectra are known. For an iron meteorite this has been done by Arnold et al. (1961) for two different depths, 10-g/cm^2 and 100-g/cm^2 iron. These authors used the primary and secondary spectra discussed above and constructed the necessary excitation functions from experimental cross sections and from a semiempirical formula given by Rudstam (1956) (see Section V,C,1). In this way, Arnold et al. (1961) obtained absolute production rates for a great number of radionuclides. The results are given in Table III. It is common to express absolute production rates by giving the absolute production rate for ^{36}Cl and the relative production rates of all other nuclides normalized to ^{36}Cl.

A similar approach to calculate radionuclide production in iron meteorites has been made by Goel (1962). This author obtained the flux and the energy distribution inside of an iron meteorite and the total isobaric cross section for each mass number from the primary spectrum, the total inelastic cross section, and the star size distributions in nuclear emulsions. To split the total isobaric yields into independent yields he used Rudstam's formula (Section V,C,1). Goel's calculated production rates for the center of a iron meteorite with 20-cm diameter are also listed in Table III.

For the chemically more complex stone meteorites, Eq. (7) was inte-

TABLE III

PRODUCTION RATES OF RADIONUCLIDES IN IRON METEORITES, NORMALIZED FOR ^{36}Cl = 1.0

		By integration of Eq. (7)				
		Arnold et al. (1961)			Goel (1962)	Measured[a]
		100 g/cm²	10 g/cm²	Surface	80 g/cm²	Aroos
Nuclide ^{36}Cl	Half-life 3×10^5 y	23 atoms min kg	34 atoms min kg	35 atoms min kg	16.2 atoms min kg	14 atoms min kg
^{3}H	12.3 y	12	10	8	10.4	
^{3}He	Stable	27	23	17	20	34
^{4}He	Stable	135	108	63	71	128
^{7}Be	54 d	0.69	0.74	1.0		
^{10}Be	2.7×10^6 y	0.31	0.34	0.62		0.29
^{14}C	5770 y	0.13	0.15	0.27		
^{20,21,22}Ne	Stable	0.73	0.81	1.3		1.2
^{22}Na	2.6 y	0.11	0.12	0.18		0.15
^{26}Al	7.4×10^5 y	0.10	0.11	0.15		0.13
^{32}Si	~700 y	0.05	0.06	0.06		0.06
^{32}P	14.3 d	0.41	0.44	0.49		
^{33}P	25 d	0.30	0.32	0.35		
^{36}Cl	3.1×10^5 y	=1.0	=1.0	=1.0	=1.0	=1.0
^{36}Ar	Stable	1.2	1.2	1.2	1.1	1.2
^{37}Ar	34 d	0.65	0.65	0.61	0.48	1.3
^{38}Ar	Stable	1.6	1.5	1.4	2.0	1.9
^{39}Ar	269 y	0.90	0.85	0.74	1.7	1.0
$^{40+}$Ar	Stable	0.38	0.35	0.31	1.0	
^{42}Ar	33 y	0.11	0.09	0.08	0.07	
^{39}K	Stable	1.8	1.7	1.5	2.3	
^{40}K	1.3×10^9 y	1.5	1.4	1.2	1.4	0.6
^{41}K	Stable	2.6	2.3	1.8	2.8	

grated by Kirsten et al. (1963). Since these authors discussed only stable reaction products, they compared the relative production rates of different stable nuclides with the measured ratios for a great number of stone meteorites, chondrites as well as achondrites. It was not necessary to know the exposure times T. Measured and calculated ratios were in reasonable agreement. Wasson (1963) applied the same method to predict the activities to be expected in cosmic dust under the influence of the relatively low-energy solar radiation. Of course no secondaries had to be considered for cosmic dust.

Lavrukhina et al. (1967) extended the method to targets of chondritic, sideritic, tektitic, granitic, and basaltic materials. They give closed expressions for the production of various radionuclides at various depths in the model targets. Time variations of the primary radiation and absorption reactions like neutron capture were taken into account.

TABLE III (continued)

Nuclide	Half-life	By integration of Eq. (7)				Measured[a] Aroos
		Arnold et al. (1961)			Goel (1962)	
		100 g/cm²	10 g/cm²	Surface	80 g/cm²	
		23 atoms min kg	34 atoms min kg	35 atoms min kg	16.2 atoms min kg	14 atoms min kg
^{36}Cl	3×10^5 y					
^{41}Ca	1.1×10^5 y	1.1	1.0	0.8	0.56	
^{42}Ca	Stable	3.1	2.6	1.9		3.5
^{43}Ca	Stable	3.6	3.0	2.0		3.9
^{44}Ca	Stable	4.2	3.5	2.1		
^{45}Ca	165 d	0.28	0.22	0.13	0.82	0.3
^{46}Ca	Stable	0.05	0.04	0.02		0.09
^{45}Sc	Stable	5.0	4.0	2.3	4.3	6.2
^{46}Sc	84 d	1.5	1.2	0.63	2.6	2.1
^{44}Ti	$\sim$200 y	0.38	0.31	0.19	0.09	0.31
^{48}V	16 d	7.3	5.2	2.2		6
^{49}V	330 d	9.6	6.7	2.6	5.0	12
V^{50}	6×10^{14} y	5.2	3.5	1.2		10
^{51}Cr	28 d	19	11	2	5.9	20
^{52}Mn	5.7 d	6.6	4.2	0.8		
^{53}Mn	$\sim$2 $\times 10^6$ y	33	18	2.5	11.4	38
^{54}Mn	300 d	38	20	2.5	7.6	34
^{55}Fe	2.6 y	220	100	6.2		120
^{60}Fe	$\geq$3 $\times 10^5$ y	0.1	0.1	0.03		
^{56}Co	77 d	4.5	2.1	0.1		4.4
^{57}Co	270 d	5.5	3.0	0.6		6.4
^{58}Co	71 d	17	7.8	0.1		4.4
^{60}Co	5.2 y	0.6	0.3	0.01		1.3
^{59}Ni	7.5×10^4 y	1.4	0.7	0.1		3.4

[a] For source of data see Arnold et al. (1961) and Arnold (1961).

Merrihue (1966) and Pepin (1966) applied the method to heavy radionuclides such as krypton and xenon isotopes produced by spallation on heavy meteoritic trace elements (e.g., strontium, selenium, zirconium, barium, cerium, and others). Ratios of measured production rates agreed with theoretical ratios. Similar results were reported by Bogard (1967), Hohenberg et al. (1967) and Munk (1967).

Another way to calculate production rates is to apply different forms of a cascade theory. In fact, a simplified cascade theory yielded the first verifications of measured concentrations and depth distributions of spallation products in iron meteorites. The method was proposed by Martin (1953). He assumed that the intensity of the primary cosmic radiation decreases exponentially inside of an iron meteorite without essential change of the energy spectrum. For the total interaction probability of primaries with iron he used the geometrical cross section. The number of secondaries produced per interaction was taken from nuclear emulsion data for primary, secondary, and higher-order interactions.

The production of evaporation particles has been estimated in the same way. In that case the absorption coefficients of primary and secondary particles serve as parameters to fit measured depth distributions of helium in large iron meteorites. Moreover, the best fit also gives us the preatmospheric radius of the meteorite. Ebert and Wänke (1957), Hoffman and Nier (1958, 1960), and Signer and Nier (1960) extended the method from He to Ne and Ar isotopes and used the few cross sections available at this time instead of nuclear emulsion data. They simplified the excitation functions by assuming two representative cross sections, one for primary and one for secondary mean energies, and were able to reproduce quantitatively the measured spallation product distribution in the Grant iron meteorite.

Another way to calculate absolute production rates makes use of the Monte Carlo method applied to the extranuclear cascade (More and Tiffany, 1962; Ebeoglu et al., 1966). This formalism allows a calculation of the propagation of an incident primary particle spectrum inside a thick target by statistical methods. As input data, Ebeoglu et al. (1966) used cross sections obtained from total interaction probabilities (proportional to the geometric cross section) and intranuclear cascade calculations (Bertini, 1963) (see Section V,C,2). Since the latter data were available only for incident energies below 400 MeV, the method is not yet applicable to meteorites. However, Ebeoglu and Wainio (1966) have calculated the expected absolute production rates on the lunar surface due to solar radiation ($E < 500$ MeV). Results of the production of ^7Be, ^{14}C, ^{22}Na, and ^{54}Mn at various depths of iron, stony meteorite, and olivine–basalt targets have been reported. Comparison with measured thick iron target data as discussed in Section V,B showed reasonable agreement.

Instead of calculating production rates, one can measure them directly in meteorites since for radioactive isotopes with a half-life considerably less than the exposure time, the production rate in space is identical with the decay rate at the time of the fall of the meteorite [see Eq. (9)]. For stable isotopes one has to divide the measured concentration by the exposure time T, assuming an irradiation constant in time. T can be found by methods discussed in Section VII,B,3.

There are about a hundred papers reporting measured concentrations of cosmic ray-produced isotopes in meteorites. Generally, each such number can be used to calculate production rates by some generalizations. Production rates obtained in this way usually refer to the total complex target composition (atoms/gram meteorite). However, if one uses meteorites with different chemical compositions, one can also calculate the yields from particular elements like silicon, magnesium, aluminum, calcium, etc. (Hintenberger *et al.*, 1964a; Begemann, 1965; Heymann and Anders, 1967).

The comparison of calculated and measured production rates shows a general agreement in most cases. To give an example we have listed the nuclides measured in the iron meteorite Aroos in Table III together with the calculated production rates. However, in some cases, striking differences exist. They reflect the fact that simplifications or assumptions made in the calculations are not valid in that particular case. If one considers which modifications one has to make in order to yield agreement (e.g., use of a modified energy spectrum, different shape of the meteorite), one obtains a powerful method for learning facts about the history of the radiation and the targets. A detailed discussion of such applications will be given in Section VII.

The availability of measured production rates allows one to introduce a general method for the prediction of production rates in particular meteorites with unique composition and radiation history. Stauffer and Honda (1962) deduce, from measurements of radionuclides in iron meteorites, the empirical correlation

$$P_i = k_1(A_t - A_p)^{-k_2} \tag{11}$$

(valid for $6 \leq A_t - A_p \leq 36$; k_1, k_2: constants A_t, mass number of the target element; A_p, mass number of the radionuclide).

The equation holds true for radionuclides which accumulate the total isobaric yield of their mass number. In other cases one has to multiply the right side of the equation with a factor

$$\gamma = \sigma_{\text{included isobars}}/\sigma_{\text{total isobar}} < 1.$$

Equation (11) enables us to calculate all production rates by measuring only two. Two measured rates determine the constants k_1 and k_2 in the meteorite involved. The values for k_1 vary from 2×10^{15} to 2×10^{16} atoms/g and for k_2 from 1.5 to 2.5 for different meteorites and different depths.

It is interesting to note (Geiss *et al.*, 1962), that for iron meteorites, a spallation product distribution of the form given in Eq. (11) agrees with a closed solution of Eq. (7) when one uses Rudstam's spallation formula for the excitation functions and an energy spectrum of the form

$$f(E)\,dE = f_0 E^{-\alpha}\,dE.$$

It is often important to know which part of the energy spectrum is mainly responsible for the production of a particular isobar. Low-energy particles lead overwhelmingly to the production of isotopes close to the target element while nuclei further away will be formed mainly by high-energy radiation.

The energy E_{max} at which the production of a particular isobar is maximal under natural radiation conditions may be calculated by

$$E_{\text{max}} = 5.7(\Delta A)^{3/2} \qquad (E \text{ in MeV})$$

(Geiss *et al.*, 1962).

The direct measurement of production rates in thick targets by means of accelerators will be discussed in Section V,B.

V. NUCLEAR DATA

To interpret the cosmic ray interaction products in meteorites a detailed knowledge of the excitation functions involved is necessary. There are three main sources of such information:

(1) Cross sections and excitation functions measured in thin targets by means of accelerators.

(2) Cross sections measured in thick targets, including the effects of secondary particles.

(3) Theoretical or semiempirical methods for the construction of excitation functions.

A. Thin-Target Cross Sections

During the last ten years numerous laboratories have contributed many nuclear reaction cross sections, with the main goal of studying nuclear systematics and nuclear structure. The large number of cross sections obtained in this work represent the major source of information for our needs. The cross sections were obtained primarily for other purposes, and as a result, there is often a lack of information about reactions which are not so important in nuclear chemistry but which are of primary interest in meteoritics. For example, in nuclear chemistry independent yields receive the most attention. In meteoritics one is more interested in total chain yields. In addition, in nuclear chemistry one is mainly interested in thin targets, while thick targets are of more importance in meteoritics. To overcome this situation, there was a feedback from meteoritics to nuclear chemistry. Programs for cross-section measurements were carried out for the particular needs of meteoritics. This was especially necessary for the determination of yields for stable isotopes which received only little attention by radiochemists.

The first cross sections for the production of stable isotopes were measured mass spectrometrically by Fireman and Zähringer (1957) and Schaeffer and Zähringer (1958). The availability of large accelerators for such experiments was greatly appreciated by cosmochemists. A survey is given in Table IV for those thin-target cross sections most important in meteoritics. The data included in this table were selected as follows:

Incident Particles. Here protons are considered with energies above 50 MeV. The table contains cross sections for the production of nuclei far away from the target nucleus as well as for nuclei produced by reactions such as (p, n) or (p, pn).

The very few known cross sections for neutron- and α-induced reactions above 50 MeV are also included in the table. Fortunately, from the nature of the spallation mechanism, one does not expect that the yields depend strongly on the properties of the incident particles, as long as the energy is high enough. Thus, proton cross sections are fairly representative for n- or α-induced reactions. This was experimentally confirmed by Koterling (1962) and by Davis *et al.* (1963).

The energy range below 50 MeV plays a role for secondary neutrons, especially when the excitation functions for special reactions such as (n, p) or (n, α) rise to very high resonance values. If needed, cross sections for such reactions might be found explicitly in the compilation by Jessen *et al.* (1966) or they might be calculated from theory of fast neutron reactions as summarized by Cindro (1966).

Target Elements. Only the main meteoritic elements, Fe, Ni, Si, Mg, Al, Na, O and S, are included in Table IV. Data for calcium do not exist. For spallation reactions on trace elements cross sections might be found in the compilations by Bruninx (1961, 1962, 1964). Cross sections for energies below 425 MeV are compiled by Bertini *et al.* (1966). In addition, cross sections for rare gas isotope production can be found for barium by Funk and Rowe (1967) and for Cu, Ag, Au, and U by Hudis *et al.* (1968).

Reaction Products. Only isotopes for which many measurements of their abundances in meteorites exist are included. These are ^{3}H, ^{3}He, ^{4}He, ^{7}Be, ^{10}Be, ^{14}C, ^{20}Ne, ^{21}Ne, ^{22}Ne, ^{22}Na, ^{26}Al, ^{32}Si, ^{32}p, ^{33}p, ^{36}Cl, ^{36}Ar, ^{37}Ar, ^{38}Ar, ^{39}Ar, ^{42}Ar, ^{39}K, ^{40}K, ^{41}K, ^{44}Ti, ^{45}Ca, ^{46}Sc, ^{48}V, ^{49}V, ^{51}Cr, ^{52}Mn, ^{54}Mn, ^{55}Fe, and ^{56}Co. Their half-lives can be found in Table III.

Arrangement. The data are arranged according to increasing mass numbers of the reaction products, increasing target element numbers and increasing incident energies. Differences between the values given in Table IV and the original papers are due to a recalculation for improved monitor-reaction cross sections, carried out by Bruninx (1961, 1964). The code for the reference numbers is given at the end of the table. Cross sections are cumulative for precursors with half-lives shorter than 10 hr; precursors with longer half-lives are not included. Exceptions are noted.

TABLE IV

MEASURED HIGH-ENERGY REACTION CROSS SECTIONS ON THIN
TARGETS FOR APPLICATIONS IN COSMOCHEMISTRY

Nuclide	Target	Energy (MeV)	Cross section (mb)		Ref.[d]
^{3}H	O	450	38	$\pm$ 5	10
	O	2050	34.6	$\pm$ 5	10
	O	2050	36		25
	O	2200	33	$\pm$ 4	15
	O	6200	38		25
	^{24}Mg	300	19	$\pm$ 6	13
	^{24}Mg	450	30	$\pm$ 9	13
	^{24}Mg	550	26	$\pm$ 8	13
	Mg	600	15	$\pm$ 4	1
	^{24}Mg	660	43	$\pm$ 13	13
	Mg	2050	35	$\pm$ 2	10
	Mg	2050	36		25
	Al	80	10	$\pm$ 2	12
	Al	120	16	$\pm$ 5	13
	Al	120	8.0		25
	Al	150	12	$\pm$ 1.5	12
	Al	200	18	$\pm$ 6	13
	Al	300	25	$\pm$ 7	13
	Al	450	24	$\pm$ 7	13
	Al	450	23	$\pm$ 3	10
	Al	500	37	$\pm$ 11	13
	Al	550	33	$\pm$ 10	13
	Al	600	44	$\pm$ 13	13
	Al	600	32	$\pm$ 2	1
	Al	660	46	$\pm$ 14	13
	Al	2050	44		25
	Al	2050	43	$\pm$ 5	10
	Al	2200	75		1
	Al	5700	45		5
	Al	5700	50		25
	Al	25,000	71		1
	Al	25,000	67		28
	Si	600	18	$\pm$ 3	1[a]
	Fe	50	4.2		6
	Fe	75	4.3	$\pm$ 0.8	6
	Fe	100	4.8	$\pm$ 0.9	6
	Fe	135	6.4	$\pm$ 1.2	6
	Fe	150	6.1	$\pm$ 1.1	6
	Fe	150	17	$\pm$ 2	12
	Fe	175	6.6	$\pm$ 1.2	6
	Fe	450	28	$\pm$ 5	10
	Fe	600	48	$\pm$ 5	1
	Fe	2050	61	$\pm$ 9	10
	Fe	2050	64		25
	Fe	2200	66	$\pm$ 10	1

TABLE IV (*continued*)

Nuclide	Target	Energy (MeV)	Cross section (mb)		Ref.[d]
³H	Fe	2200	72	± 8	11
(*cont'd*)	Fe	6200	57	± 11	25
	Fe	25,000	104	± 10	1
	Fe	25,000	92		28
	Ni	450	22	± 3	10
	Ni	600	41	± 15	1
	Ni	2050	87	± 14	10
	Ni	2050	90		25
	Ni	2200	135	± 9	1
³He	Mg	540	23		30
	Mg	600	24		1
	Al	540	27		30
	Al	600	27		1
	Al	2200	72		1
	Si	600	34		1
	Si	2200	56		1
	Fe	160	11		7
	Fe	430	45		7
	Fe	540	35		30
	Fe	600	34		1
	Fe	2200	63		1
	Fe	3000	240		7
	Fe	25,000	133		1
	Fe	25,000	113		28
	Ni	540	42		30
	Ni	600	42		1
	Ni	2200	58		1
⁴He	Mg	540	250		30
	Mg	600	260		1
	Al	540	300		30
	Al	600	268		1
	Al	2200	410		1
	Si	600	302		1
	Si	2200	330		1
	Fe	160	120		7
	Fe	430	450		7
	Fe	540	360		30
	Fe	600	336		1
	Fe	2200	410		1
	Fe	3000	1300		7
	Fe	25,000	750		1
	Fe	25,000	670		28
	Ni	540	190		30
	Ni	600	396		1
	Ni	2200	470		1

TABLE IV (*continued*)

Nuclide	Target	Energy (MeV)	Cross section (mb)			Ref.[d]
[7]Be	[16]O	52	2	±	0.5	41
	[16]O	110	5	±	1.2	41
	O	130	7.85	±	2	31
	[16]O	150	3.8	±	1	41
	[16]O	155	4.5	±	1	42
	[16]O	155	5	±	1.5	43
	O	210	4.9	±	1.2	31
	O	300	6.5	±	1.6	31
	O	400	7.5	±	1.9	31
	O	500	8.2			24
	O	800	8.7			24
	O	1500	8.4			24
	O	2200	10.9			24
	[16]O	2200	8.8	±	0.4	50
	O	2900	10.2			24
	O	5700	10	±	5	5
	Na	5700	13	±	2	5
	Mg	130	2.66	±	0.7	31
	Mg	210	2.31	±	0.6	31
	Mg	300	2.48	±	0.6	31
	Mg	400	3.25	±	0.9	31
	Mg	500	7.3			24
	Mg	800	9.0			24
	Mg	1500	9.6			24
	Mg	2200	10.8			24
	Mg	2900	12.1			24
	Al	120	1.63			38
	Al	155	1.0	±	0.2	49
	Al	335	1.6			18
	Al	660	4.2			38
	Al	1000	7.6			19
	Al	1000	7.5			20
	Al	1400	8.3			19
	Al	1400	8.0			20
	Al	1800	12.6			19
	Al	2200	9.1			19
	Al	2200	12.1			20
	Al	2900	7.5			39
	Al	3000	11.7			19
	Al	3000	10.8			20
	Al	5700	8.3	±	0.9	5
	Al	28,000	7.89	±	0.29	17
	Si	130	1.07	±	0.3	31
	Si	210	0.76	±	0.2	31
	Si	300	1.69	±	0.3	31
	Si	400	2.02	±	0.4	31
	Si	500	2.4			24

TABLE IV (*continued*)

Nuclide	Target	Energy (MeV)	Cross section (mb)	Ref.[d]
[7]Be	Si	800	2.6	24
(*cont'd*)	Si	1500	2.8	24
	Si	2200	3.0	24
	Si	2900	4.1	24
	Fe	130	0.18 ± 0.04	31
	Fe	150	0.23 ± 0.03	33
	Fe	210	0.31 ± 0.05	31
	Fe	300	0.47 ± 0.07	31
	Fe	400	0.57 ± 0.09	31
	Fe	500	1.21	24
	Fe	660	2 ± 0.3	33
	Fe	730	2.75	29
	Fe	730	3.0	29
	Fe	730	2.9	3
	Fe	800	2.1	24
	Fe	1500	3.2	24
	Fe	2200	5.0	24
	Fe	2900	5.7	24
	Fe	24,000	7.0	26
	Ni	130	0.21 ± 0.04	31
	Ni	210	0.38 ± 0.07	31
	Ni	300	0.59 ± 0.09	31
	Ni	400	0.81 ± 0.12	31
	Ni	500	1.62	24
	Ni	800	2.4	24
	Ni	1500	4.9	24
	Ni	2200	6.5	24
	Ni	2900	7.2	24
[10]Be	Fe	730	1.0	2
	Fe	730	1.1 ± 0.3	29
	Fe	730	1.2 ± 0.4	29
[14]C	O	52	2.2	8
	O	73	2.6	8
	O	100	2.8	8
	O	150	1.8	8
	O	2700	1.9	8
[20]Ne	Mg	540	25	30
	Mg	600	20.6	1
	Al	540	27	30
	Al	600	20.2	1
	Al	2200	17	1
	Si	600	22.3	1
	Fe	600	0.75	1
	Ni	600	0.72	1

TABLE IV (*continued*)

Nuclide	Target	Energy (MeV)	Cross section (mb)	Ref.[d]
^{21}Ne	Mg	540	26	30
	Mg	600	20.2	1
	Al	540	28	30
	Al	600	20.2	1
	Al	2200	16.7	1
	Si	600	19.4	1
	Si	2200	15.0	1
	Fe	430	0.1	32
	Fe	540	0.8	30
	Fe	600	0.71	1
	Fe	2200	2.4	1
	Fe	25,000	6.9	1
	Fe	25,000	5.7	28
	Ni	540	0.51	30
	Ni	600	0.67	1
	Ni	2200	1.9	1
^{22}Ne	Mg	540	17	30
	Mg	600	9.9	1
	Al	540	15	30
	Al	600	11.2	1
	Al	2200	6.9	1
	Si	600	7.6	1
	Si	2200	7.0	1
	Fe	540	0.4	30
	Fe	600	0.4	1
	Fe	2200	1.2	1
	Fe	25,000	3.4	1
	Fe	25,000	3.0	28
	Ni	540	0.38	30
	Ni	600	0.34	1
^{22}Na	Na	50	73.6	14[a]
	Na	51.5	94 ± 10	44[a]
	Na	75	55.5	14
	Na	77	67 ± 7	44
	Na	88	64 ± 7	44
	Na	110	63.7 ± 7	44
	Na	120	41.2 ± 2	14
	Na	134	47.5 ± 5	44
	Na	152	47 ± 5	44
	Na	155	45 ± 5	45
	Na	5700	31 ± 5	5
	^{25}Mg	50	38.5	46[a]
	Mg	50	50.2	37
	Mg	60	35.7	37
	Mg	70	30	37

TABLE IV (*continued*)

Nuclide	Target	Energy (MeV)	Cross section (mb)	Ref.[d]
^{22}Na	Mg	80	29	37
(*cont'd*)	Mg	90	28.6	37
	Mg	95	28.8	37
	^{25}Mg	100	43	46
	Mg	130	28.3 ± 4.2	31
	Mg	210	21.6 ± 3.2	31
	Mg	300	23.8 ± 3.6	31
	Mg	400	22.2 ± 3.3	31
	Mg	500	11.5	24
	Mg	800	11.9	24
	Mg	1500	12.3	24
	Mg	2200	15.3	24
	Mg	2900	15.4	24
	Al	50	40 ± 2	35
	Al	50.6	36.4 ± 3.6	47
	Al	60	36 ± 2	35
	Al	70	28 ± 1.5	35
	Al	70	29	27
	Al	80	24 ± 1.5	35
	Al	90	22 ± 1	35
	Al	100	21.5 ± 1	35
	Al	110	20 ± 1	35
	Al	110	28	27
	Al	120	19 ± 1	35
	Al	120	22.1	38
	Al	130	19 ± 1	35
	Al	150	18 ± 1	35
	Al	150	1.2 × σ(660)	21
	Al	155	17.5 ± 0.6	51
	Al	260	1.0 × σ(660)	21
	Al	335	13.6	18
	Al	350	0.96 × σ(660)	21
	Al	450	1 × σ(660)	21
	Al	560	1 × σ(660)	21
	Al	590	1.0	16
	Al	600	19.2	20
	Al	660	16.5	38
	Al	1000	16.7	20
	Al	1400	19.4	20
	Al	1600	14.2	20
	Al	1800	18.6	20
	Al	2000	12.3	39
	Al	2200	14	20
	Al	2900	11.1	39
	Al	3000	16.6	20
	Al	5700	17 ± 3	5
	Al	28,000	9.85 ± 0.3	17

TABLE IV (*continued*)

Nuclide	Target	Energy (MeV)	Cross section (mb)	Ref.[d]	Remarks
^{22}Na	Si	56	20	52	
(*cont'd*)	Si	60	21.8	52	
	Si	130	10.4 ± 1.1	31	
	Si	210	10.3 ± 1	31	
	Si	300	13.1 ± 1.3	31	
	Si	400	12.5 ± 1.3	31	
	Si	500	8.6	24	
	Si	800	8.2	24	
	Si	1500	7.8	24	
	Si	2200	8.4	24	
	Si	2900	9.1	24	
	Fe	150	0.03 ± 0.01	33	
	Fe	340	0.02	9	
	Fe	660	0.36 ± 0.03	33	
	Fe	730	0.36	3	
	Fe	24,000	1.6	26	
	Fe	29,000	3.4 ± 0.3	40	
^{26}Al	Fe	660	0.36	38	
	Fe	730	0.43	3	
^{32}Si	Fe	660	0.28	38	
	Fe	730	0.3	2	
	Fe	730	0.46	29	
^{32}P	S	n: 90	44	36	Neutrons
	Fe	130	0.02 ± 0.004	31	
	Fe	150	0.2 ± 0.1	33	
	Fe	210	0.09 ± 0.013	31	
	Fe	300	0.32 ± 0.05	31	
	Fe	340	0.04	9	
	Fe	400	0.61 ± 0.09	31	
	Fe	500	1.28	24	
	Fe	660	2.3 ± 0.2	33	
	Fe	800	3.6	24	
	Fe	1500	5.4	24	
	Fe	2200	6.1	24	
	Fe	2900	6.1	24	
	Ni	130	0.007 ± 0.001	31	
	Ni	210	0.025 ± 0.004	31	
	Ni	300	0.13 ± 0.02	31	
	Ni	400	0.27 ± 0.04	31	
	Ni	500	0.89	24	
	Ni	800	2.1	24	
	Ni	1500	3.4	24	
	Ni	2200	4.5	24	
	Ni	2900	4.6	24	

TABLE IV (*continued*)

Nuclide	Target	Energy (MeV)	Cross section (mb)	Ref.[d]	Remarks
^{33}P	Fe	150	0.065 ± 0.03	33	
	Fe	660	1.2 ± 0.2	33	
^{36}Cl	Fe	660	3.7	38	
	Fe	730	6.8	3	
^{36}Ar	Fe	430	1.0	32	
	Fe	540	1.1	30	
	Fe	600	1.13	1	
	Fe	600	4.4	38	
	Fe	2200	1.37	1	
	Fe	25,000	1.84	1	
	Fe	25,000	1.64	28	
	Fe/Cr/Ni	660	0.9 ± 0.1	23	b
	Fe/S/O	660	1.2 ± 0.6	23	c
	Ni	540	0.9	30	
	Ni	600	1.14	1	
	Ni	2200	1.3	1	
^{37}Ar	Fe	100	0.014	34	
	Fe	150	0.21	34	
	Fe	200	0.56	34	
	Fe	380	3.2	34	
	Fe	430	3.3	32	
	Fe	500	5.1	34	
	Fe	600	4.2 ± 0.5	1	
	Fe	600	4.65	1	
	Fe	660	2.8	38	
	Fe	750	6.5	34	
	Fe	α: 880	6.0	34	α particles
	Fe	1000	8.2	34	
	Fe	1500	8.1	34	
	Fe	2000	8.1	34	
	Fe	2900	7.2	34	
	Fe	25,000	6.1 ± 0.3	1	
	Fe	25,000	7.1	1	
	Fe	25,000	5.7	28	
	Fe	28,000	6.4	34	
	Ni	100	0.007	34	
	Ni	200	0.89	34	
	Ni	380	2.0	34	
	Ni	500	3.5	34	
	Ni	600	2.9 ± 0.7	1	
	Ni	600	3.9	1	
	Ni	750	6.5	34	
	Ni	α: 880	5.6	34	α particles
	Ni	1000	7.9	34	

TABLE IV (*continued*)

Nuclide	Target	Energy (MeV)	Cross section (mb)	Ref.[d]	Remarks
^{37}Ar	Ni	1500	9.2	34	
(*cont'd*)	Ni	2000	8.7	34	
	Ni	2900	9.2	34	
	Ni	28,000	6.7	34	
^{38}Ar	Fe	430	8.0	32	
	Fe	540	10	30	
	Fe	600	9.8	1	
	Fe	2200	12.6	1	
	Fe	25,000	16.3	1	
	Fe	25,000	13.6	28	
	Fe/Cr/Ni	660	8.2 ± 0.8	23	[b]
	Fe/S/O	660	7.4 ± 1.5	23	[c]
	Ni	540	6.4	30	
	Ni	600	7.15	1	
	Ni	2200	7.4	1	
^{39}Ar	Fe	430	4.1	32	
	Fe	540	5.1	30	
	Fe	600	5.1	1	
	Fe	α: 880	6.4	34	α particles
	Fe	2000	9.3	34	
	Fe	2200	6.2	1	
	Fe	25,000	10.9 ± 1	1	
	Fe	25,000	9	1	
	Fe	25,000	7.7	28	
	Fe/Cr/Ni	660	3.9 ± 0.3	23	[b]
	Fe/S/O	660	4.3 ± 0.5	23	[c]
	Ni	540	3.0	30	
	Ni	600	2.97	1	
	Ni	α: 880	4.6	34	α particles
	Ni	2000	6.6	34	
	Ni	2200	4.2 ± 0.8	1	
	Ni	2200	2.7	1	
^{39}K	Fe	575	6.7 ± 0.7	22	
	Fe	590	1.97	4	Estimated
	Fe	18,300	2.13	4	Estimated
	Ni	590	1.71	4	Estimated
	Ni	18,300	2.48	4	Estimated
^{40}K	Fe	575	9.2 ± 0.8	22	
	Fe	590	3.5	4	Estimated
	Fe	730	8.0	29	
	Fe	18,300	3.65	4	Estimated
	Ni	590	2.50	4	Estimated
	Ni	18,300	3.60	4	Estimated

TABLE IV (*continued*)

Nuclide	Target	Energy (MeV)	Cross section (mb)	Ref.[d]	Remarks
^{41}K	Fe	575	8.0 ± 0.7	22	
	Fe	590	4.22	4	Estimated
	Fe	18,300	4.15	4	Estimated
	Ni	590	2.25	4	Estimated
	Ni	18,300	3.25	4	Estimated
^{42}Ar	Fe	730	0.04	29	
^{44}Ti	Fe	660	0.97	38	
	Fe	730	2.0	2	
	Fe	730	3.6	29	
^{45}Ca	Fe	150	0.36 ± 0.06	33	
	Fe	340	0.56	19	
	Fe	660	1.2 ± 0.13	33	
	Fe	730	1.4	29	
	Fe	24,000	0.67	26	
^{46}Sc	Fe	150	3.0 ± 0.6	33	
	Fe	340	3.2	9	
	Fe	660	5.8 ± 0.9	33	
	Fe	730	6.4	29	
	Fe	24,000	4.0	26	
^{48}V	Fe	130	≥5.9	31	Cumulative
	Fe	150	15 ± 2	33	
	Fe	210	≥6.4	31	Cumulative
	Fe	300	≥8.5	31	Cumulative
	Fe	340	10.3	9	
	Fe	400	≥9.5	31	Cumulative
	Fe	500	10.7	24	Cumulative
	Fe	660	15 ± 2	33	
	Fe	660	16.1	38	
	Fe	800	6.8	24	Cumulative
	Fe	1500	6.3	24	Cumulative
	Fe	2200	6.3	24	Cumulative
	Fe	2900	4.9	24	Cumulative
	Fe	24,000	12	26	
	Ni	130	≥3.5	31	Cumulative
	Ni	210	≥4.6	31	Cumulative
	Ni	300	≥6.5	31	Cumulative
	Ni	400	≥7.5	31	Cumulative
	Ni	500	6.1	24	Cumulative
	Ni	800	7.1	24	Cumulative
	Ni	1500	7.9	24	Cumulative
	Ni	2200	8.8	24	Cumulative
	Ni	2900	8.5	24	Cumulative

TABLE IV (*continued*)

Nuclide	Target	Energy (MeV)	Cross section (mb)		Ref.[d]	Remarks
[49]V	Fe	150	33	± 5	33	Independent
	Fe	340	31		9	Independent
	Fe	660	25	± 4	33	Independent
	Fe	660	32.4		38	Independent
	Fe	730	30		29	
	Fe	24,000	22		26	Independent
[51]Cr	Fe	130	30	± 6.0	31	
	Fe	150	63	± 19	33	Independent
	Fe	210	22	± 3.0	31	
	Fe	300	25	± 3.8	31	
	Fe	340	41		9	Independent
	Fe	400	25	± 3.8	31	
	Fe	500	21.4		24	
	Fe	660	35	± 6.2	33	Independent
	Fe	660	41.3		38	Independent
	Fe	730	27		29	
	Fe	800	20.1		24	
	Fe	1500	17.3		24	
	Fe	2200	14.6		24	
	Fe	2900	15.8		24	
	Fe	24,000	21		26	Independent
	Ni	130	20.0	± 4.0	31	
	Ni	210	19.0	± 2.8	31	
	Ni	300	19.2	± 2.8	31	
	Ni	400	18.0	± 2.7	31	
	Ni	500	18.2		24	
	Ni	800	15.8		24	
	Ni	1500	14.1		24	
	Ni	2200	14.7		24	
	Ni	2900	11.6		24	
[52]Mn	Fe	130	68.6	± 10.3	31	
	Fe	150	14	± 3	33	Independent
	Fe	210	49	± 7.4	31	
	Fe	300	33.7	± 5	31	
	Fe	340	12.9		9	Independent
	Fe	400	24.2	± 4	31	
	Fe	500	14.7		24	
	Fe	500	16		3	Independent
	Fe	660	16	± 2	33	Independent
	Fe	730	13		3	Independent
	Fe	800	14.1		24	
	Fe	1500	10.3		24	
	Fe	2200	10.4		24	
	Fe	2900	10.1		24	
	Fe	24,000	5.6		26	Independent

TABLE IV (*continued*)

Nuclide	Target	Energy (MeV)	Cross section (mb)	Ref.[d]
^{52}Mn	Ni	130	39.5 ± 6	31
(*cont'd*)	Ni	210	33.5 ± 6	31
	Ni	300	38.5 ± 6	31
	Ni	400	28.0 ± 4	31
	Ni	500	18.7	24
	Ni	800	12.0	24
	Ni	1500	12.4	24
	Ni	2200	12.2	24
	Ni	2900	8.9	24
^{54}Mn	Fe	130	38.2 ± 7.6	31
	Fe	150	36 ± 16	33
	Fe	210	27.7 ± 4.2	31
	Fe	300	21.4 ± 3.2	31
	Fe	340	12	9
	Fe	400	18.6 ± 2.8	31
	Fe	500	28	3
	Fe	500	21.3	24
	Fe	660	34 ± 10	33
	Fe	730	33	3
	Fe	730	32	29
	Fe	800	18.5	24
	Fe	1500	19.0	24
	Fe	2200	16.6	24
	Fe	2900	11.5	24
	Fe	24,000	17	26
	Ni	130	10.5 ± 2	31
	Ni	210	8.5 ± 1.3	31
	Ni	300	9.8 ± 1.4	31
	Ni	400	6.95 ± 1	31
	Ni	500	6.6	24
	Ni	800	4.5	24
	Ni	1500	4.9	24
	Ni	2200	4.9	24
	Ni	2900	4.2	24
^{55}Fe	Fe	150	110 ± 10	33
	^{56}Fe	370	63.9 ± 3.8	48
	Fe	660	60 ± 20	33
	Fe	660	60.8	38
	Fe	24,000	36	26
^{56}Co	Fe	130	4.25 ± 0.85	31
	Fe	150	1.6 ± 0.3	33
	Fe	210	2.26 ± 0.3	31
	Fe	300	1.5 ± 0.2	31
	Fe	340	0.24	9

TABLE IV (*continued*)

Nuclide	Target	Energy (MeV)	Cross section (mb)	Ref.[d]
^{56}Co	^{56}Fe	370	0.91	48
(*cont'd*)	^{56}Fe	370	0.95	48
	^{56}Fe	370	0.90	48
	Fe	400	0.89 ± 0.13	31
	Fe	500	0.94	24
	Fe	660	1.3 ± 0.3	33
	Fe	800	0.81	24
	Fe	1500	0.82	24
	Fe	2200	0.67	24
	Fe	2900	1.01	24
	Ni	130	90	31
	Ni	210	59	31
	Ni	300	62	31
	Ni	400	45	31
	Ni	500	32	24
	Ni	800	25	24
	Ni	1500	25	24
	Ni	2200	28	24
	Ni	2900	21	24

[a] For fine structure of this energy range see original paper.
[b] Target composition 75% Fe, 15% Cr, 10% Ni.
[c] Target composition 64.5% Fe, 25.1% S, 10.4% O.
[d] Reference code for Table IV: 1, Goebel *et al.* (1964); 2, Arnold *et al.* (1961); 3, Honda and Lal (1960); 4, Chackett (1965); 5, Benioff (1960); 6, Goebel (1958); 7, Schaeffer and Zähringer (1958); 8, Murry *et al.* (1961); 9, Rudstam *et al.* (1952); 10, Curry *et al.* (1956); 11, Fireman (1955); 12, Brun *et al.* (1962); 13, Kutznetsov and Mekhodov (1958); 14, Meadows and Holt (1951a); 15, Fireman and Rowland (1955); 16, Goebel and Schultes (1960); 17, Cumming *et al.* (1962a); 18, Marquez and Perlman (1951); 19, Baker *et al.* (1958); 20, Friedländer *et al.* (1955); 21, Prokoskin and Tiapkin (1957); 22, Lämmerzahl and Zähringer (1966); 23, Gerling and Levskii (1961); 24, Rayudu (1964a); 25, Curry (1959); 26, Estrup (1963); 27, Hintz and Ramsay (1952); 28, Goebel and Zähringer (1961); 29, Honda and Lal (1964); 30, Bieri and Rutsch (1962) [corrected according to Goebel *et al.* (1964)]; 31, Rayudu (1964b); 32, Schaeffer and Zähringer (1959); 33, Lavrukhina *et al.* (1963); 34, Davis *et al.* (1963); 35, Gauvin *et al.* (1962); 36, Knox (1949); 37, Meadows and Holt (1951b); 38, Lavrukhina *et al.* (1964); 39, Cummings *et al.* (1962b); 40, Porile and Tanaka (1964); 41, Albouy *et al.* (1962); 42, Valentin *et al.* (1963); 43, Gradsztajn (1960); 44, Gusakov (1962); 45, Valentin (1965); 46, Valentin *et al.* (1964); 47, Cummings (1963); 48, Remsberg and Miller (1963); 49, Ligonniere *et al.* (1964); 50, Reeder (1965); 51, Nguyen Long Den (1961); 52, Sheffey (1968).

As evident from the data, typical excitation functions undergo an increase in the energy range between 100 MeV and 1 GeV but reach saturation above 1 GeV. Except for evaporation particles (^{3}H, ^{3}He, ^{4}He), the cross sections are negligible for energies below 100 MeV if ΔA exceeds 4–5 mass units. In more detail, the general pattern of excitation functions is well represented by Eq. (24) (Section V,C,1).

B. Thick-Target Measurements

The multiplication of the combined primary and secondary energy spectrum by the excitation functions and by the element concentrations yields production rates in an analytical way [Eq. (7)]. A more direct method to obtain such information involves bombardment of thick targets, thus simulating the cosmic ray bombardment of a whole meteorite. If one measures the activities in thin foils which are placed inside of a thick absorber at various depths, one can calculate effective cross sections from the relation

$$\sigma_{\mathrm{eff}}(d,\,E_{\mathrm{inc}}) = DA_{\mathrm{t}}/(\lambda FN\,\Delta d), \tag{12}$$

where

$$D = \text{decay rate at the end of the bombardment,}$$
$$A_{\mathrm{t}} = \text{target mass number,}$$
$$\lambda = \text{decay constant of the radionuclide,}$$
$$F = \text{total proton flux,}$$
$$\Delta d = \text{thickness of the foil,}$$
$$d = \text{depth at which the foil is placed,}$$
$$N = 6 \times 10^{23},$$
$$E_{\mathrm{inc}} = \text{energy of the incident particles.}$$

These effective cross sections include the contributions from secondary particles such as protons, neutrons or π mesons. If, as is usually the case, a collimated beam of protons is used, then the radionuclide production depends not only on the depth but also on the lateral distances from the beam axis. To describe this dependency one defines a differential cross section, $\sigma(d, s)$, for a small ring concentric to the beam axis and normalized in such a way that

$$\int \sigma'(d,\,s) = \sigma(s). \tag{13}$$

Averaging over the total area inside of a circle with radius s one defines

$$\sigma(d,\,s) = \int_0^s \sigma'(d,\,s)\,ds. \tag{14}$$

Sometimes a spherical geometry is more useful. Then one integrates over a hemisphere of constant distance ρ from the point where the beam hits the target and defines

$$\sigma(\rho) = \int_0^{\pi/2} \sigma'(\rho,\,\alpha)\rho\,d\alpha, \tag{15}$$

where α is the angle of spread centered in the beam entry point (Honda 1962). If one measures $\sigma(d)$ or $\sigma(\rho)$ for different targets, distances, and incident energies, the resulting information can be applied to predict either the production rates at a certain point inside of a meteorite or the concentration profiles which reflect the depth distribution of radionuclides. Such results can then be compared either with measured concentration profiles in large meteorites or with results of extra nuclear cascade calculations as discussed in Section IV,B.

The depth variations of radionuclide concentrations in meteorites were first proposed by Bauer (1947), quantitatively predicted by Martin (1953), and detected by Paneth et al. (1953). Following these pioneering studies, further investigations were carried out by Vinogradov et al. (1957), Fireman (1958b), Hoffman and Nier (1958, 1959, 1960), Fechtig et al. (1960), and Signer and Nier (1960, 1962). (For a more detailed discussion see Section VII,B,2.)

The first thick-target experiment was performed by Fireman and Zähringer (1957). They irradiated a stack of iron blocks of 23-cm maximal length and 9-cm² cross section with 0.16-, 1.0-, 3.0-, and 6.2-GeV protons and measured the production of tritium and ^{37}Ar as a function of depth. The tritium and ^{37}Ar production at 6.2 GeV was found to increase with depth for the first 5 cm below the surface and then to decrease exponentially for greater depths. At lower energies no such buildup occurred.

For further discussion it is convenient to introduce a few definitions:

Let

$\quad \sigma_0 = $ effective cross section at the surface,

$\quad \sigma_m = $ effective cross section at the depth where σ reaches its maximum,

$\quad \sigma_{m/e} = $ effective cross section at the depth where $\sigma_{\text{eff}} = (1/e)\sigma_m$ $(e = 2.71)$,

$\quad d_m = $ depth where $\sigma_{\text{eff}} = \sigma_m$,

$\quad d_{m/e} = $ depth where $\sigma_{\text{eff}} = \sigma_{m/e}$,

$\quad s = $ lateral distance from the beam axis,

$\quad A_t = $ target mass number, and

$\quad A_p = $ mass number of the reaction product.

Then let

$$\zeta = (\sigma_m/\sigma_0), \text{ the buildup factor,} \tag{16}$$

$$d_m = \text{the maximum position,} \tag{17}$$

$$d_\mu = d_{m/e} - d_m, \text{ the mean thickness,} \tag{18}$$

$$s = \text{the lateral distance, and} \tag{19}$$

$$\Delta A = A_t - A_p, \text{ the mass difference.} \tag{20}$$

Fireman and Zähringer concluded from their results that the medium *incident* energy of the *primary* cosmic radiation must have been a few gigavolts. They came to this conclusion by comparing their results to the measured depth variation in the Carbo iron meteorite.

Stoenner *et al.* (1960) irradiated a sample of the chondrite Hamlet and samples of calcium fluoride, potassium fluoride, and stainless steel behind a 10-cm iron shield with protons of 3-GeV energy. The $^{37}Ar/^{39}Ar$ ratio measured in the irradiated Hamlet sample agreed with the ratio found in the freshly fallen meteorite. In addition, it was also possible to reproduce the measured ratio by combining the ^{37}Ar and ^{39}Ar yields in steel, calcium, and potassium with the abundance of these elements in Hamlet. In another experiment, Davis and Stoenner (1962) found that the ^{37}Ar concentrations produced by 3-GeV protons in a thick iron target go through a broad maximum at depths between 2 and 20 cm and decrease exponentially after that.

Systematic studies have been carried out by Kohman and Goel (1963) and by Honda (1962). Honda measured a great number of reaction products varying from ^7Bc to ^{55}Fe produced by 3-GeV protons in iron and nickel foils, which were located at different depths in a thick iron absorber. In addition to depth variations, he also measured the lateral spread of the radionuclides in the foils. It was found that nuclides with large mass differences were produced almost entirely along the beam axis, while for small mass differences there was a large lateral spread. For some products with small ΔA the lateral target dimensions (30×30 cm^2) were not sufficient to trap all the reaction products. The result can easily be explained by multiple scattering and by the fact that the secondary particles are not collimated. Since secondaries have relatively low energies, they mainly form products with small mass differences.

Honda studied the depth variation and found that the buildup factor increases with decreasing mass difference. The maximum position increases slightly with increasing lateral distance. The mean thickness varies between 10 and 35 cm and increases for one nuclide with increasing lateral distance.

In order to apply the measured differential cross sections to meteorites, Honda calculated the effective cross sections $\sigma(\rho)$ by integration over a hemisphere according to Eq. (15). Thus, effective cross sections were obtained for different radii ρ. The energy of 3 GeV is fairly close to the mean energy of the primary cosmic radiation. Therefore, values $\sigma(\rho)$ allow us to predict the activities to be expected in the *center* of spherical bodies with any radius ρ by means of Eq. (12). In particular the parameter k_2 from Eq. (11) can be deduced from the data. Once the k_2 values for specific meteorites are deduced from radionuclide measurements, the corresponding depth can be calculated. Agreement between the thick-target data and meteorite data exists in so far as the range of k_2 values for different meteorites corresponds to reasonable depths. *A priori,* one does not know the original (preatmospheric) sizes of the meteorites. However, one can reverse the procedure and calculate such preatmospheric depths (see Section VII,B,2). To calculate the expected activities at any point of a meteorite or for non-spherical shape or for a more detailed incident energy spectrum, some mathematical treatment is necessary.

Attempts to systematize this task have been made by Bender and Kohman (1964). They introduced a set of ten parameters which describe the measured data in a unique way in order to simplify interpolations, extrapolations, and calculations. At least some of these parameters seem to have physical meaning. However, these considerations are still in a preliminary state.

Further thick-target measurements have been carried out by the Carnegie group. Shedlovsky and Rayudu (1964) and Shen (1962) irradiated iron foils lodged at different depths in a thick iron absorber with 1 and 3 GeV protons. For 3 GeV protons, Honda's results were confirmed. At 1 GeV, they found that the buildup factors and maximum positions are smaller than for 3 GeV. Again the side spread was more pronounced at greater depths and for small mass differences. The mean thickness for iron was 13 cm at 1 GeV and 22 cm at 3 GeV.

From the ^{22}Na and ^{24}Na differential cross sections an attempt was made to reproduce the depth variation of neon isotopes in the Grant iron meteorite as measured by Signer and Nier (1960). They found that the mean thickness of the meteorites is larger than that of the target. This might be explained by higher primary radiation energies incident on the Grant iron meteorite.

The same trends as mentioned above have been confirmed by further measurements at Carnegie for 100-MeV, 450-MeV, and 6.1-GeV energy (Rayudu, 1963a, b, 1964). To simulate stone meteorites, glass targets have also been used (Rayudu and Shedlovsky, 1963). The low-energy data might be used for measurements made on space probes exposed to the solar radiation.

C. Theoretical Cross-Section Data

In spite of the many measured thin-target cross sections given in Table III, the data are still incomplete for the integration of Eq. (8). Fortunately, methods have been developed to predict spallation cross sections theoretically. The accuracy of such excitation functions is not as good as for the most experimentally determined cross sections and an inaccuracy of a factor 2 is the rule. Nevertheless, the calculations are extremely helpful when no measurements exist. Two different ways have been tried:

(1) Generalization of trends deduced from available data combined with the application of some known facts about nuclear systematics and distribution laws.

(2) Monte Carlo calculations for the intranuclear cascade based on a statistical method.

1. Semiempirical Method

This method has been developed by Rudstam (1955, 1956). Based on the assumption of an exponential mass yield distribution and a Gaussian charge distribution, Rudstam gave a formula which could describe the behavior of measured excitation functions for medium-weight target elements within a fairly wide range of reaction products. In order to fit the data quantitatively he introduced four parameters P, Q, R, and S. Then, the *individual* cross section for the production of a nuclide with the atomic number Z and the mass number A became

$$\sigma(Z, A) = \exp[PA - Q - R(Z - SA)^2]. \tag{21}$$

The parameter P reflects the influence of the incident energy. Q arises essentially from a proportionality constant of the form e^{-Q} and is therefore related with the total inelastic interaction cross section σ_i. R and S describe the charge distribution. In particular SA defines the position of the charge distribution maximum Zp. P depends on the energy as well as the target mass number A_t and so does Q. R and S are independent of the radiation conditions; however, R depends on A_t.

Honda and Lal (1960, 1964) proved that Eq. (21) holds true in the important case of iron spallation for nuclides with mass numbers between 20 and 45. To fit the data best in this particular case, a slight change of Rudstam's original values for the parameters was necessary. The values are

$$P = 10.2E^{-0.63} \quad (E \text{ in MeV}), \qquad R = 1.9, \qquad S = 0.474.$$

Q depends on the energy according to

$$Q = \ln \frac{\exp(PA_t) - \exp(PA_{t/2})}{P\sigma_i(R/\pi)^{1/2}},$$

where σ_i is the total inelastic interaction cross section. For 730 MeV, Q was found to be 4.0 (the units of σ are millibars).

The usefulness of the formula at this point depended on the question whether or not a few measured data were available for each target element. If so, one could estimate the most suitable parameter values and then calculate the whole excitation function.

In meteorites, total isobaric yields are often more important than independent yields. To get an analytical expression for this case, Geiss *et al.* (1962) summed Eq. (21) for all independent yields of one isobar and derived the formula

$$\sigma_{tot}(Z, A) = \sigma_i c E^{-2/3} \exp(-c \, \Delta A E^{-2/3}), \tag{22}$$

where σ_{tot} is the total isobaric yield, $\sigma_i = \sigma_o A_t^{2/3}$ and is the total inelastic or geometrical cross section, $c = $ constant, and $\Delta A = A_t - A$.

The most recent values for the constant c were given by Schwarz and Oeschger (1967). Using their data, the explicit formula for σ_{tot} becomes

$$\sigma_{tot}(\Delta A, E, A_t) = \frac{711(A_t/E)^{2/3}}{A_t + 50} \exp\left(-\frac{11.85 \, \Delta A}{E^{2/3}(A_t + 50)}\right), \tag{23}$$

where E is given in gigavolts and σ_{tot} in millibars.

A breakthrough was achieved when Rudstam (1966) used an increased number of experimental data in order to deduce equations for the variation of the parameters P, Q, R, and S, valid for almost all target masses and energies.

For this extension of Eq. (21) two essential changes had to be made.

1. Evidently, for an extended mass range the charge distribution maximum Zp is no longer linearly proportional to A. Thus, for this term to become flexible a fifth parameter T has been introduced, and SA is substituted with $SA - TA^2$.

2. The Gaussian charge distribution $\sigma \sim (Z - Zp)^2$ is replaced by the analytical expression $\sigma \sim (Z - Zp)^{3/2}$ in which the distribution decreases to its limits less steeply.

Thus, Eq. (21) becomes

$$\sigma(Z, A) = \exp[PA - Q - R(Z - SA + TA^2)^{3/2}]. \tag{24}$$

Making use of the increased number of experimental data, the following relations for the parameters P, Q, R, S, and T were obtained:

$$P = \begin{cases} (20 \pm 7)E^{-(0.77 \pm 0.06)} & \text{for } E < 2.1 \text{ GeV,} \\ 0.056 \pm 0.003 & \text{for } E > 2.1 \text{ GeV,} \end{cases} \tag{25}$$

where the units of E are in megavolts. This relation is valid for medium weight nuclei. For heavy nuclei, P decreases:

$$R = (11.8 \pm 3)A^{-(0.45 \pm 0.07)}, \tag{26}$$
$$S = 0.486 \pm 0.001, \tag{27}$$
$$T = 0.00038 \pm 0.00002, \tag{28}$$

$$Q = \ln \frac{1.79\left[e^{PA_i}\left(1 - \frac{2}{3}\frac{(0.45 \pm 0.07)}{PA_t}\right) - 1 + \frac{2}{3}(0.45 \pm 0.07) + \frac{2}{3}\frac{(0.45 \pm 0.07)}{PA_t}\right]}{\sigma_i P(11.8 \pm 3)^{2/3}A_t^{-2/3(0.45\pm0.07)}}. \tag{29}$$

The total inelastic cross section σ_i is in first approximation

$$^{(1)}\sigma_i = \pi(1.26A_t^{1/3} \times 10^{-13})^2 \quad \text{cm}^2, \tag{30}$$

and in second approximation

$$^{(2)}\sigma_i = {}^{(1)}\sigma_i \exp\left[-(0.25 \pm 0.12) + (0.0074 \pm 0.0016)A_t\right]f_2, \tag{31}$$

where

$$f_2 = \begin{cases} \exp\left[(1.73 \pm 0.35) - (0.0071 \pm 0.0018)E\right] & \text{for } E < 240 \text{ MeV}, \\ 1 & \text{for } E > 240 \text{ MeV}. \end{cases} \tag{32}$$

One has to keep in mind that the parameters are chosen in such a way that the best fit is obtained when one considers all kinds of targets. For special cases, the fit can be improved if one deduced slightly different values from a least square fit of the data for only this case. Goebel *et al.* (1964) did so for iron [they knew Rudstam's Eq. (24) from a private communication].

The parameters were as follows:

$$P = 0.19 \text{ (for 600 MeV)}, \quad Q = 4.88 \text{ } (\sigma \text{ in mb}),$$
$$R = 2.0, \quad S = 0.495, \quad T = 0.0005.$$

The accuracy of the cross sections obtained from Eq. (24) lies within a factor of 3. Since the total variation of the absolute cross sections has a range of 10^6, the approximation is quite satisfactory.

2. *Intranuclear Cascade Calculations*

The interaction mechanism described in Section IV,A is the starting point for a statistical treatment of the path of single nucleons in the nuclei (Monte Carlo method). The knock-on phase can be treated in terms of single particle–particle collisions within the nucleus because the wavelengths of both the incoming particles and of the subsequent collision products are shorter than, or at least comparable to, the average nucleon–nucleon distance in the nucleus. Thus, interaction cross-section and angular distribution data for free particle reactions such as p, p; n, p; π^+, p can be used. In a three-dimensional model one can trace the path of each single nucleon coming into existence by means of relativistic kinematics in conjunction with the computerized statistical treatment of the parameters. For instance, type and place of collision, momentum of the struck nucleon, and scattering angle are such parameters.

The method was proposed by Serber (1947), first applied by Goldberger (1948), and cast into its present form by Metropolis *et al.* (1957a, b). To carry out the calculation, an agreement concerning the nuclear properties had to be made. Metropolis *et al.* (1957a, b) chose a constant nucleon density and a Fermi-gas energy distribution for the nucleus. Furthermore, the potential of π mesons within the nucleus was assumed to be zero. Calculations have been made for the production of ^{64}Cu and ^{27}Al from some heavy target nuclei using incident protons at energies between 82 and 365 MeV without pion participation (Metropolis *et al.*, 1957a). For proton energies between 450 and 1840 MeV the pion production was taken into account by Metropolis *et al.* (1957b). Using a modified model for the nuclear density which makes use of Hofstadter's electron scattering data, Bertini (1963) carried out similar calculations for ten elements ranging from carbon to uranium and for energies between 25 and 400 MeV.

Since we are interested here mainly in the final spallation yields, the evaporation phase has also to be taken into account. This was done by Metropolis *et al.* (1957b). They assumed that no further mass loss occurs if the excitation energy drops below 10 MeV and that for each successive 17 MeV increase one mass unit gets lost. The authors were able to present calculated mass yield curves for 360 and 1840 MeV protons on ^{64}Cu and found a rather good agreement with the measured radiochemical data, except for reactions with $\Delta A = 1$.

A further advance was made in the work of Dostrovsky *et al.* (1958). Using the Metropolis results for the knock-on phase as input data, these authors introduced the statistical handling of the evaporation phase. Based on Weisskopf's formula for the emission probability as a function of excitation energy and of nuclear properties, they traced the development of "evaporation cascades" with the Monte Carlo method and obtained as a result the average number and energy of emitted particles of each sort as a function of target mass number and excitation energy (e.g., for neutrons, protons, deuterons, tritons, ^{3}He and ^{4}He-nuclei). These data can be translated into mass yield distributions and thus cross sections can be predicted. Calculations were carried out for average *excitation* energies between 100 and 700 MeV for target masses between 49 and 239 (Dostrovsky *et al.*, 1958) and for average *excitation* energies below 50 MeV for target masses between 50 and 74 (Dostrovsky *et al.*, 1959). Average excitation energies between 25 and 50 MeV correspond to initial energies below 400 MeV (Metropolis *et al.*, 1957b). This energy range is important for the secondary radiation in large objects and also for objects exposed to the solar radiation.

A great number of Monte Carlo calculations for special reactions were performed and have demonstrated the good agreement of the theory with experimental values (see Ref. 5 in Porile and Tanaka 1964). However, available data concern mainly reactions not so important from our point of view. To overcome this shortcoming, many cross sections not explicitly given in the literature could be obtained by making use of the code and programs already worked out. Other cross sections have already been calculated but they have not been published. They may be obtained from Bertini (1963) and Bertini and Dresner (1963).

VI. MEASUREMENT OF INTERACTION PRODUCTS

In this section we will outline the methods which have been applied to study nuclear interaction products in extraterrestrial objects. Because of the very small abundances of these products classical methods of microchemical analysis are usually ineffective even in their most sophisticated form. Moreover, the method has to yield concentrations of isotopes rather than of elements. Thus, mass spectrometry, counting methods and neutron activation are applicable.

Stable or long-lived isotopes are not completely cosmogenic. They are partially primordial, radiogenic or contamination originated. We will not discuss here how to divide the measured quantities into these components, though to do this is sometimes more difficult than the measurement itself. In principle, one uses the mass spectrometer to measure abundances of stable isotopes. Counting techniques are used to detect radioactive isotopes. However, long lived isotopes such as ^{40}K or ^{81}Kr have also been measured mass spectrometrically (Voshage and Hintenberger, 1963; Marti 1967).

A. Mass Spectrometry

Different techniques have been developed depending upon whether the isotopes to be detected are gaseous or solid. The first discovery of a cosmogenic isotope (He^3 in iron meteorites, Paneth et al., 1952) was performed by gas mass spectrometry. Since then rare gas isotopes have been a major source of information. In fact, gas mass spectrometry applied to cosmochemistry still deals exclusively with noble gases. Since Paneth, the method has been refined by numerous authors with particularly important contributions by Reynolds (1956), Gentner and Zähringer (1955, 1957) and Schaeffer (1959). A recent review of the technique is given by Kirsten (1966).

The rare gases are extracted by melting of the meteorite sample in a high vacuum furnace. Active gases are removed with hot titanium getters. One then introduces the gas into a staticly operated Nier-type mass spectrometer, equipped with an electron impact ion source and an electron multiplier. Calibration occurs by means of known standard samples or isotope dilution. In a sophisticated system, gas amounts as small as 10^{-13} cm³ STP can be detected.

For solid isotopes one concentrates the element to be detected on the filament of the thermion-source of a solid source spectrometer by either chemical or physical methods. Calibration is usually done by isotopic dilution. Stable solid cosmogenic isotopes have first been measured by Voshage and Hintenberger (1959). They discovered cosmogenic ^{39}K, ^{40}K, and ^{41}K in iron meteorites. Subsequently, cosmogenic ^{42}Ca, ^{43}Ca, ^{44}Ca, ^{46}Ca, ^{50}V, ^{54}Cr, ^{6}Li, and ^{7}Li were detected mass spectrometrically (Stauffer and Honda, 1961; Hintenberger et al., 1965a). Detailed descriptions of the method are given by Voshage and Hintenberger (1961) and by Brunnee and Voshage (1964).

B. Low-Level β-Counting

Radioactive nuclides of cosmogenic origin were first detected by Begemann et al. (1957) and Fireman and Schwarzer (1957). They measured meteoritic tritium by means of β-counters. Since that time a great number of β-active isotopes with half-lives ranging from 15 hr up to 1.3×10^9 years have been measured in many laboratories. In β-counting it is necessary to extract the active components by chemical or physical methods. Geiger counters, proportional counters and end window counters have been used. Great efforts were made to increase the sensitivity of detection. This can be achieved by background reduction through shielding and by use of coincidence techniques (low-level counting). Details and progresses in low-level counting applied to cosmochemistry are treated extensively in the "Proceedings" (1967) of a recent meeting on this matter in Monaco.

C. γ-Spectroscopy

Anders (1960) placed a 70-g piece of the Plainview Chondrite between two thallium doped NaI-scintillation crystals. He then measured the ^{26}Al activity by counting the coincident emission of the two 511-keV positron-annihilation γ rays and the 1.83 MeV γ ray from the transition between excited state and ground state. This γ–γ coincidence method is nondestructive but requires relatively large specimens. It can be applied to all positron emitters (e.g., ^{22}Na). Details may be found in Gfeller et al. (1961), Biswas *et al.* (1963), and Heymann and Anders (1967). The advantage of three coincident γ rays does not exist for electron emitters accompanied by a single γ ray. Nevertheless, activity measurements by means of single γ-spectroscopy were also successful. For details see Van Dilla *et al.* (1960), Rowe *et al.* (1963), and "Proceedings" (1967).

D. Neutron Activation

Small concentrations of stable or long lived isotopes exposed to pile neutrons will undergo nuclear reactions such as (n, p), (n, γ), or (n, α). If the reaction product is radioactive one can measure its decay by counting. Then, the decay rate is a measure of the concentration of the stable isotope involved. Neutron activation in general is described e.g. by Schulze (1962). Neutron activation was first applied to cosmochemistry by Fireman and Schwarzer (1957). They measured meteoritic ^{3}He by means of the reaction ^{3}He (n, p)^{3}H $\xrightarrow{\beta^-}$ ^{3}He. Among the cosmogenic isotopes measured in this way are ^{45}Sc and ^{41}K (Wänke 1960, 1961).

VII. INFORMATION DERIVED FROM HIGH-ENERGY INTERACTION PRODUCTS

A. Information about Radiation

1. *Cosmic Ray Variations in Time and Space*

The parent bodies of meteorites are thought to be quite large, having diameters well exceeding the average penetrating depth of cosmic rays in matter. Thus, meteoritic matter was shielded in the early stage of its evolution. Sometimes, the parent bodies broke into meter-sized parts due to collisions. Small pieces thus were exposed to cosmic radiation, and high-energy interaction products were accumulated.

The concentration level buildup in the interaction depends on the half-life of the reaction product. A stable isotope is accumulated during the whole exposure time and it acts as a monitor for the total flux. On the other hand, a radioactive isotope reaches an equilibrium concentration after about five half-lives. Then, the decay rate becomes equal to the production rate as long as the cosmic ray intensity remains stable. Irradiation is interrupted when the meteorite falls to the earth and is shielded by the earth's atmosphere. After the fall, the radioactive isotopes decay while the stable reaction products remain in the meteorite.

If the cosmic ray flux changes during the journey of the meteorite in space, either in time or along the orbit of the meteorite, the activity level reached during equilibrium will also change. If the characteristic time of such variations is small compared to the half-life of the radioactive isotope, then the activity level will reflect an averaged intensity. If the half-life is smaller than the variation period, the activity will follow these changes. Thus, we have a method to detect cosmic ray flux variations over either time or space. All one has to do is to compare the activity ratio of a long- and a short-lived isotope in a meteorite at the time of fall to the relative production rates of these two isotopes as obtained from nuclear data.

The selection of different pairs of isotopes with different half-lives allows us to study short-term variations as well as long-term variations. The task becomes greatly simplified if one uses pairs of isotopes of similar mass numbers. It is well known that the shapes of the excitation functions for such pairs are very similar. Thus, the ratio of production rates of the two isotopes becomes independent of the energy, and one needs to know the cross section ratio for one energy only. If the measured activity ratio of two isotopes A_1 and A_2 with half-lives T_1 and T_2 ($T_1 \gg T_2$) agrees with the laboratory cross section ratio, one can conclude that the cosmic ray intensity averaged over approximately the last $3T_1$ years was essentially the same as when averaged over the last $3T_2$ years. In the following, this will be expressed by the simplified phrase "The intensity has been constant during the last $3T_1$ years." More accurately, the activity ratio reflects the ratio of the time-averaged cosmic ray fluxes $I(t)$ weighted exponentially by the time elapsed because of the decay of the radioactivity:

$$\frac{D_1}{D_2} = \frac{T_2 \int_0^\infty I(t) P_1{}^* \exp(-\lambda_1 t)\, dt}{T_1 \int_0^\infty I(t) P_2{}^* \exp(-\lambda_2 t)\, dt}, \tag{33}$$

where D_i are the measured decay rates and $P_i{}^*$ are the production rates per flux unit.

As mentioned, the selection of particular isotopes to be used depends upon the time scale for which information is sought. The 34 day-half-life of $^{37}\mathrm{Ar}$ is comparable to the time during which the meteoroid was near the earth's orbit before it entered the atmosphere. Thus, $^{37}\mathrm{Ar}$ reflects the cosmic ray intensity at ~ 1 a.u. immediately before the fall of the meteorite. The 2.6-year half-life of $^{22}\mathrm{Na}$ is comparable to typical revolution periods of objects on eccentric orbits reaching the outer asteroid belt at 3–4 a.u. It is believed that many meteorites circled the sun on such paths. Indeed, stereoscopic photography of the fall of the Pribram meteorite in Czechoslovakia in 1959 yielded such an orbit and a corresponding revolution period of 4 years (Ceplecha, 1960). Kepler's second law requires that objects in eccentric orbits spend more time near aphelion than near perihelion. Hence, the ^{22}Na-activity buildup during the whole journey of the meteoroid reflects, mainly, the average cosmic ray intensities at 3–4-a.u. distance from the sun. ^{22}Na may also serve to study variations due to the 11-year cyclic modulation of galactic cosmic rays at such distances.

The isotope ^{39}Ar (half-life 269 years) yields average cosmic ray intensities with regard to orbit and 11-year variations. It might also reflect secular intensity variations on a 100-year time scale, as indicated by ^{14}C data (De Vries, 1958; Suess, 1965). ^{36}Cl (half-life 3.1×10^5 years), ^{26}Al (7.4×10^5 years), and ^{10}Be (2.7×10^6 years) average over the last 10^6–10^7 years. ^{40}K (1.3×10^9 years) covers the whole exposure history of iron meteorites. The same is true for stable isotopes.

Work on the problem discussed here depends upon the availability of freshly fallen meteorites because isotopes with relatively short half-lives have to be detected. In addition, one needs to know the date of fall of the meteorite in order to correct for the activity already decayed at the time of the measurements. First results were obtained by Sprenkel et al. (1959), Vilcsek and Wänke (1961), and Heymann and Schaeffer (1962). It was found that the ^{39}Ar/^{36}Cl ratios in iron meteorites are in general agreement with the cross-section ratio and it was concluded that average cosmic ray intensities have probably been constant within 10%, at least for the last 100,000 years. From similar measurements of ^{26}Al/^{10}Be, ^{10}Be/^{36}Cl, and ^{26}Al/^{36}Cl ratios in iron meteorites, it was concluded that the cosmic ray intensity has been constant within a factor of 2 during the last 10^7 years (Heymann and Schaeffer, 1961; Honda and Arnold, 1961; Honda et al., 1961; Sammet and Herr, 1963; Crevecoeur and Schaeffer, 1963).

Since Arnold et al. (1961) showed that there was a general agreement between measured activities in the Aroos meteorite and calculated production rates (based on present flux rates, see Section IV,B), it is clear that even during the whole exposure time of iron meteorites no dramatic intensity changes have occurred. Voshage (1962, 1967) studied this time interval by means of ^{40}K in more detail. His latest data led him to state that cosmic ray intensity has been stable within 25% during the last 10^9 years.

Local variations have been studied by means of the pairs ^{37}Ar–^{39}Ar and ^{53}Mn–^{54}Mn in freshly fallen stone and iron meteorites (Fireman and DeFelice, 1960a; Stoenner et al., 1960; Davis et al., 1963; Cobb et al., 1963) and Shedlovsky et al. (1967). To a first approximation ti was found that cosmic ray intensities at 1 a.u. and 4 a.u. are equal. More accurate measurements have indicated a slight heliocentric radial intensity gradient (Schaeffer et al., 1963). Recently, Fireman (1967a) obtained an increase of 20% per a.u. distance from the sun, independent on the phase of the solar cycle. A 10%-increase per atomic unit was directly measured between 1 and 1.5 a.u. by O'Gallagher and Simpson (1967). Fireman (1967a) measured the ^{22}Na/^{26}Al ratios in six stony meteorites, which fell between 1959 and 1967. He concluded that the 11-year modulation at 3–4 a.u. is less intense than it is in the neighborhood of the earth's orbit, as revealed by the ^{22}Na changes.

Yokoyama and Mabuchi (1967) pointed out that some chondrites contain excess amounts of isotopes with half-lives shorter than half a year (^{45}Ca, ^{51}Cr). This may be due to the capture of thermalized solar neutrons by ^{44}Ca and ^{50}Cr. The required neutron flux should also cause a detectable excess of the 5.2-year ^{60}Co, but no such excess is observed. This might be an indication of either a rapid decrease of the solar radiation intensity at 3–4 a.u. or of very eccentric meteorite orbits.

The methods described above presume that the shape of the meteorite involved has remained unaltered. If the meteorite suffered a major change in size by a collision, the fresh surface will be exposed to cosmic rays so that it is possible that some of the longer-lived isotopes may not be at equilibrium with the cosmic ray flux. This question will be discussed in more detail in Section VII,B.

2. *Cosmic Ray Energy Spectrum and Effective Energy in the Past*

In Section IV,B it was mentioned that a spallation product distribution of the form

$$P(\Delta A) = k_1(\Delta A)^{-k_2} \tag{11}$$

is a closed analytical solution of the integral in Eq. (7).

Considering only iron meteorites (neglecting the changes due to different behavior of nickel or minor constituents), this can be expressed by a relation

$$P^*(\Delta A) = \int f(E)\sigma(\Delta A, E)\, dE = k_1(\Delta A)^{-k_2}, \tag{34}$$

where P^* is the production rate per iron atom and $\sigma(\Delta A, E)$ is taken from Eq. (22). By means of Eq. (34) it is possible to deduce the energy spectrum, $f(E)\, dE$, from the measured distribution $k_1(\Delta A)^{-k_2}$. An analytical treatment has been presented by Geiss *et al.* (1962). The spectrum obtained by this method is the effective spectrum present in the meteorite and might differ significantly from the primary spectrum of large meteorites.

To calculate k_1 and k_2 from Eq. (11) one uses results from stable isotope measurements since they represent mostly the total isobaric yield of their mass number. From stable isotope abundances measured in the Grant iron meteorite, the constant k_2, responsible for the energy spectrum, is found to be 2.4. The resulting energy spectrum is then

$$f(E)\,dE = f_0 E^{-1.93}\,dE \tag{35}$$

(Geiss *et al.,* 1962).

For energies below 500 MeV, the spallation product distributions are not sensitive to varying spectra. This is because low-energy particles become partly absorbed by different mechanisms, such as ionization. However, for energies above 800 MeV, the spectrum of Eq. (35) agrees within 20% with the observed present cosmic ray spectrum according to Eq. (1). Therefore, for energies above 1 GeV one can conclude that the average energy spectrum received by the Grant meteorite has been similar to the present spectrum.

It might be useful to define an effective energy in such a way that the production rates of certain isotopes derived by use of the cross sections valid at this energy agree with the production rates resulting from the energy spectrum actually present in the meteorite, including the secondary spectrum. Kirsten *et al.* (1963) calculated effective energies by comparison of the average ^{3}He/^{21}Ne ratio in chondrites with the cross-section ratios at different energies. In stone meteorites, ^{21}Ne is mainly produced by Si, Mg, and Al. The cross sections for these reactions are not very energy dependent. However, the ^{3}He cross sections increase to higher energies. Hence, the ^{3}He/^{21}Ne ratio is sensitive to changes of the effective energy. It was found that the effective energy inside the chondrites, averaged over the whole exposure time and including secondaries, has been ~300 MeV.

3. *Composition of Cosmic Radiation*

Inelastic interactions depend only very little on the nature of the bombarding particles. Therefore, spallation reaction products are not applicable to the study of the chemical composition of cosmic radiation in the past. However, recently Cantelaube *et al.* (1967) reported the possibility of detecting tracks caused by primary heavy ions ($Z > 25$) in meteorites. The radiation damage caused by these ions can be made visible in the microscope by etching techniques. It can be distinguished from similar but isotropically distributed tracks due to fission phenomena by its anisotrophic distribution. This method might yield more information in the future.

B. Information about Irradiated Objects

1. *Terrestrial Ages of Meteorites*

The terrestrial age Θ of a meteorite is defined as the time elapsed since its fall to the earth. If the fall is not observed, it is possible to calculate Θ by studying the decaying radionuclides in the meteorite because their production was interrupted at the time of fall. Equations (9) and (10) yield the relation

$$D_i(\Theta)/D_k(\Theta) = (P_i \exp -\lambda_i\Theta)/(P_k \exp -\lambda_K\Theta). \qquad (36)$$

D_i, $k(\Theta)$ are the decay rates measured at the time Θ elapsed since the fall of the meteorite. Equation (36) is valid for isotopes that have half-lives which are short compared to the total exposure time. From Eq. (36) we obtain

$$\Theta = \frac{\ln (D_i(\Theta)P_k/D_k(\Theta)P_i)}{\lambda_K - \lambda_i}. \qquad (37)$$

The ratio P_k/P_i can be measured in meteorites with a known date of fall. This ratio is essentially the same for all meteorites of similar composition and size. Another possibility is to use pairs of isotopes having similar mass numbers. In this case the excitation functions behave similarly and one has $P_k/P_i = \sigma_k/\sigma_i$ with $\sigma_{i,k}$ taken at any energy. Θ can then be obtained from the laboratory cross sections and from the measured decay rates $D_{i,k}(\Theta)$. The choice of isotopes to be used depends on the presumed terrestrial age. The method is most sensitive if Θ is comparable to $1/\lambda_k$ and $1/\lambda_i \gg \Theta$.

Terrestrial ages have been determined for many stone and iron meteorites by means of the ratios $^{39}Ar/^{36}Cl$, $^{36}Cl/^{10}Be$, $^{14}C/^{26}Al$, $^{14}C/^{10}Be$, and $^{26}Al/^{36}Cl$ (Fisher and Schaeffer, 1960; Honda et al., 1961; Vilcsek and Wänke, 1961; Goel and Kohman, 1962; Suess and Wänke, 1962; Kohman and Goel, 1963; Vilcsek and Wänke, 1963). The results are reviewed by Anders (1964) and recently by Wänke (1966). A surprising result obtained by these methods was the discovery of unexpectedly high terrestrial ages. This indicates that the destruction by weathering may be much slower than previously assumed. The Potter stone meteorite has been on earth longer than 21,000 years (Goel and Kohman, 1962), and the terrestrial age of the iron meteorite Tamarugal is higher than 1.5×10^6 years (Vilcsek and Wänke, 1963). The method has also been used to determine the age of meteorite craters on the earth (Goel and Kohman, 1962).

The knowledge of terrestrial meteorite ages allows in some cases a determination of previously unknown half-lives for certain isotopes. The method is obvious from Eq. (37). The half-lives for ^{44}Ti and ^{32}Si have been found in this way (Lipschutz, 1964).

2. Ablation

During its flight through the atmosphere, the surface of the meteorite melts and a considerable part of its mass is lost by ablation. The depth dependency of radionuclide production may serve as a tool to calculate the original size of the meteorite that was exposed to radiation before ablation took place. If the $^3He/^4He$ or the $^3He/^{21}Ne$ ratio, measured directly below the surface of the meteorite, corresponds to the theoretical production ratio at a depth of 10 cm, one can conclude that the ablation affected the outer 10 cm of the meteorite. The method has been applied to iron meteorites by Ebert and Wänke (1957), Hoffman and Nier (1958, 1959, 1960), Fireman (1958b), and Signer and Nier (1960). Typical relative mass losses range from 40 to 90% of the original mass. The variation is not surprising since the ablation depends on the geocentric velocity and on the total mass of the entering objects. Furthermore, ablation may well occur in an asymmetric manner. This has been proved by Fireman (1959), Hoffman and Nier (1959), and Signer and Nier (1960). These authors reconstructed the original shape of iron meteorites by measuring helium concentrations at various depths. The obtained profiles of constant helium concentration are necessarily parallel to the original surface. They are, however, by no means parallel to the present surface of the meteorite. In the same way one can recognize if a meteorite has been broken in pieces during its flight through the atmosphere. (Breakup on the earth's surface is obvious form the partial lack of a melting crust.)

Begemann and Vilcsek (1965) measured the preatmospheric size of stony meteorites in a way which has been suggested in principle by Van Dilla *et al.* (1960). They extracted the chlorine contained in three stony meteorites without affecting the iron and calcium phases which contain the spallation-produced ^{36}Cl. The ^{36}Cl contained in the dissolved phase is produced mainly by thermal and epithermal neutron capture by primordial ^{35}Cl. The yield of the reaction is strongly depth dependent since the secondary neutrons can be slowed down only if the body exceeds a certain minimum size. Applying the calculations of Eberhardt *et al.* (1963) for the thermalization of neutrons inside of the meteorites, Begemann and Vilcsek found ablations between 3 and 16 cm for the three meteorites. An even lower ablation (~1 cm) has been reported by Price *et al.* (1967) and by Cantelaube *et al.* (1967) for a mesosiderite and an amphoterite. This was done by comparing measured and calculated radiation damage trace densities of heavy cosmic ray nuclei immediately below the meteorite surfaces.

The existence of meteorites with such surprisingly low ablation losses may give us an opportunity to study the interaction products caused by the low-energy solar radiation within the uppermost layers of recovered meteorites (Lal *et al.*, 1967; Yokoyama and Mabuchi, 1967).

It should be mentioned that the results of this section would be altered if a slow and continuous size alteration in space occurs. However, the present experimental evidence does not tend to confirm the presence of appreciable space erosion. The matter will be discussed in Section VII,B,4.

3. Meteorite Exposure Ages

In Eq. (8), we defined $Q_i(d)$ as the total number of atoms of a stable radionuclide formed during the exposure time T in one gram of meteorite material taken at the depth d. The results of Section VII,A,1 justify, as a first approximation, the assumption of a time independent radiation intensity during T. Therefore, the production rate of the stable isotope becomes

$$P_s(d) = Q_s(d)/T. \tag{38}$$

According to Eq. (9), the production rate of a radioactive isotope with a half-life small compared to T is

$$P_r(d) = D_r(d), \tag{9}$$

where $D_r(d)$ is the decay rate of the active isotope in decays per gram and second at the fall time of the meteorite and measured at the same depth d.

Nuclides with similar mass numbers are formed in reactions with a fairly similar shape of the excitation functions. This simplifies the expression for the production rates as given in Eq. (7), and one has for the ratio of a stable and a radioactive isotope

$$P_s/P_r = \sigma_s/\sigma_r \tag{39}$$

(σ_s, σ_r taken at any energy).

Then, from Eqs. (38), (9), and (39) we have

$$T = (\sigma_r/\sigma_s)Q/D. \tag{40}$$

This equation enables us to calculate the exposure time T (also called radiation age or exposure age) of the meteorite from one ratio of nuclear reaction cross sections, the concentration of a stable isotope, and the decay rate of a radioactive isotope. Roughly, T measures the time during which the meteorite was exposed to radiation as a body whose dimensions were small compared to the average penetrating depth of the cosmic radiation.

The first radiation age of an iron meteorite based on the method described was obtained by Fireman and Schwarzer (1957). The first radiation age of a stony meteorite was obtained by Begemann et al. (1957). In both cases the isobaric pair ^3He–^3H was used. Since then exposure ages of a great number of stony and iron meteorites have been measured by numerous authors. Frequently used pairs of nuclides are ^3He–^3H, ^{36}Ar–^{36}Cl, ^{38}Ar–^{39}Ar, ^{22}Ne–^{22}Na, ^{21}Ne–^{26}Al, and ^{21}Ne–^{36}Cl.

It is striking to note that the stable isotope is always a rare gas. This is by no means accidental. It is brought about by the necessity of using isotopes which have very low primordial concentrations. Otherwise, one would be unable to distinguish the small cosmogenic contribution from the primordial background. Since meteorites have been highly fractionated with regard to rare gases during their formation, rare gases fit the requirement better than any other element.

For isobaric pairs such as ^{3}He–^{3}H, ^{22}Ne–^{22}Na, or ^{36}Ar–^{36}Cl one has to take into account the fact that the active isotope decays into the stable one. If the half-life of the active isotope is short compared to the radiation time, the cumulative cross section $\sigma_s + \sigma_r$, rather than just σ_s, has to be used to calculate exposure ages, according to Eq. (40). If the radioactive isotope has a very long half-life, it is possible to replace the activity measurement by mass spectrometric measurement. This has been done by Voshage (1967) using ^{41}K–^{40}K and, recently, by Marti (1967) using ^{78}Kr–^{81}Kr.

As an exception, the ^{40}K–^{41}K method does not use a rare gas isotope and it is indeed a major problem to distinguish the cosmogenic from the primordial potassium. However, the primordial potassium concentrations of iron meteorites are very low and usually comparable to the cosmogenic amounts. Hence, the distinction is made possible by means of the very different isotopic composition of cosmogenic and primordial potassium. Stone meteorites have much higher primordial potassium concentrations than iron meteorites. This is one of the reasons why the ^{40}K–^{41}K method cannot be applied to stony meteorites. The ^{40}K–^{41}K method is unique in that the active isotope is not in radioactive equilibrium. This is the case even for those objects which have the longest known radiation ages. This can be taken into account by a simple mathematical correction, with the advantage that both isotopes used for the exposure age calculation record the actual radiation history during the whole exposure time.

In the following we will summarize the results of reported radiation ages in general. Details may be found in many comprehensive papers (Schaeffer, 1962; Kirsten et al., 1963; Vilcsek and Wänke, 1963; Anders, 1964; Hintenberger et al., 1964b; Heymann, 1965, 1967; Vinogradov and Zadorozhny, 1965; Eberhardt et al., 1966; Zähringer, 1966; Wänke, 1966; Voshage, 1967; Heymann and Mazor, 1967; Zähringer, 1968).

The interpretation of the experimental data will be discussed in Section VII,B,4.

a. Exposure Ages of Stone Meteorites. A fairly large number of exposure ages based on activity measurements have been obtained. The main contributions were made by Begemann et al. (1957, 1959), Goebel et al. (1959), Fireman and DeFelice (1960b, 1960c, 1961), Geiss et al. (1960a, 1960b), Goebel and Schmidlin (1960), Stoenner et al. (1960), St. Charalambus and Goebel (1962), Fireman et al. (1963), Rowe et al. (1963).

Cobb (1963), Vilcsek and Wänke (1963), Fechtig and Gentner (1965), Begemann (1966), Heymann and Anders (1967), and Shedlovsky *et al.* (1967). The most important result was the discovery of a large gap between typical exposure ages with a mean value of a few million years and the age of the meteorites as measured by radioactive dating methods (about 4.5 billion years). Hence it was evident that exposure age and radiogenic age date two quite distinct events.

Despite the importance and accuracy of the data obtained by means of actual activity measurements, the data have not been sufficiently abundant to study regularities among the different meteorite classes statistically. This is, at least partly, due to the requirement of relatively large samples of meteorites with known date of fall. Fortunately, a somewhat simplified procedure can be applied to many stone meteorites for which stable rare gases are measured but radioactivities are not measured. In these cases it is possible to assume that the average production rate is the same for all stone meteorites. Due to the relatively small size of the recovered parts of stone meteorites, the measured tritium decay rates decline only rarely by more than 30% from the average of 0.5 dpm/g* at the time of fall (Urey 1959; Bainbridge *et al.,* 1962). Despite the inaccuracy of the single exposure age derived from average production rates, it yields very valuable information if one considers the exposure age distribution of meteorites of a particular meteorite class statistically. The individual error is compensated for by the gain of statistical reliability due to the increase in the number of available exposure ages.

To calculate ^{3}He ages one usually uses a cross section ratio $\sigma(^3H)/\sigma(^3He) = 1$. This ratio is somewhat uncertain since no cross section data for oxygen, a major constituent of stone meteorites, are available. However, Goebel *et al.* (1964) radiated the Ramsdorf chondrite and obtained, in a direct measurement, the same value for the ratio $\sigma(^3H)/\sigma(^3He)$. Using the averaged data as discussed, one arrived at a ^{3}H production rate in chondrites of $\sim 2 \times 10^{-8}$ cm^3(STP)/g-10^6 years (Kirsten *et al.,* 1963; Anders, 1964). Average production rates for ^{21}Ne and ^{38}Ar can be calculated by multiplying the above ^{3}H production rate with the average values of the ^{21}Ne/^{3}He and ^{38}Ar/^{3}He ratios measured in many cases for each chemically similar group of meteorites. This results in a ^{21}Ne range from 0.2 to 0.43 × 10^{-8} cm^3 ^{21}Ne/g-10^6 years for the different meteorite classes (Anders, 1964). The ^{38}Ar production depends on the calcium contents of the meteorites which vary considerably within all meteorite classes except bronzite and hypersthene chondrites. Thus, only for the latter classes is an average production rate of ^{38}Ar meaningful (0.05 × 10^{-8} cm^3 ^{38}Ar/ g-10^6 years).

* Disintegrations per minute per gram.

Cobb (1963) measured the ^{38}Ar–^{39}Ar exposure ages of the separated iron phase of a number of chondrites and calculated the production rates for ^3He, ^{21}Ne, and ^{38}Ar in the whole meteorite from their whole rock rare gas concentrations. Thus, he overcame the uncertainty of the unknown $He^3/$ ^{3}-production cross section in oxygen. The results show general agreement with the values cited above.

By means of the production rates given above one can convert the measured stable cosmogenic rare gas concentrations directly to exposure ages. This has been done for the different classes by various authors. The most recent compilation of all available exposure ages of stony meteorites has been given by Zähringer (1968). The results are compiled in Figs. 1 and 2. Most of the rare gas data compiled in these figures are taken from Kirsten et al. (1963), Hintenberger et al. (1964b, 1965b), Vinogradov and Zadorozhny (1965), Kaiser and Zähringer (1965), Eberhardt et al. (1966), Heymann (1967), Heymann and Mazor (1967, 1968), Zähringer (1968).

The numbers included in Figs. 1 and 2 refer to the morphological type of the chondrites according to a classification given by Van Schmus and Wood (1967). Increasing numbers correspond to an increasing degree of metamorphism as observed in microscopical studies of the texture of the meteorites. The numbers may well be correlated to some genetic relationships. The main features of the exposure age distributions may be reviewed as follows:

1. Most of the stony meteorites have ages below 100 million years. A concentration exists for very young ages below 10 million years. Extreme values are ~20,000 years for the Farmington meteorite (Heymann and Anders, 1967) and 220 million years for Norton County (Begemann et al., 1957).

2. The distribution shows marked differences among the different meteorite classes (this was first assumed by Geiss et al., 1960b).

3. Bronzites have a distinct peak at about 4 million years. The further distribution decreases nearly exponentially at higher ages.

4. Hypersthenes show a rather broad distribution with a nearly linear decrease at higher ages. The statistical significance of some broad maxima is still in question (see, e.g., Tanenbaum, 1967; Zähringer, 1968).

5. Carbonaceous chondrites are concentrated at ages below 15 million years (Heymann and Mazor, 1967).

6. The relative abundance of exposure ages above 50 million years is higher for achondrites than for chondrites. Distinct accumulations exist for aubrites at 50 million years (Kirsten et al., 1963; Eberhardt et al., 1965), and for diogenites at 20 million years (Eberhardt et al., 1966).

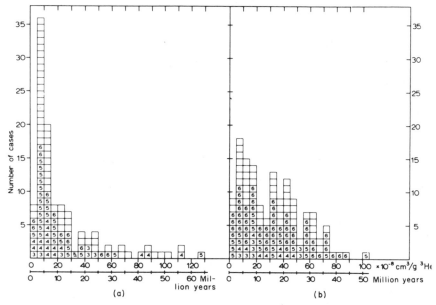

FIG. 1. Exposure age distribution of ordinary chondrites based on the ³He content: (a) bronzite chondrites (115 cases); (b) hypersthene chondrites (142 cases). The numbers 3–6 refer to the morphological type according to the classification by Van Schmus and Wood (1967). [Taken from Zähringer (1968).]

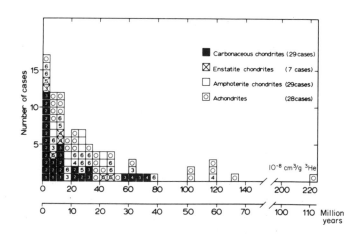

FIG. 2. Exposure age distribution of nonordinary stony meteorites based on the ³He content. The numbers 1–6 refer to the morphological type according to the classification by Van Schmus and Wood (1967). [Taken from Zähringer (1968).]

7. Clustering exists for the bronzites of texture type 5 (Zähringer, 1968) and for carbonaceous chondrites of texture type 2 (Heymann and Mazor, 1967).

8. The amphoterite distribution is similar to that of the hypersthenes (Kaiser and Zähringer, 1965; Heymann, 1965). Exposure ages of eucrites and howardites are distributed more like bronzites.

9. For at least one meteorite, New Concord, two distinct exposure ages for different pieces have been observed (2 and 36 million years, Zähringer, 1966), indicating multiple collisions in space.

10. There is some indication from lowered $^3He/^{21}Ne$ ratios and lowered radiogenic ^4He content in bronzites and hypersthenes that a part (10%, Zähringer, 1966) of the helium of these meteorites underwent diffusion losses. This does not change the exposure age histogram if it is based on ^{21}Ne.

Before we interpret the observed features we will also summarize the exposure age distribution of iron meteorites.

b. Exposure Ages of Iron Meteorites. Exposure age determinations for iron meteorites based on activity measurements yielded radiation ages averaging more than one order of magnitude longer than those obtained for stony meteorites (Fireman, 1958a; Wänke, 1958, 1960; Vilcsek and Wänke, 1961, 1963; Fisher and Schaeffer, 1960; Van Dilla *et al.,* 1960; Honda and Arnold, 1961, 1964; Honda *et al.,* 1961; Heymann and Schaeffer, 1961, 1962; Signer and Nier, 1962; Goel and Kohman, 1963; Bauer, 1963; Hintenberger and Wänke, 1964; Lipschutz *et al.,* 1965; Begemann, 1965; Schaeffer and Heymann, 1965; Cobb, 1966). Just as in the case of stones, the amount of available results did not achieve statistical significance. (The $^{40}K-^{41}K$ method yielded a large number of data but requires additional information to take shielding effects into account.) Again, one would like to calculate exposure ages from the much more abundant stable isotope data alone. However, for the large iron meteorites it is necessary to take into account the depth variation of the radionuclide production. Shielding lowers the total flux. In addition, secondary particles lower the mean energy and thus change the effective production cross section at larger depths. Fortunately, it is possible to decide from stable rare gas data how deep the sample was located. To do this one uses the energy dependency of the $^4He/^{21}Ne$ ratio. ^{21}Ne from iron is produced almost exclusively by the highly energetic primary radiation. ^4He is also produced by secondaries abundant at greater depths. If one knows for *one* large iron meteorite the radiation age from an activity measurement and also measures the variation of the ^{21}Ne concentration and of the $^4He/^{21}Ne$ ratio with depth, then one can obtain the relation between the ^{4}He/^{21}Ne ratio and ^{21}Ne production rates at corresponding depths, valid for all iron meteorites. These production rates may then be applied to calculate exposure ages of any iron meteorite whose ^{21}Ne and ^{4}He concentrations are known (Signer and Nier, 1962; Wänke, 1966; Schultz and Hintenberger, 1967).

As was already mentioned, the exposure ages measured by the ^{40}K–^{41}K method have also to be corrected for depth dependent production rates. This is because the active isotope is not in radioactive equilibrium and thus does not give the individual production rate at a certain depth. The necessary correction may also be made by means of the ^{4}He/^{21}Ne ratio.

Figure 3 shows the distribution of exposure ages for the different meteorite classes based on the ^{40}K/^{41}K–^{4}He/^{21}Ne method (Voshage, 1967) and based on the ^{21}Ne–^{4}He/^{21}Ne method (Wänke, 1966).

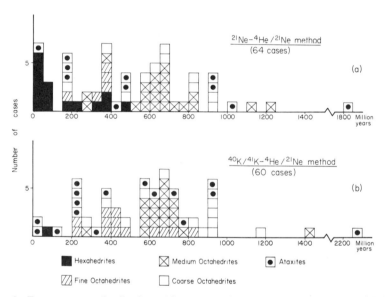

FIG. 3. Exposure age distribution of iron meteorites based on (a) the ^{21}Ne–^{4}He/^{21}Ne method and (b) the ^{40}K/^{41}K–^{4}He/^{21}Ne method. [Data taken from Wänke (1966) and Voshage (1967).]

The results may be summarized as follows:

1. The mean exposure age of iron meteorites is more than ten times longer than the mean exposure age for stony meteorites. Many irons have exposure ages around 700 million years. Extreme values are 4 million years for the Pitts meteorite (Cobb, 1966) and 2.3 billion years for the Deep Springs meteorite (Voshage, 1967).

2. Marked differences in the distribution exist for different types of iron meteorites.

3. All hexahedrites have radiation ages below 500 million years. Most of them have very low exposure ages.

4. Most fine octahedrites have exposure ages of about 400 million years.

5. Exposure ages between 600 and 700 million years are typical for normal octahedrites. Coarse octahedrites have typical exposure ages around 900 million years.

6. Voshage (1967) claims distinct exposure age clusters for subgroups of octahedrites according to their gallium and germanium content (which might reflect genetic similarities).

7. In some cases (Sikhote Alin, Arispe, Odessa) very different exposure ages have been measured on different pieces from the same meteorite fall (Vilcsek and Wänke, 1963; Heymann *et al.*, 1966).

8. ^{41}K–^{40}K ages tend to be about 30% higher than ^{36}Ar–^{36}Cl or ^{38}Ar–^{39}Ar ages.

9. The measured tritium activities in some of the iron meteorites are much lower than expected.

10. Meteorites with low exposure ages often have low $^{3}He/^{4}He$ ratios (Voshage, 1967; Hintenberger *et al.*, 1967).

4. *Origin and History of Meteorites*

It has been found from the exposure age determinations that the exposure ages are generally much shorter than the ages of the meteorites themselves and that they are distributed over a large time interval. It is clear therefore that exposure ages do not determine the lifetime of the primary meteoritic parent bodies. It has been shown that cosmic rays had roughly their present intensity for at least the last 2×10^9 years (Section VII,A,1). Thus, meteorites must have either lost their cosmic ray products or they must have been shielded from cosmic rays inside a larger body.

Heating may lead to a partial loss of gaseous products and indeed does so to a certain extent. However, this cannot be the explanation for the low exposure ages. Many meteorites with low exposure ages do still retain the ^{40}Ar produced by radioactive decay of ^{40}K during their whole lifetime which is ~4.5 billion years. In addition, there are exposure ages not based on rare gases, such as ^{40}K–^{41}K ages. It must be concluded then that the meteorites have altered size and shape considerably during their lifetime and exposure history. This may have occurred either in sudden events (e.g., collisions, breakups, and explosions) or in a rather slow process such as space erosion due to the continual impact of dust and ions on the surface of the meteorites.

a. Space Erosion. If a meteorite is continuously eroded during its exposure time by the impact of high-velocity interplanetary dust and ions, a specimen located at d cm depth below the (ablation-corrected) surface has spent its lifetime on the average at significantly greater depths. Since the production rate of a stable isotope such as ^{3}He decreases exponentially with the distance from the surface, the ^{3}He amount integrated over the whole exposure time will be much lower than that expected for a depth of d cm. However, the production rate of a radioactive isotope is unaffected if its half-life is such that very little erosion takes place during several lifetimes. As a result, radiation ages will be lowered significantly and will no longer reflect the true exposure time (Whipple and Fireman, 1959); Fireman and DeFelice, 1960b; Fisher, 1966).

This process has been proposed to explain the exposure age distribution in terms of different rates of erosion for individual meteorites. Different erosion rates could result either from varying dust densities along the different orbits or from differences in the mechanical strength of the meteorites. The latter argument seemed to be an easy explanation for the large difference between the mean exposure ages of irons and stones.

Another indication of the influence of space erosion was provided by the fact that ^{40}K–^{41}K ages tended to be some 30% higher than exposure ages based on relatively short half-lives such as ^{36}Cl–^{36}Ar or ^{3}H–^{3}He ages. An erosion rate of 10^{-7} cm/year would not affect the buildup of ^{36}Cl activity within $\sim$1 million years. However, it would affect the ^{40}K activity during some 10^8 years by providing additional shielding of more than 10 cm. Space erosion could then account for the 30% difference between ^{40}K–^{41}K and ^{36}Cl–^{36}Ar ages.

There is no question that the process of space erosion takes place. The controversial question is whether or not the erosion rates are high enough to affect the exposure ages appreciably. If they do, they could be calculated according to Whipple and Fireman (1959) by the relation

$$K = L/T, \tag{41}$$

where K is the erosion rate in centimeters per million years, L is the mean absorption length of the cosmic radiation, and T is the apparent exposure age in millions of years. Clusters of exposure ages for stones and irons are found at 5×10^6 years and 5×10^8 years, respectively. If one takes these values as being representative, these groups of meteorites would have undergone erosion according to Eq. (41) at a rate of 10^{-5} cm/years (stones) and 5×10^{-8} cm/years (irons).

In the following we will outline some arguments against such high erosion rates:

1. The size distribution of stony meteorites cannot be understood if erosion rates are high. Most of the small stones should have disappeared (Urey, 1959).

2. The mechanical strength of iron and stone is not very different if small mineral grains are considered. This applies to the problem since the impacting particles are very small (Anders, 1964). Anders' conclusions have been questioned recently by Comerford (1967) and by Fisher (1966).

3. Clustering can not be explained since the erosion rates for meteorites in different orbits should be different (Anders, 1964).

4. If stone meteorites shrink continuously at high erosion rates, they all had to pass through a size range of about 50-cm radius. For bodies of this size the flux of thermalized neutrons in the center becomes that high that an excess of ^{36}Ar should be formed by neutron capture in meteoritic chlorine [$^{35}Cl(n, \gamma)^{36}Cl \xrightarrow{\beta^-} {}^{36}Ar$]. However, such excess ^{36}Ar has not been found (see Eberhardt et al., 1963; Begemann and Vilcsek, 1965). Fireman (1966) came to a contradictory conclusion using similar considerations.

5. The difference between $^{40}K-^{41}K$ and $^{36}Cl-^{36}Ar$ ages in iron meteorites does not exist for all meteorites (Voshage, 1967). $^{26}Al-^{21}Ne$ ages should also be affected but they are not (Lipschutz et al., 1965).

6. The actual dust densities as measured from zodiacal light observation (Van de Hulst, 1947), combined with the meteor impact theory (Öpik, 1958), lead to erosion rates not in agreement with the apparent exposure age distributions (Anders, 1964). Comerford (1967) questioned the applicability of Öpik's impact theory for our case and Fireman (1966) found agreement between zodiacal light density and space erosion. Price et al. (1967) pointed out that the zodiacal light measurements would correspond to erosion rates below 10^{-7} cm/year.

7. Price et al. (1967) studied the tracks due to radiation damage caused by heavy cosmic radiation particles in the outer surface of the Patwar meteorite. At erosion rates of more than 10^{-7} cm/year all tracks should have disappeared, but this is not the case. Altogether, it seems that typical values for the actual erosion rates converge to a value $<10^{-7}$ cm/year for stones (Price et al., 1967) and about 10^{-8} cm/year for iron (Fisher, 1967). At these erosion rates, the observed exposure ages of stones should be >500 million years, and those of iron about 2 billion years. This is contradictory to the actual observation. The conclusion is that space erosion does not essentially affect the exposure ages. This leads to the statement that measured exposure ages must agree with the actual radiation time, that is, with the lifetime of the meteorites as bodies having sizes of no more than a few meters.

b. Collisions and the History of Meteorites. Exposure ages are much shorter than the radiogenic ages since meteorites have been shielded for a long time within relatively large parent bodies. In that space erosion was found to be unimportant, meteorites had to be brought to the surface late in their history in a catastrophic event. The spontaneous explosion of a single body is very unlikely since there is simply no energy source which could account for explosions or for volcanic eruptions in relatively small and solid parent bodies. Ejecta from the solid outer layers of a large body with a molten core and volcanic activity may exist as curiosities but this can not explain the exposure ages in general. The only remaining possibility is that collisions between different parent bodies occurred. Collisional breakups between planetary objects have been suggested by Begemann *et al.* (1957), Geiss (1957), and Eberhardt and Hess (1960). It is now widely accepted that the present size distribution of meteorites is the result of single or multiple collisions in space. The observation of different exposure ages for different parts of the same meteorites (see Section VII,B,3) is a direct experimental proof for breakups.

It is clear now that the exposure age is a measure of the time between the last breakup and the capture of a fragment by the earth. Two cases have to be considered. The first case is one in which the probability for collisions leading to a break up in the locality proposed for the origin of a certain group of meteorites is high. The average time interval T_{co} between two subsequent collisions is much shorter than the average lifetime T_{capt} of the fragments against capture by the earth. The average time between two collisions rather than the capture time determines the exposure age distribution.

The second case is one in which the mean lifetime, T_{capt}, is short compared to the mean time between two subsequent breakups. Therefore, the exposure age distribution is determined by T_{capt}.

It is evident that exposure age distributions provide a method to test models for the origin of meteorites. This is made possible since calculations allow the prediction of both average lifetimes up until planetary capture and collision (and/or breakup) probabilities. Models proposed to explain the origin of meteorites have, of course, to consider many aspects other than exposure ages. This is discussed in review articles by Anders (1964) and Wänke (1966). In this chapter we will discuss evidence from only high-energy interaction products in relation to the origin of meteorites. A proper model should be consistent with the distinct exposure age distributions for the different meteorite classes. This leads, among other things, to the important conclusions that meteorites originated in more than one parent body.

Calculations for the mean lifetime of planetary objects against capture by the earth have been carried out by Öpik (1951) and subsequently refined by Arnold (1963, 1965a, b) and Wetherill (1967). Five cases have been considered.

In the first model, the successive perturbation of the orbits of ring asteroids leads, finally, to capture by the Earth. The model is rejected since impacts between asteroids are not sufficient to produce the necessary initial escape velocity.

In the second model one considers the transfer of ring asteroids into Mars like orbits in the first step. The further perturbation of the orbit by Mars leads to the capture by Earth (Arnold, 1963). The resulting mean life times are comparable to the exposure ages of octahedrites.

Collisions between Earth crossing Mars asteroids are considered in the third model. This model is favored by Anders (1964) as an account of the origin of stone meteorites despite the fact that the predicted capture times are about fifty times higher than the actual exposure ages of chondrites.

Another model involves the capture of apollo asteroids. They had initialy highly eccentric orbits which extend from the asteroid belt to the orbits of Mars *and* Earth. The resulting capture times are too short to explain the exposure ages of octahedrites. Exposure ages of hexahedrites and stone meteorites may be in agreement with this model, especially if subsequent collisions played a role.

A final proposal concerns itself with the capture of debris in impacts of large meteorites or comets on the moon or on mars. This debris is ejected in circular orbits similar to that of Earth. The very short T_{capt} Values are comparable to exposure ages of stone metorites but much too short to account for exposure ages of iron meteorites.

The present opinions concerning the origin of different types of meteorites are controversial. They may be reviewed as follows.

RING ASTEROID FRAGMENTS DIRECTLY EJECTED INTO EARTH-CROSSING ORBITS. The time between two collisions of a member of the asteroidical belt is of the order of 10 years (Kuiper, 1950). Collisions mainly destroy stones. They rarely lead to break up of irons, due to their higher macroscopic strength. The fragments are collected randomly by the earth. One expects continuous exposure age distributions and the difference between irons and stones is understandable. This model is rejected because, among other reasons, it can not explain the different clusters found in exposure age distributions.

RING ASTEROID FRAGMENTS ARRIVED ON EARTH VIA MARS. This model is accepted to account for the origin of octahedrites and ataxites because it is the only one which can explain their long exposure ages.

MARS ASTEROIDS DEFLECTED TO THE EARTH. If one considers only exposure ages, there is not much infavor of this model. However, the exposure ages could be lowered by subsequent collisions $(T_{co} < T_{capt})$. It is also possible that many fragments were produced in a few relatively recent events. In spite of their long mean capture time, the very young end of their expected life time probability distribution may account for the observed exposure ages. Outstanding peaks in the exposure age distribution could be related to single major events.

APOLLO ASTEROIDS DIRECTLY CAPTURED BY THE EARTH. According to their exposure age distribution, stone meteorites as well as hexahedrites could be explained by this model.

LUNAR OR MARTIAN ORIGIN OF STONY METEORITES. A lunar origin of stony meteorites was first suggested by Urey (1959). Recently, the model has come to be favored especially for bronzite chondrites (Urey, 1965; Arnold, 1965a; Zähringer, 1966; Wänke, 1966). It requires the ejection of material on impact out of the Earth-Moon system. This model finds support in the fact that the mean capture time for objects in earth like orbits is in very good agreement with the exposure age distribution of chondrites. Significant peaks in the exposure age distribution, as, for example, the 4-million-year peak for bronzites, are explained as the result of distinct large impacts yielding very many fragments. The 4-million-year peak could be outstanding simply because it is the most recent one (Zähringer, 1968). Meteorites other than bronzites may have been more abundant at earlier times (e.g., aubrites, 50 million years). Differences between the different meteorite groups could be due to their production in different layers of the moon (Zähringer, 1966) or to the origin from different places (e.g., Mars). A similar origin has been proposed for hexahedrites because of their relatively low exposure ages.

It should be mentioned that this model may explain the unusually low $^3He/^{21}He$ ratios in some 10% of the bronzite and hypersthene chondrites and in most of the hexahedrites. Half of the ^3He is produced by the decay of tritium which tends to diffuse at quite low temperatures (Fechtig and Gentner, 1965). This might cause us to believe that the orbits of these meteorites came so close to the sun that the temperature was sufficient to cause tritium losses during the exposure time of the meteorites. Orbits so close to the sun are possible only for meteorites of lunar or martian origin.

The chemical composition of the moon as determined from the Apollo Lunar Missions rules out any appreciable production of meteorites from lunar material.

5. *Early Planetesimal Radiation*

A major problem for nucleosynthesis models is to explain the existence of Li, Be, and B. Fowler *et al.* (1962) suggested that these elements were produced during the formation of the solar system in solid planetesimals by spallation reactions as a result of an intense and highly energetic proton radiation from the sun. The planetesimals are thought to be located in the region of the present planets. In average, they may have been no larger than some 10 m in diameter and no smaller than 50 cm. Otherwise, the Li, Be, and B production would have been insufficient, or the isotopic composition of these elements could no longer be understood by the effect of neutrons thermalized inside of the planetesimals. It is assumed that the planetary bodies have been formed out of the planetesimals. This requires a through mixture of irradiated and unirradiated material. However, there is no *a priori* reason why meteoritic parent bodies formed in the asteroidical belt should have the same percentage of radiation products as the earth. Rather, the ratio of irradiated to unirradiated material and the total flux received could well have been different. Therefore, the actual existence of the proposed high-energy proton radiation may be detected by isotopic anomalies in meteorites. Involved are those spallation products which can not be explained by radiation obtained after the last break up. Krummenacher *et al.* (1962) found ^{124}Xe and ^{126}Xe excess possibly due to spallation of barium and rare earth isotopes in carbonaceous chondrites.

From the ^{124}Xe anomaly, the amount of heavy elements, and the corresponding cross sections, one can calculate that a total proton flux of 10^{18} p/cm^2 could account for the anomaly.

To a smaller extent, a similar effect has been found by Merrihue (1963) in chondrules of the Bruderheim hypersthene chondrite. However, in the latter case this may be due to recent radiation after the last breakup (Merrihue, 1966).

As we have seen, a proton flux of 10^{18} p/cm^2 leads to a detectable amount of ^{124}Xe since the primordial concentrations of rare gases are very low as compared to nonvolatile elements. Corresponding anomalies caused by 10^{18} p/cm^2 in nonvolatile elements lead only to relatively much smaller anomalies and do not exceed the present limits of detection, except perhaps in the case of potassium, as discussed below.

Reported isotope anomalies in nonvolatile elements due to early radiation correspond to required total proton fluxes of 10^{21} p/cm^2 or more (e.g., Ba, Umemoto, 1962; Cr, Shima and Honda, 1966). These seem unlikely if one considers the limit set by the ^{124}Xe anomaly in carbonaceous chondrites. No effects have been found for lithium (Krankowsky and Müller, 1964; Dews, 1966), ^{36}S (Hulston and Thode, 1965), ^{54}Cr (Murthy and Sandoval, 1965; Shima and Honda, 1966), and tin (DeLaeter and Jeffery, 1965).

The most recent study concerning ^{40}K in stone and iron meteorites was made by Burnett et al. (1966a) [note correction in Burnett et al. (1966b)]. The authors set up experiments in which they would have been able to see a ^{40}K enrichment corresponding to a total proton radiation larger than 3×10^{18} protons. However, no difference between terrestrial and meteoritic material has been found. The finding is of particular importance with respect to silicate inclusions from the Weekeroo Station octahedrite. This is the case since one could argue that no differences are to be expected if the meteorites come from the moon. However, we have seen that at least octahedrites are not of lunar origin (Section VII,B,4). As a result one may state that a possible early planetesimal radiation is limited to a total flux of some 10^{18} p/cm^2.

6. Cosmic Dust

The detection of radionuclides in cosmic dust requires sample weights of the order of 1 g. Dust collection in space can not provide such quantities in the foreseeable future, but one can work on dust collected by and accumulated on the earth over long time periods. Most promising are studies on ocean sediments and snow cores from the polar regions. Both should contain appreciable amounts of cosmic dust. Unfortunately, the following difficulties arise:

1. The sampling mechanism is not well understood. The residence time of small particles in the interplanetary space, in the atmosphere, and in the oceans and the sedimentation or precipitation rates are not uniform but depend on changing parameters. Only rough estimates are available. Educated guesses for the time intervals involved are: for the residence time of dust in the interplanetary space, $\sim 10^5$ years; for the residence time of dust in the atmosphere, a few days; for the residence time of vaporized material in the atmosphere, a few years; and for the residence time of dissolved material in the oceans, a few thousand years.

2. Cosmic dust becomes diluted by the bulk of inactive terrestrial dust during settlement. This aggravates the detection of weak radioactivities. Also, apparent production rates will be lowered. This problem is more serious for ocean sediments than for samples collected in arctic snow and ice.

3. Deposited dust might become altered or destroyed by geochemical and geophysical processes.

4. Terrestrial radioactivities are added to the cosmic activities. Such contaminants may be produced by cosmic rays in the atmosphere, by cosmic ray secondaries on the earth's surface, by reactions caused by natural radioactivities, by nuclear explosions and radioactive waste, and by low-energy solar wind particles captured in the atmosphere. Atmospheric activities are abundant only for isotopes below mass number 40 since the heaviest component of air with appreciable abundance is ^{40}Ar. Nuclear weapon tests do not affect isotopes with long half-lives which are detected in relatively old sediments or snow cores.

Easiest to detect are radioactivities induced by the solar radiation since the flux of solar protons is much higher than the galactic flux. If dust has chondritic composition, one would expect to find in it mainly ^{56}Co, ^{55}Fe, ^{26}Al, and ^{53}Mn (Wasson, 1963). ^{56}Co and ^{55}Fe have short half-lives and cannot be used in samples accumulated for hundreds of years. Hence, the isotopse most suitable for detection are ^{26}AI and ^{53}Mn.

Even if activities have been detected and identified as being cosmic, their interpretation faces the following problems:

1. For long-lived isotopes, the residence time of dust in regions of intense solar radiation might not have been sufficient to build up the activities to steady state equilibrium.

2. Interaction products formed in small particles might escape due to recoil.

3. Low-energy solar wind particles stopped in the dust will simulate interaction products.

Lal and Venkatavaradan (1966) and Wasson et al (1967) reported ^{26}Al activities in ocean sediments exceeding the contaminants by factors of 5 and 10. On the other hand, Schaeffer et al. (1964) did not find any ^{26}Al in excess of the atmospheric contamination in Pacific red clays. The same result was arrived at by studies of dissolved and undissolved materials in 200-year-old Greenland ice (Fireman and Langway, 1965; McCorkell et al., 1967).

If one makes certain assumptions about the residence time of dust in the solar system in order to account for the undersaturation and assumes reasonable and stable average solar proton fluxes, it is possible to calculate the expected ^{26}Al activity in pure cosmic dust of assumed chondritic composition by means of Eq. (7). Comparison of this result with the ^{26}Al activity actually measured yields the percentage of cosmic material in the sediment. Since the total sedimentation rates are available from oceanographic studies (e.g., the ionium method), one arrives, finally, at an influx rate of cosmic dust. The ^{26}Al data obtained by the authors mentioned above correspond to influx rates of about $10^{6 \pm 0.5}$ tons/year over the whole earth.

Schaeffer *et al.* (1964) studied ^{36}Cl in dust extracted from Pacific red clays. ^{36}Cl is mainly produced by galactic protons interacting with Ca and Fe. The calculation of influx rates from ^{36}Cl activities is easier to perform, since the average flux of galactic protons is much better known than the average solar flux. Two different samples yielded influx rates of 0.4 and 2×10^6 tons/year over the earth's surface (Schaeffer *et al.,* 1964).

Fireman (1967b) found ^{60}Co activity in Greenland ice. It is not yet decided if this is the result of fallout contamination or of interactions in nickel-rich cosmic dust. Slight indications of spallogenic ^{21}Ne and ^{3}He in Pacific red clays have been found by Merrihue (1964). He also found ^{36}Ar and ^{38}Ar excesses. This argon may be a primordial gas residue which was included while the dust particles were formed. It may also be low-energy solar wind particles captured by the dust. Latter findings were confirmed by Tilles (1966, 1967).

The quantitative interpretation of radionuclides in cosmic dust is still somewhat ambiguous. Qualitatively, however, cosmogenic isotopes are the only proof of the existence of terrestrial dust at the present time.

7. *The Lunar Surface*

The lunar surface is not shielded by an atmosphere. Also the uppermost layers probably do not get lost as in the case for meteorites falling to earth. Hence, a lunar surface sample should show a complete record of both radiation components, solar and galactic. Since the total solar proton flux is about 100 times higher than the flux of galactic protons, one expects, in the upper few centimeters of the lunar surface, reaction products mainly due to low-energy radiation. The deeper layers should reflect the effects due to primary and to secondary galactic radiation.

Most of the γ activity of the lunar surface, as measured by the Luna 10 satellite, is induced by cosmic radiation. Only a small part comes from the natural radioactivity of lunar material. Most of the induced γ radiation comes from the instantaneous deexcitation of the stable or radioactive interaction products while they are formed. Interaction products may be identified by the recorded γ energies and by a consideration of the energy levels off possible reaction products. The measured spectrum indicates the existence of cosmogenic ^{26}Al, ^{20}Ne, ^{24}Mg, ^{14}N, and other isotopes. Another part of the radiation is a result of the decay of radionuclides with very short half-lives such as ^{14}O (72 sec), ^{19}O (27 sec), or ^{20}F (10.7 sec) (Vinogradov *et al.,* 1966).

A theoretical approach to the prediction of the production rates of cosmogenic isotopes such as ^{54}Mn, ^{22}Na, ^{14}C, and ^{7}Be in the uppermost layers of the lunar surface under the influence of the solar radiation has been made by Ebeoglu and Wainio (1966). They used Monte Carlo calculations for the intra- and extranuclear cascade and an assumed basaltic composition mixed with 5% meteoritic material. In the future it will be possible to compare these predicted patterns with the actual distribution of the reaction products. This may well reveal some surprises about the history of the solar cosmic radiation.

8. *Tektites*

Tektites are found in a few geographically well defined areas. These areas include the Philippines, Australia, the Ivory Coast, and Czechoslovakia. Tektites are black glassy objects, generally smaller than an egg. Often tektites seem to be aerodynamically shaped. Chemically they are similar to very acidic silicates. The SiO_2 content is about 80%.

For a long time it was believed that tektites are an unusual type of meteorite. Today it seems clear that they are formed as ejecta of molten material during impacts of very large meteorites on the earth. There is still some discussion of a possible lunar origin. This is not the place to discuss all the evidence for and against each of these arguments (see O'Keefe, 1963; Zähringer, 1963; Gentner *et al.,* 1967). However, we do want to touch upon one piece of evidence which is contributed by applied nuclear chemistry; namely, the exposure ages of tektites. If tektite material was exposed to cosmic radiation, either as "meteoroid," or on the lunar surface, the detection of radioactivies due to spallation processes would prove their extraterrestrial origin. ^{26}Al is a most suitable isotope to look for since it is formed with very high yield from silicon, a major element in tektites. To interpret the results of a search for ^{26}Al we must first know how long the tektites have been on the earth. No tektite "fall" has been observed, but the terrestrial age of tektites is known from radioactive dating methods. Tektites from Southeast Asia, Australia, and the Ivory Coast have K–Ar and fission track ages between 0.7 and 1.3 million years (Zähringer, 1963; Wagner 1966; Gentner *et al.,* 1967). After about 1.3 million years, about 30% of the ^{26}Al activity corresponding to radioactive equilibrium in space should still be present in the tektite. However, no cosmogenic isotopes have been found (Ehmann and Kohman, 1958; Reynolds, 1960; Anders, 1960; Viste and Anders, 1962; Rowe *et al.,* 1963). This could mean that ^{26}Al was not in radioactive equilibrium because tektites were shielded at least until 100,000 years before their fall on earth. It is more likely, though, that tektites are not meteorite like objects at all. From the lack of ^{21}Ne, Reynolds (1960) found an upper limit for the tektite exposure age of 28,000 years. Recently, the limit has been lowered to less than 300 years by the fission track method (Fleischer et al., 1965).

9. Space Probes

Quite a few parts of satellites and rockets have been recovered and radionuclides have been studied in these materials. Due to effects caused by the geomagnetical field, satellites are irradiated in a much more modulated manner than meteorites. In a satellite, the record of radioactivities depends very much on its orbit. Polar orbits receive more radiation than equatorial orbits because of the geomagnetic cutoff. The flight times of satellites are mostly very short. The solar radiation received depends on the solar activity during this short time interval. A part of the isotropic radiation is shielded by the earth's mass. Also, the effective flux will change if the orbit extends into the inner radiation belt of the earth where magnetically trapped protons add to the total flux. Finally, solar flare particles may be stopped in the satellites without interaction and simulate higher fluxes.

Interesting results concerning the solar proton flux and spectrum, the short term variations in the environment of the earth, and the composition of solar flare particles were obtained from interaction products as well as from stopped particles.

Most investigations were carried out on parts of Discoverer satellites. Discoverers have been placed in nearly polar orbits for only a few days, with apogees extending into the inner radiation belt of the earth.

The most interesting results were obtained on Discoverer 17. This satellite was exposed to a 3^+ solar flare on November 12, 1960. Radionuclides were studied in iron battery cases, lead ballast, aluminum flanges and silver bromide emulsions from Discoverers 17, 18, 25, 29, and 36. Activity measurements have been carried out for ^3H, ^{37}Ar, and ^{57}Co in iron (Fireman et al., 1961, 1963; Stoenner and Davis, 1961; Tiles et al., 1961; Wasson, 1962), for ^3He, ^3H, and ^4He in aluminum (Tiles et al., 1961; Fireman et al., v961, 1963; Schaeffer and Zähringer, 1962), for ^3He, ^{37}Ar, ^{127}Xe, and ^{205}Bi in lead (Dodson et al., 1961; Fireman et al., 1961, 1963; Tiles et al., 1961; Schaeffer and Zähringer, 1962; Keith and Turkevich, 1962); and for ^{106}Ag in silver bromide (Wasson, 1961).

The presence of ^{37}Ar and ^{127}Xe can be explained by considering protons trapped in the Van Allen belt which interact with the satellite. Radionuclide production due to galactic cosmic rays did not contribute to the measured activities because of the short exposure time.

The isotopes ^{57}Co from steel, ^{106}Ag from AgBr, and ^{205}Bi from Pb are also explainable if one considers the interaction of solar flare protons. The activities in Discoverer 17 correspond to a total proton flux of about 10^8

p/cm^2 (Wasson, 1961, 1962; Keith and Turkevich, 1962). From the depth variation of ^{205}Bi in thick lead layers, an energy spectrum for solar flare particles proportional to $\sim E^{-5}$ has been deduced (Keith and Turkevich, 1962).

The observed amounts of ^{3}H, ^{3}He, and ^{4}He cannot be accounted for as interaction products. They are due to stopped solar flare particles, and for stopped particles with energies between 85 and 150 MeV it was found that the triton:proton ratio is about 1:200 (Fireman et al., 1963). The ^{3}He/^{4}He ratio for flare particles with energies above $\sim$200 MeV varies between 0.02 and 0.2 (Schaeffer and Zähringer, 1962). On the surface of the sun, the ^{3}He/^{4}He ratio is $\leq$0.02, while for low-energy cosmic radiation (200–400 MeV/nucleon) a ^{3}He/^{4}He ratio of 0.7 has been observed (Rao 1961). Schaeffer and Zähringer therefore concluded that ^{3}He is produced during the flare, probably by the chain reaction H(n, γ)^{2}H(p, γ)^{3}He. Stopped tritium has also been detected in the aluminum flange of a rocket which was fired to a height of more than 60 km during the same flare (Fireman et al., 1963).

Two months later Discoverer 18 was placed in orbit. Although there were no flares at this time, a large excess of tritium has also been found in steel of Discoverer 18. This might indicate that tritons from the November flare were still trapped in the Van Allen belt at this time, corresponding to lifetimes of a few months for particles trapped in the Van Allen belt. This explanation is confirmed by the fact that much lower tritium excesses exist in Discoverers 26 and 36 and no helium was detectable in Discoverers 25 and 29. Nevertheless, the tritium concentrations in Discoverers 26 and 36 are still about fifty times higher than could be accounted for by the interaction with Van Allen particles or by direct radiation. The excess tritium is probably due to trapped tritons. Argon-37 and ^{127}Xe activities in these satellites do agree with expected production rates.

A completely different situation exists for the observations made on iron from Sputnik 4. This satellite was in orbit for 2.3 years with an initial apogee of 690 km. Only minor solar activity took place while Sputnik 4 was in orbit. In this case it makes sense to compare the measured production rates with those measured in iron meteorites, since the radiation time was sufficient to build up detectable amounts of radionuclides formed in the interaction with the galactic radiation.

We find that radioisotopes like ^{3}H, ^{32}P, ^{37}Ar, ^{46}Sc, ^{48}V, ^{49}V, ^{51}Cr, ^{54}Cr, ^{54}Mn, ^{55}Fe, ^{57}Co, ^{59}Fe, and ^{60}Co are produced in abundances corresponding to about 20–50% of the production rates measured in the Aroos iron meteorite. The reduction is caused by the geomagnetic shielding of a part of the galactic radiation (Kammerer et al., 1963; DeFelice et al., 1963; Shedlovsky and Kaye, 1963; Wasson, 1964).

The ratio of the production rate observed in Sputnik to the one measured in Aroos is about 0.2 for ^{37}Ar but 0.7 for ^{55}Fe. This reflects the influence of the low-energy solar radiation for the production of isotopes with small ΔA. The tritium concentrations in Sputnik 4 are very low. Probably a large part of the tritium was lost by diffusion when the satellite reentered the atmosphere.

Note: After completion two related publications came to our attention (Shen, 1967; Sitle, 1967). The chapters written by T. P. Kohlman and M. L. Bender in the former and by M. Monda and J. R. Arnold, as well as W. R. Webber in the latter are closely related to the topic of this chapter.

REFERENCES

Albouy, G., Cohen, J. P., Gusakow, M., Poffe, N., Sergolle, H., and Valentin, L. (1962). Spallation de l'oxygène par des protons de 20 à 150 MeV, *Phys. Lett.* **2,** 306–307.

Anders, E. (1960). The record in the meteorites. II. On the presence of aluminum-26 in meteorites and tektites, *Geochim. Cosmochim. Acta* **19,** 53–62.

Anders, E. (1962). Meteorite ages; *Rev. Mod. Phys.* **34,** 287–325.

Anders, E. (1964). Origin, age, and composition of meteorites, *Space Sci. Rev.* **3,** 583–714.

Anderson, K. A., Arnoldy, R., Hoffman, R., Petersen, L., and Winckler, J. R. (1959). Observations of low-energy solar cosmic rays from the flare of 22, August 1958, *J. Geophys. Res.* **64,** 1133–1147.

Arnold, J. R. (1961). Nuclear effects of cosmic rays in meteorites, *Ann. Rev. Nucl. Sci.* **11,** 349–370.

Arnold, J. R. (1963). "The Origin of Meteorites as Small Bodies, Isotopic and Cosmic Chemistry," pp. 374–365. North-Holland Publ., Amsterdam.

Arnold, J. R. (1965a). The origin of meteorites as small bodies II. The model, *Astrophys. J.* **141,** 1536–1547.

Arnold, J. R. (1965b). The origin of meteorites as small bodies III. General considerations, *Astrophys. J.* **141,** 1548–1556.

Arnold, J. R., Honda, M., and Lal, D. (1961). Record of cosmic ray intensity in the meteorites, *J. Geophys. Res.* **66,** 3519–3531.

Bailey, D. K. (1959). Abnormal ionization in the lower ionosphere associated with cosmic-ray flux enhancements, *Proc. Inst. Radio Engrs.* **47,** 255–266.

Bailey, D. K. (1962). Time variations of the energy spectrum of solar cosmic rays in relation to the radiation hazard in space, *J. Geophys. Res.* **67,** 391–396.

Bainbridge, A. E., Suess, H. E., and Wänke, H. (1962). The tritium content of three stony meteorites and one iron meteorite, *Geochim. Cosmochim. Acta* **26,** 471–473.

Baker, E., Friedländer, G., and Hudis, J. (1958). Formation of Be^7 in interactions of various nuclei with high energy protons, *Phys. Rev.* **112,** 1319–1321.

Bauer, C. A. (1947). Production of helium in meteorites by cosmic radiation, *Phys. Rev.* **72,** 354–355.

Bauer, C. A. (1963). The helium contents of metallic meteorites, *J. Geophys. Res.* **68,** 6043–6057.

Begemann, F. (1965). Edelgasmessungen an Eisenmeteoriten und deren Einschlüssen, *Z. Naturforsch.* **20a**, 950–960.

Begemann, F. (1966). Tritium content of two chondrites, *Earth Planet. Sci. Lett.* **1**, 148–150.

Begemann, F., and Vilcsek, E. (1965). Durch Spallationsreaktionen und Neutroneneinfang erzeugtes Cl^{36} in Meteoriten und die Prä-Atmosphärische Größe von Steinmeteoriten, *Z. Naturforsch.* **20a**, 533–540.

Begemann, F., Geiss, J., and Hess, D. C. (1957). Radiation age of a meteorite from cosmic-ray-produced He^3 and H^3, *Phys. Rev.* **107**, 540–542.

Begemann, F., Eberhardt, P., and Hess, D. C. (1959). He^3–H^3-Strahlungsalter eines Steinmeteoriten, *Z. Naturforsch.* **14a**, 500–503.

Bender, M. L., and Kohman, T. P. (1964). Interpretation of thick target bombardments, *Nucl. Chem. Progr. Rep. Carnegie Inst. Techn.,* 20ff.

Benioff, P. A. (1960). Nuclear reactions of Low-Z elements with 5.7 BeV protons, *Phys. Rev.* **119**, 316–324.

Bertini, H. W. (1963). Low-energy intranuclear cascade calculations, *Phys. Rev.* **131**, 1801–1821.

Bertini, H. W., and Dresner, L. (1963). Compilation of reactions calculated for particles with energies from about 50 to 350 MeV, *Neutron Phys. Div. Ann. Prog. Rep.* ORNL-3360, Oak Ridge.

Bertini, H. W., Guthrie, M. P., Pickell, E. H., and Bishop, B. L. (1966). Literature survey of radiochemical cross section data below 425 MeV, ORNL Rep. 3884, Oak Ridge.

Bieri, R. H., and Rutsch, W. (1962). Erzeugungsquerschnitte für Edelgase aus Mg, Al, Fe, Ni, Cu und Ag bei Bestrahlung mit 540 MeV Protonen, *Helv. Phys. Acta* **35**, 553–554.

Birnbaum, M., Shapiro, M. M., Stiller, B., and O'Dell, F. W. (1952). Shape of cosmic ray star size distribution in nuclear emulsions, *Phys. Rev.* **86**, 86–89.

Biswas, F., and Fichtel, C. E. (1965). Composition of solar cosmic rays, *Space Sci. Rev.* **4**, 709–736.

Biswas, M. M., Mayer-Böricke, C., and Gentner, W. (1963). Cosmic ray produced Na^{22} and Al^{26} activities in chondrites, "Earth Science and Meteoritics," pp. 207–218. North-Holland Publ., Amsterdam.

Bogard, D. D. (1967). Krypton anomalies in achondritic meteorites, *J. Geophys. Res.* **72**, 1299–1309.

Brocas, J., and Picciotto, E. (1967). Nickel content of antarctic snow: Implications of the influx rate of extraterrestrial dust, *J. Geophys. Res.* **72**, 2229–2236.

Brun, C., Lefort, M., and Tarago, X. (1962). Contribution à l'étude du double pick-up indirect measure de la production de tritium par des protons de 82 et 150 MeV dans diverses cibles, *J. Phys. Rad.* **23**, 167ff.

Bruninx, E. (1961). High energy nuclear reaction cross sections, CERN-Rep. 61–1, Geneva.

Bruninx, E. (1962). High energy nuclear reaction cross sections, CERN-Rep. 62–9, Geneva.

Bruninx, E. (1964). High energy nuclear reaction cross sections, CERN-Rep. 64–17, Geneva.

Brunnee, C., and Voshage, H. (1964). "Massenspektrometrie." Thiemig, Munich.

Burnett, D. S., Lippolt, H. J., and Wasserburg, G. C. (1966a). The relative isotopic abundance of ^{40}K in terrestrial and meteoritic samples, *J. Geophys. Res.* **71**, 1249–1269.

Burnett, D. S., Lippolt, H. J., and Wasserburg, G. J. (1966b). Correction to Burnett et al., 1966a, J. Geophys. Res. **71**, 3609.

Cantelaube, Y., Maurette, M., and Pellas, P. (1967). Traces d'ions lourds dans les mineraux de la chondrite de Saint Severin, In "Radioactive Dating and Methods of Low-Level Counting," pp. 215–229. IAEA, Vienna.

Ceplecha, Z. (1960). New Czechoslovak meteorite luhy, Bull. Astron. Inst. Czechoslovakia **12**, 21ff.

Chakett, K. F. (1965). Yields of potassium isotopes on high energy bombardments of vanadium, iron, cobalt, nickel, copper and zinc, J. Inorg. Nucl. Chem. **27**, 2493–2505.

Charalambus, S., and Goebel, K. (1962). Tritium and argon in the Bruderheim meteorite, Geochim. Cosmochim. Acta **26**, 659–663.

Cindro, N. (1966). A survey of fast neutron reactions, Rev. Mod. Phys. **38**, 391–446.

Cobb, J. C. (1963). Exposure ages of some chondrites from Ar^{39} and Ar^{38} measurements, Brookhaven Nat. Lab. Rep. 8915.

Cobb, J. C. (1966). Iron meteorites with low cosmic-ray exposure ages, Science **151**, 1524ff.

Cobb, J. C., Davis, R., Schaeffer, O. A., Stoenner, R. W., and Thompson, S. (1963). Cosmogenic products in the Bogou meteorite, Trans. Amer. Geophys. Union **44**, 1, 89.

Comerford, M. F. (1967). Comparative erosion rates of stone and iron meteorites under small particles bombardment, Geochim. Cosmochim. Acta **31**, 1457–1471.

Crevecoeur, E. H., and Schaeffer, O. A. (1963). Separation et Mesures de Al^{26} et Be^{10} dans les méteorites, In "Radioactive Dating," pp. 335–340. IAEA, Vienna.

Cumming, J. B. (1963). Absolute cross section for the $^{12}C(p, pn)^{11}C$ reaction at 50 MeV, Nucl. Phys. **49**, 417–423.

Cumming, J. B., Friedländer, G., Hudis, J., and Poskanzer, A. M. (1962a). Spallation of aluminum by 28-GeV protons, Phys. Rev. **127**, 950–954.

Cumming, J. B., Hudis, J., Poskanzer, A. M., and Kaufman, S. (1962b). $Al^{27}(p, 3pn)$ $Na^{24}/C^{12}(p, pn)C^{11}$ cross section ratio in the GeV Region, Phys. Rev. **128**, 2392–2397.

Currie, L. A. (1959). Tritium production by 6-BeV protons, Phys. Rev. **114**, 878–880.

Currie, L. A., Libby, W. F., and Wolfgang, R. L. (1956). Tritium production by high energy protons, Phys. Rev. **101**, 1557–1563.

D'Arcy, R. G. (1960). Balloon observations of solar protons in the auroral zone, July 1959, Ph.D. thesis, Univ. of California, Berkeley.

Davis, L. R., Fichtel, C. E., Geiss, D. E., and Ogilvie, K. W. (1961). Rocket observations of solar protons on September 3, 1960, Phys. Rev. Lett. **6**, 492–494.

Davis, R., and Stoenner, R. W. (1962). A study of the production of Ar^{37} by 3 BeV protons in a thick sample of iron, J. Geophys. Res. **67**, 3552ff.

Davis, R., Stoenner, R. W., and Schaeffer, O. A. (1963). Cosmic ray produced Ar^{37} and Ar^{39} activities in recently fallen meteorites, In "Radioactive Dating," pp. 355–364. IAEA, Vienna.

DeFelice, J., Fireman, E. L., and Tilles, D. (1963). Tritium, argon-37 and manganese-54 radioactivities in a fragment of Sputnik 4, J. Geophys. Res. **68**, 5289–5296.

DeLaeter, J. R., and Jeffery, P. M. (1965). The isotopic composition of terrestrial and meteoritic tin, J. Geophys. Res. **70**, 2895–2903.

De Vries, H. (1958). Variation in concentration of radiocarbon with time and location on earth, Koninkl. Ned. Akad. Wetenschap. Proc. **B61**, 94ff.

Dews, J. R. (1966). The isotopic composition of lithium in chondrules, J. Geophys. Res. **71**, 4011–4020.

Dodson, R. W., Friedländer, G., Hudis, J., and Schaeffer, O. A. (1961). Induced radioactivities in Pb carried on Discoverer XVII, *Bull. Amer. Phys. Soc.* **6**, 277.

Dostrovsky, J., Rabinowitz, P., and Bivins, R. (1958). Monte Carlo calculations of high energy nuclear interactions I. Systematic of nuclear evaporation, *Phys. Rev.* **111**, 1659–1677.

Dostrovsky, J., Fraenkel, Z., and Friedländer, G. (1959). Monte Carlo calculations of nuclear evaporation processes III. Applications to low energy reactions, *Phys. Rev.* **116**, 683–702.

Ebeoglu, D. B., and Wainio, K. M. (1966). Solar proton activation of the lunar surface, *J. Geophys. Res.* **71**, 5863–5872.

Ebeoglu, D. B., Wainio, K. M., More, K., and Tiffany, O. L. (1966). Monte Carlo calculation of radionuclide production in iron targets bombarded with 400 MeV protons, *J. Geophys. Res.* **71**, 1445–1451.

Eberhardt, P., and Hess, D. C. (1960). He in stone meteorites, *Astrophys. J.* **131**, 38–46.

Eberhardt, P., Geiss, J., and Lutz, H. (1963). "Neutrons in Meteorites; Earth Science and Meteoritics" pp. 143–168. North-Holland Publ., Amsterdam.

Eberhardt, P., Eugster, O., and Geiss, J. (1965). Radiation ages of aubrites, *J. Geophys. Res.* **70**, 4427–4434.

Eberhardt, P., Eugster, O., Geiss, J., and Marti, K. (1966). Rare gas measurements in 30 stone meteorites, *Z. Naturforsch.* **21a**, 414–426.

Ebert, K. H., and Wänke, H. (1957). Über die Einwirkung der Höhenstrahlung auf Eisenmeteorite, *Z. Naturforsch.* **12a**, 766–773.

Ehmann, W. D., and Kohman, T. P. (1958). Cosmic-ray-induced radioactivities in meteorites. II. Al^{26}, Be^{10}, and Co^{60} in aerolites, siderites and tektites, *Geochim. Cosmochim. Acta* **14**, 364–379.

Estrup, P. J. (1963). Spallation yields from iron with 24 GeV protons, *Geochim. Cosmochim. Acta* **27**, 891–895.

Fechtig, H., and Gentner, W. (1965). Tritium diffusionssungen an vier Steinmeteoriten, *Z. Naturforsch.* **20a**, 1686–1691.

Fechtig, H., Gentner, W., and Kistner, G. (1960). Räumliche Verteilung der Edelgasisotope im Eisenmeteoriten Treysa, *Geochim. Cosmochim. Acta* **18**, 72–80.

Fireman, E. L. (1955). Tritium production by 2.2 GeV protons on iron and its relation to cosmic radiation, *Phys. Rev.* **97**, 1303–1304.

Fireman, E. L. (1958a). Argon-39 in the Sikhote Alin meteorite fall, *Nature* **181**, 1613–1614.

Fireman, E. L. (1958b). Distribution of helium-3 in the Carbo meteorite, *Nature* **181**, 1725.

Fireman, E. L. (1959). The distribution of helium-3 in the Grant meteorite and a determination of the original Mass, *Planet. Space Sci.* **1**, 66–70.

Fireman, E. L. (1966). Neutron exposure ages of meteorites, *Z. Naturforsch.* **21a**, 1138–1146.

Fireman, E. L. (1967a). Radioactivities in meteorites and cosmic ray variations, *Geochim. Cosmochim. Acta* **31**, 1691–1700.

Fireman, E. L. (1967b). Evidence for extraterrestrial particles in polar ice, *Smithsonian Contrib. Astrophys.* **11**, 373–379.

Fireman, E. L., and DeFelice, J. (1960a). Argon radioactivity in a recently fallen meteorite—a deep space probe for cosmic rays, *J. Geophys. Res.* **65**, 2489–2490.

Fireman, E. L., and DeFelice, J. (1960b). Argon-39 and tritium in meteorites, *Geochim. Cosmochim. Acta* **18**, 183–192.

Fireman, E. L., and DeFelice, J. (1960c). Argon-37, argon-39, and tritium in meteorites and the spatial constancy of cosmic rays, *J. Geophys. Res.* **65**, 3035–3041.

Fireman, E. L., and DeFelice, J. (1961). Tritium, argon-37 and argon-39 in the Bruderheim meteorite, *J. Geophys. Res.* **66**, 3547–3551.

Fireman, E. L., and Langway, C. C. (1965). Search for aluminum-26 in dust from the Greenland ice sheet, *Geochim. Cosmochim. Acta* **29**, 21–27.

Fireman, E. L., and Rowland, F. (1955). Tritium and neutron production by 2.2 BeV protons on nitrogen and oxygen, *Phys. Rev.* **97**, 780–782.

Fireman, E. L., and Schwarzer, D. (1957). Measurement of Li⁶, He³, and H³ in meteorites and its relation to cosmic radiation, *Geochim. Cosmochim. Acta* **11**, 252–262.

Fireman, E. L., and Zähringer, J. (1957). Depth variation of tritium and argon-37 produced by high energy protons in iron, *Phys. Rev.* **107**, 1695–1698.

Fireman, E. L., DeFelice, J., and Tilles, D. (1961). Solar flare tritium in a recovered satellite, *Phys. Rev.* **123**, 1935–1936.

Fireman, E. L., DeFelice, J., and Tilles, T. (1963). Tritium and radioactive argon and xenon in meteorites and satellites, "Radioactive Dating," pp. 323–333. IAEA, Vienna.

Fisher, D. E. (1966). The origin of meteorites: Space erosion and cosmic radiation ages, *J. Geophys. Res.* **71**, 3251–3259.

Fisher, D. E. (1967). Cosmic radiation ages and space erosion II. The iron meteorites. "Radioactive Dating and Methods of Low-Level Counting," pp. 269–280. IAEA, Vienna.

Fisher, D. E., and Schaeffer, O. A. (1960). Cosmogenic nuclear reactions in iron meteorites, *Geochim. Cosmochim. Acta* **20**, 5–14.

Fleischer, R. L., Naeser, C. W., Price, P. B., Walker, R. M., and Maurette, M. (1965). Cosmic ray exposure ages of tektites by the fission-track technique, *J. Geophys. Res.* **70**, 1491–1496.

Fowler, W. A., Greenstein, J. L., and Hoyle, F. (1962). Nucleosynthesis during the early history of the solar system, *Geophys. J.* **6**, 148–220.

Friedländer, G., Hudis, J., and Wolfgang, R. (1955). Disintegrations of aluminum by protons in the energy range 0.4 to 3 BeV, *Phys. Rev.* **99**, 263–268.

Funk, H., and Rowe, M. W. (1967). Spallation yield of xenon from 730 MeV proton irradiation of barium, *Earth Planet. Sci. Lett.* **2**, 215–219.

Gauvin, H., Lefort, M., and Tarrago, X. (1962). Emission d'helions dans les réactions de spallation, *Nucl. Phys.* **39**, 447–463.

Geiss, J. (1957). Über die Geschichte der Meteorite aus Isotopenmessungen, *Chimia* **12**, 349–363.

Geiss, J., Hirt, B., and Oeschger, H. (1960a). Tritium and helium concentration in meteorites, *Helv. Phys. Acta* **33**, 590–593.

Geiss, J., Oeschger, H., and Signer, P. (1960b). Radiation ages of chondrites, *Z. Naturforsch.* **15a**, 1016–1017.

Geiss, J., Oeschger, H., and Schwarz, M. (1962). The history of cosmic radiation as revealed by isotopic changes in the meteorites and on the earth; *Space Sci. Rev.* **1**, 197–223.

Gentner, W., and Zähringer, J. (1955). Argon-und-Heliumbestimmungen an Eisenmeteoriten, *Z. Naturforsch.* **102**, 498–499.

Gentner, W., and Zähringer, J. (1957). Argon und Helium als Kernreaktionsprodukte in Meteoriten, *Geochim. Cosmochim. Acta* **11**, 60–71.

Gentner, W., and Zähringer, J. (1959). Kalium-Argon-Alter einiger Tektite, *Z. Naturforsch.* **14a**, 686–687.

Gentner, W., Kleinmann, B., and Wagner, G. A. (1967). New K–Ar and fission track Ages of impact glasses and tektites, *Earth Planet. Sci. Lett.* **2**, 83–86.

Gerling, E. K., and Levskii, L. K. (1961). Yield of argon isotopes from targets irradiated with 660 MeV protons (in Russian), *Radiochimia* **3**, 97ff.

Gerloff, U., Weihrauch, J. H., and Fechtig, H. (1967). Electron microscope and microprobe measurements on luster—flight samples, *Space Res.* **7**, 1412–1420.

Gfeller, Ch., Houtermans, F. G., Oeschger, H., and Schwarz, U. (1961). γ-γ-Koinzidenzmessung zur zerstörungsfreien Messung des Gehaltes von Meteoriten an Positronenstrahlern und γ-aktiven Isotopen, *Helv. Phys. Acta* **32**, 466–469.

Goebel, K. (1958). Tritium production in iron by protons at energies between 50 and 177 MeV, CERN Rep. 58–2, Geneva.

Goebel, K., and Schmidlin, P. (1960). Tritium-Messungen an Steinmeteoriten, *Z. Naturforsch.* **15a**, 79–82.

Goebel, K., and Schultes, H. (1960). The cross section of the reaction $^{27}Al(p, 3pn)$ ^{24}Na at 590 MeV, CERN Rep. 60–63, Geneva.

Goebel, K., and Zähringer, J. (1961). Erzeugung von Tritium und Edelgasisotopen bei Behandlung von Fe und Cu mit Protonen von 25 GeV Energie, *Z. Naturforsch.* **16a**, 231–236.

Goebel, K., Schmidlin, P., and Zähringer, J. (1959). Das Tritium-Helium- und das Kalium-Argon-Alter des Meteoriten Ramsdorf, *Z. Naturforsch.* **14a**, 996–998.

Goebel, K., Schultes, H., and Zähringer, J. (1964). Production cross sections of tritium and rare gases in various target elements, CERN Rep. 64–12, Geneva.

Goel, P. S. (1962). Calculation of production rates of specific nuclides in iron meteorites, *In* "Researches on Meteorites," pp. 36–67. Wiley, New York.

Goel, P. S., and Kohman, T. P. (1962). Cosmogenic carbon-14 in meteorites and terrestrial ages of "finds" and craters, *Science* **136**, 875–876.

Goel, P. S., and Kohman, T. P. (1963). Cosmic-ray exposure history of meteorites from cosmogenic Cl^{36}, "Radioactive Dating," pp. 413–432. IAEA, Vienna.

Goldberger, M. L. (1948). The interaction of high energy neutrons and heavy nuclei, *Phys. Rev.* **74**, 1269–1277.

Gradsztajn, E. (1960). Determination par spectronomie de masse des sections éfficaces de production de fragments louds par des protons de 155 MeV sur O^{16}, *J. Phys. Radium,* **21**, 761ff.

Grjebine, T., Lalou, C., Ros, J., and Capitant, M. (1964). Study of magnetic spherules in three cores of the occidental basin of the Mediterranean Sea; *Ann. N.Y. Acad. Sci.* **119**, 143–165.

Gusakow, M. (1962). Contribution à l'étude des reactions (p, pn) à moyenne energie, *Ann. Phys. (Paris)* **7**, 67ff.

Heymann, D. (1965). Cosmogenic and radiogenic helium, neon and argon in amphoterite chondrites, *J. Geophys. Res.* **70**, 3735–3743.

Heymann, D. (1967). On the origin of hypersthene chondrites: Ages and shock-effects of black chondrites, *Geochim. Cosmochim. Acta* **31**, 1793–1809.

Heymann, D., and Anders, E. (1967). Meteorites with short cosmic ray exposure ages, as determined from their ^{26}Al content, *Geochim. Cosmochim. Acta* **31**, 1793–1809.

Heymann, D., and Mazor, E. (1967). Radiation ages and gas-retention ages of carbonaceous chondrites and of unequilibrated ordinary chondrites, *In* "Radioactive Dating and Methods of Low-Level Counting," pp. 239–258. IAEA, Monaco.

Heymann, D., and Mazor, E. (1968). Noble gases in unequilibrated ordinary chondrites, *Geochim. Cosmochim. Acta* **32**, 1–19.

Heymann, D., and Schaeffer, O. A. (1961). Exposure ages of some iron meteorites, *J. Geophys. Res.* **66**, 2535–2536.

Heymann, D., and Schaeffer, O. A. (1962). Constancy of cosmic rays in time, *Physica* **28**, 1318–1323.

Heymann, D., Lipschutz, M. E., Nielson, B., and Anders, E. (1966). Canyon Diablo meteorite: Metallographic and mass spectrometric study of 56 fragments, *J. Geophys. Res.* **71**, 619–641.

Hintenberger, H., and Wänke, H. (1964). Helium und Neonisotope in Eisenmeteoriten, *Z. Naturforsch.* **19a**, 210–218.

Hintenberger, H., König, H., Schultz, L., Wänke, H., and Wlotzka, F. (1964a). Die relativen Produktionsquerschnitte für He^3 und Ne^{21} aus Mg, Si, S und Fe in Steinmeteoriten, *Z. Naturforsch.* **19a**, 88–91.

Hintenberger, H., König, H., Schultz, L., und Wänke, H. (1964b). Radiogene, Spallogene und Primordiale Edelgase in Steinmeteoriten, *Z. Naturforsch.* **19a**, 327–341.

Hintenberger, H., Voshage, H., and Sarkar, H. (1965a). Durch die Kosmische Strahlung Produziertes Lithium und Calcium in Eisenmeteoriten, *Z. Naturforsch.* **20a**, 965–967.

Hintenberger, H., König, H., Schultz, L., and Wänke, H. (1965b). Radiogene, Spallogene und Primordiale Edelgase in Steinmeteoriten. III., *Z. Naturforsch.* **20a**, 983–989.

Hintenberger, H., Schultz, L., Wänke, H., and Weber, H. (1967). Helium- und Neonisotope in Eisenmeteoriten und der Tritiumverlust in Hexaedriten, *Z. Naturforsch.* **22a**, 780–878.

Hintz, N. M., and Ramsey, N. F. (1952). Excitation functions to 100 MeV, *Phys. Rev.* **88**, 19–27.

Hoffman, J. H., and Nier, A. O. (1958). Production of helium in iron meteorites by the action of cosmic rays, *Phys. Rev.* **112**, 2112–2117.

Hoffman, J. H., and Nier, A. O. (1959). The cosmogenic He^3 and He^4 distribution in the meteorite Carbo, *Geochim. Cosmochim. Acta* **17**, 32–36.

Hoffman, J. H., and Nier, A. O. (1960). Cosmic-ray-produced helium in the Keen Mountain and Casas Grandes meteorites, *J. Geophys. Res.* **65**, 1063–1068.

Hohenberg, C. M., Munk, M. N., and Reynolds, J. H. (1967). Spallation and fissiogenic xenon and krypton from stepwise heating of the Pasamonte achondrite; The case for extinct ^{244}Pu in meteorites, *J. Geophys. Res.* **72**, 3139–3177.

Honda, M. (1962). Spallation products distributed in a thick iron target bombarded by 3 BeV protons, *J. Geophys. Res.* **67**, 4847–4848.

Honda, M., and Arnold, J. R. (1961). Radioactive species produced by cosmic rays in the Aroos iron meteorite, *Geochim. Cosmochim. Acta* **213**, 219–232.

Honda, M., and Arnold, J. R. (1964). Effects of cosmic rays on meteorites, *Science* **143**, 203–212.

Honda, M., and Lal, D. (1960). Some cross sections for the production of radionuclides in the bombardment of C, N, O, and Fe by medium energy protons, *Phys. Rev.* **118**, 1618–1625.

Honda, M., and Lal, D. (1964). Spallation cross sections for long lived radionuclides in iron and light nuclei, *Nucl. Phys.* **51**, 363–368.

Honda, M., Shedlovsky, J. P., and Arnold, J. R. (1961). Radioactive species produced by cosmic rays in iron meteorites, *Geochim. Cosmochim. Acta* **22**, 133–154.

Hudis, J., Kirsten, T., Schaeffer, O. A., and Stoenner, R. W. (1970). Yields of stable and radioactive isotopes formed by 3 and 29 GeV proton bombardment of copper, silver, gold and uranium targets, *Phys. Rev.* **C1**, 2019–2030.

Hulston, J. R., and Thode, H. G. (1965). Variations in the S^{33}, S^{34}, and S^{36} contents of meteorites and their relation to chemical and nuclear effects, *J. Geophys. Res.* **70**, 3475–3484.

Jessen, P., Borman, M., Dreyer, F., and Neuer, H. (1966). Experimental Excitation Functions for (n, p), (n, t), (n, α), (n, 2n), (n, np) and (n, nα) reactions, *Nucl. Data.* **1,** 103ff.

Kaiser, W., and Zähringer, J. (1965). Kalium Analyse von Amphoteritchondriten und deren K-A-Alter, *Z. Naturforsch.* **20a,** 963–965.

Kammerer, O. F., Sadofsky, J., and Gurinsky, D. H. (1963). Metallographic studies of Sputnik 4 fragment, *J. Geophys. Res.* **68,** 5049–5057.

Keith, J. E., and Turkevich, A. L. (1962). Radioactivity induced in Discoverer 17 by solar flare protons, *J. Geophys. Res.* **67,** 4525–4532.

Kirsten, T. (1966). "Determination of radiogenic argon, "Potassium Argon Dating," pp. 7–39. Springer, Berlin.

Kirsten, T., Krankowsky, D., and Zähringer, J. (1963) Edelgas- und Kaliom-Bestimmungen an einer größeren Zahl von Steinmeteoriten *Geochim. Cosmochim. Acta* **27,** 13–42.

Knox, W. J. (1949). Relative cross sections for nuclear reactions induced by high energy neutrons in light elements, *Phys. Rev.* **75,** 537–541.

Kohman, T. P., and Goel, P. S. (1963). Terrestrial ages of meteorites from cosmogenic C^{14}, "Radioactive Dating," pp. 395–411. IAEA, Vienna.

Koterling, R. G. (1962). High energy nuclear reactions of niobium with incident protins and helium ions, Ph.D. thesis, Univ. of California, Berkeley; UCRL-10461.

Krankowsky, D., and Müller, O. (1964). Isotopic content and concentration of lithium in stone meteorites, *Geochim. Cosmochim. Acta* **28,** 1625–1630.

Krummenacher, D., Merrihue, C. M., Pepin, R. O., and Reynolds, J. H. (1962). Meteoritic krypton and barium versus the general isotopic anomalies in meteoritic xenon, *Geochim. Cosmochim. Acta* **26,** 231–249.

Kuiper, G. P. (1950). On the origin of asteroids, *Astron. J.* **55,** 164.

Kutznetzov, V., and Mekhedov, V. (1958). Tritium production by 120–660 MeV protons in metals (in Russian), Joint Inst. Nucl. Res. Moscow, NP 6901.

Lämmerzahl, P., and Zähringer, J. (1966). K–Ar-Altersbestimmungen an Eisenmeteoriten. II. Spallogenes Ar^{40} and Ar^{40}–Ar^{38}-Bestrahlungsalter, *Geochim. Cosmochim. Acta* **30,** 1059–1074.

Lal, D., and Venkatavaradan, V. S. (1966). Low energy protons: Average flux in interplanetary space during the last 100,000 years, *Science* **151,** 1381–1383.

Lal, D., Rajan, R. S., and Venkatavaradan, V. S. (1967). Nuclear effects of solar and galactic cosmic-ray particles in near-surface regions of meteorites, *Geochim. Cosmochim. Acta* **31,** 1859–1869.

Lavrukhina, A. K., Revina, L. D., Malyshev, V. V., and Satarova, L. M. (1963). Reactions of deep spallation of Fe nuclei by 150 MeV protons, *Sov. Phys. JETP* **17,** 960–964.

Lavrukhina, A. K., Kuznetsova, R., and Satarova, L. M. (1964). The formation rate of radioactive isotopes in chondrites by cosmic rays, *Geokhimiya* **12,** 1219–1227.

Lavrukhina, A. K., Ustinova, G. K., and Ibraev, T. A. (1967). Effects of nuclear reactions with cosmic rays in cosmic bodies of different size and composition, *Geokhimiya* **11,** 1350–1360.

Ligonniere, M., Vassent, B., and Bernas, R. (1964). Cross sections for production of beryllium-7 in aluminum, vanadium, tantalum and gold by protons of 155 and 550 MeV, *C. R. Acad. Sci. Paris* **259,** 1406ff.

Lipschutz, M. E. (1964). Estimation of half-lives from meteoritic data, *Nature* **204,** 1289–1290.

Lipschutz, M. E., Signer, P., and Anders, E. (1965). Cosmic ray exposure ages of iron meteorites by the Ne^{21}/Al^{26} method, *J. Geophys. Res.* **70,** 1473–1489.

McCorkell, R. H., Fireman, E. L., and Langway, C. C. (1967). Dissolved iron, nickel, and cobalt in Greenland ice, *Trans. Amer. Geophys. Union* **48**, 158.

McDonald, F. B. (1959). Primary cosmic-ray intensity near solar maximum, *Phys. Rev.* **116**, 462–463.

Marquez, L., and Perlman, J. (1952). Observations on lithium and beryllium nuclei ejected from heavy nuclei by high energy particles, *Phys. Rev.* **81**, 953–957.

Marti, K. (1967). Mass-spectrometric detection of cosmic-ray produced ^{81}Kr in meteorites and the possibility of Kr–Kr dating, *Phys. Rev. Lett.* **18**, 264–266.

Martin, G. R. (1953). Recent studies on iron meteorites IV. The origin of meteoritic helium and the age of meteorites, *Geochim. Cosmochim. Acta* **3**, 288–309.

Mason, B. (1962). "Meteorites." Wiley, New York.

Meadows, J. W., and Holt, R. B. (1951a). Excitation functions for proton reactions with sodium and magnesium, *Phys. Rev.* **83**, 47–49.

Meadows, J. W., and Holt, R. B. (1951b). Some excitation functions for protons on magnesium, *Phys. Rev.* **83**, 1257.

Merrihue, C. M. (1963). Excess xenon-129 in chondrules from the Bruderheim meteorite, *J. Geophys. Res.* **68**, 325–330.

Merrihue, C. M. (1964). Rare gas evidence for cosmic dust in modern Pacific red clay, *Ann. N.Y. Acad. Sci.* **119**, 351–367.

Merrihue, C. M. (1966). Xenon and krypton in the Bruderheim meteorite, *J. Geophys. Res.* **71**, 263–313.

Metropolis, N., Bivins, R., Storm, M., Turkevich, A., Miller, J. M., and Friedländer, G. (1957a). Monte Carlo calculations on intranuclear cascades. I. Low energy studies, *Phys. Rev.* **110**, 185–203.

Metropolis, N., Bivins, R., Storm, M., Miller, J. M., Friedländer, G., and Turkevich, A. (1957b). Monte Carlo calculation on intranuclear cascades. II. High energy studies and pion processes, *Phys. Rev.* **110**, 204–219.

Munk, M. N. (1967). Spallation neon, argon, krypton, and xenon in an iron meteorite, *Earth Planet. Sci. Lett.* **2**, 301–309.

Murry, M., Tamers, A., and Delibrias, G. (1961). Section éfficaces de l'oxygene 16 pour la production de carbone 14 par des protons de hautes energies, *C. R. Acad. Sci. Paris* **253**, 1202ff.

Murthy, V. R., and Sandoval, J. (1965). Chromium isotopes in meteorites, *J. Geophys. Res.* **70**, 4379–4389.

Ney, E. P., Winckler, J. R., and Freier, P. S. (1959). Protons from the Sun on May 12, 1959, *Phys. Rev. Lett.* **3**, 183–185.

Nguyen Long Den, M. (1961). Section éfficace de formation du sodium par la reaction $Al^{27}(p, 3p3n)Na^{22}$ à 155 MeV, *C. R. Acad. Sci. Paris* **253**, 2919ff.

Öpik, E. J. (1951). Collision probabilities with the planets and the distribution of interplanetary matter, *Proc. Roy. Irish Acad.* **54A**, 165–199.

Öpik, E. J. (1956). Interplanetary dust and terrestrial accretion of meteoritic matter, *Irish Astron. J.* **4**, 84–133.

Öpik, E. J. (1958). I. Meteorite impact on solid surface, II. On the catastrophic effects of collisions with celestial bodies, *Irish Astron. J.* **5**, 14–33.

O'Gallagher, J. J., and Simpson, J. A. (1967). The heliocentric intensity gradients of cosmic-ray protons and helium during minimum solar modulation, *Astrophys. J.* **147**, 819–827.

O'Keefe, O. A. (1963). "Tektites." Univ. of Chicago Press, Chicago, Illinois.

Paneth, F. A., Reasbeck, P., and Mayne, K. I. (1952). Helium-3 content and age of meteorites, *Geochim. Cosmochim. Acta* **2**, 300–303.

Paneth, F. A., Reasbeck, P., and Mayne, K. I. (1953). Production by cosmic rays of helium-3 in meteorites, *Nature* **172**, 200–201.

Park, F. R., and Reid, A. M. (1964). A comparative study of some metallic spherules, *Ann. N.Y. Acad. Sci.* **119**, 250ff.

Pepin, R. O. (1966). Heavy rare gases in silicates from the Estherville mesosiderite, *J. Geophys. Res.* **71**, 2815–2829.

Porile, N. T., and Tanaka, S. (1964). Formation of sodium and potassium nuclides in the bombardment of light elements with 29 GeV protons, *Phys. Rev.* **135B**, 122–128.

Powell, C. F., Fowler, P. H., and Perkins, D. H. (1959). "The Study of Elementary Particles by the Photographic Method." Pergamon Press, New York.

Price, P. B., Rajan, R. S., and Tamhane, A. S. (1967). On the pre-atmospheric size and maximum space erosion rate of the Patwar stony-iron meteorite, *J. Geophys. Res.* **72**, 1377–1388.

Proc. Symp. Radioactive Dating Methods Low-Level Counting, Monaco, 2–10 March, 1967. IAEA, Vienna (1967).

Prokoshkin, J., and Tiapkin, A. (1957). Investigation of the excitation functions for the reactions $C^{12}(p, pn)C^{11}$, $Al^{27}(p, 3pn)Na^{24}$ and $Al^{27}(p, 3p, 3n)Na^{22}$ in the 150–660 MeV energy range, *Sov. Phys. JETP* **5**, 148–149.

Rao, M. A. (1961). Isotopic composition of the low energy helium nuclei in the primary cosmic radiation, *Phys. Rev.* **123**, 295–300.

Rayudu, G. V. S. (1963). Simulated cosmic ray irradiation in a thick iron target, *Trans. Amer. Geophys. Union* **44**, 89.

Rayudu, G. V. S. (1964a). Formation cross sections of various radionuclides from Ni, Fe, Si, Mg, O, and C for protons of energies between 0.5 and 2.9 GeV, *J. Inorg. Nucl. Chem.* **30**, 2311–2315.

Rayudu, G. V. S. (1964b). Formation cross sections of various radionuclides from Ni, Fe, Si, Mg, O, and C for protons of energies between 130 and 400 MeV, *Canad. J. Chem.* **42**, 1149–1154.

Rayudu, G. V. S., and Shedlovsky, J. P. (1963). *Nucl. Chem. Prog. Rep. Carnegie Inst. Techn.*

Reeder, P. L. (1965). Cross sections for production of Be^7 from light targets by 2.2 GeV protons, *J. Inorg. Nucl. Chem.* **27**, 1879ff.

Remsberg, L. P., and Miller, J. M. (1963). Study of (p, pn) reactions in medium weight nuclei at 370 MeV, *Phys. Rev.* **130**, 2069–2076.

Reynolds, J. H. (1956). High sensitivity mass spectrometer for noble gas analysis, *Rev. Sci. Instrum.* **27**, 928ff.

Reynolds, J. H. (1960). Rare gases in tektites, *Geochim. Cosmochim. Acta* **20**, 101–114.

Rowe, M. W., Van Dilla, M. A., and Anderson, E. C. (1963). On the radioactivity of stone meteorites, *Geochim. Cosmochim. Acta* **27**, 983–1001.

Rudstam, G. (1955). Spallation of elements in mass range 51–75, *Phil. Mag.* **46**, 344ff.

Rudstam, G. (1956). Spallation of medium weight elements, Ph.D. thesis, Univ. of Uppsala.

Rudstam, G. (1966). Systematics of spallation yields, *Z. Naturforsch.* **21a**, 1027ff.

Rudstam, G., Stevenson, P. C., and Folger, R. L. (1952). Nuclear reactions of iron with 340 MeV protons, *Phys. Rev.* **87**, 358–365.

Sammet, F., and Herr, W. (1963). Studies on the cosmic-ray produced nuclides Be^{10}, Al^{26} and Cl^{36} in iron meteorites, *In* "Radioactive Dating," pp. 343–354. IAEA, Vienna.

Schaeffer, O. A. (1959). High sensitivity mass spectrometry of the rare gases, Brookhaven Nat. Lab. Rep. BNL 581, 12ff.

Schaeffer, O. A. (1962). Radiochemistry of meteorites, *Ann. Rev. Phys. Chem.* **13**, 151–170.

Schaeffer, O. A., and Heymann, D. (1965). Comparison of Cl[36]–Al[36] and Ar[39]–Ar[38] cosmic ray exposure ages of dated fall iron meteorites, *J. Geophys. Res.* **70**, 215–240.

Schaeffer, O. A., and Zähringer, J. (1958). Production of helium and argon in iron by high energy protons, *Z. Naturforsch.* **13a**, 346–347.

Schaeffer, O. A., and Zähringer, J. (1959). High-sensitivity mass spectrometric measurement of stable helium and argon isotopes produced by high-energy protons in iron, *Phys. Rev.* **113**, 674–678.

Schaeffer, O. A., and Zähringer, J. (1962). Solar flare helium in satellite materials, *Phys. Rev. Lett.* **8**, 389–390.

Schaeffer, O. A., Davis, R., Stoenner, R. W., and Heymann, D. (1963). The temporal and spatial variation in cosmic rays, *Int. Conf. Cosmic Rays, Jaipur* **3**, 480–486; Brookhaven Nat. Lab. Rep. 7943.

Schaeffer, O. A., Megrue, G., and Thompson, S. O. (1964). Experiments to test the presence of cosmogenic nuclides in ocean sediments; *Ann. N.Y. Acad. Sci.* **119**, 347–350.

Schultz, L., and Hintenberger, H. (1967). Edelgasmessungen an Eisenmeteoriten, *Z. Naturforsch.* **22a**, 773–779.

Schulze, W. (1962). "Neutronen-Aktivierung als Analytisches Hilfsmittel." Enke, Stuttgart.

Schwarz, M., and Oeschger, H. (1967). Spallation formula deduced from cascade and evaporation theories compared with experiments, *Z. Naturforsch.* **22a**, 972ff; *Science* **167**, 449ff (1970).

Serber, R. (1947). Nuclear reactions at high energies, *Phys. Rev.* **72**, 1114–1115.

Shedlovsky, J. P., and Kaye, J. H. (1963). Radioactive nuclides produced by cosmic rays in Sputnik 4, *J. Geophys. Res.* **68**, 5069–5078.

Shedlovsky, J. P., and Rayudu, G. V. S. (1964). Radionuclide production in thick iron targets bombarded with 1- and 3-GeV protons, *J. Geophys. Res.* **69**, 2231–2242.

Shedlovsky, J. P., Cressy, J. P., and Kohman, T. P. (1967). Cosmogenic radioactivities in the Peace River and Harleton chondrites, *J. Geophys. Res.* **72**, 5051–5058.

Sheffy, D. W. (1968). Excitation functions of radionuclides produced by 0–60 MeV proton reactions in silicon, Ph.D. thesis, ORNL-Report TM 2060, Oak Ridge (1968).

Shen, B. S. P. (1962). *Proc. Symp. Protection against Radiation Hazards in Space;* Oak Ridge Nat. Lab. TID 7652, 852–865.

Shen, B. S. P. (ed.) (1967). "High Energy Nuclear Reactions in Astrophysics." Benjamin, New York.

Shima, M., and Honda, M. (1966). Distribution of spallation produced chromium between alloys in iron meteorites, *Earth Planet. Sci. Lett.* **1**, 65–74.

Signer, P., and Nier, A. O. (1960). The distribution of cosmic ray produced rare gases in iron meteorites, *J. Geophys. Res.* **65**, 2947–2964.

Signer, P., and Nier, O. A. (1962). The measurement and interpretation of rare gas concentrations in iron meteorites, *In* "Researches on Meteorites" (C. B. Moore, ed.), pp. 7–35. Wiley, New York.

Singer, S. F. (1958). The primary cosmic radiation and its time variations, *Progr. Elementary Particle Cosmic Ray Phys.* **4**, 203ff.

Sitte, K. (ed.) (1967). "Handbuch der Physik," Vol. 46/2. Springer, Berlin.

Sprenkel, E. L., Davis, R., and Wiig, E. O. (1959). Cosmic-ray-produced Cl[36] and Ar[39] in iron meteorites, *Bull. Amer. Phys. Soc.* **4**, 223.

Stauffer, H., and Honda, M. (1961). Cosmic-ray produced V[50] and K[40] in the iron meteorite Aroos, *J. Geophys. Res.* **66**, 3584–3586.

Stauffer, H., and Honda, M. (1962). Cosmic-ray produced stable isotopes in iron meteorites, *J. Geophys. Res.* **67**, 3503–3512.

Stoenner, R. W., and Davis, R. (1961). Argon-37 produced in stainless steel from Discoverers XVII and XVIII and its relation to meteorites, *Bull. Amer. Phys. Soc.* **6**, 277

Stoenner, R. W., Schaeffer, O. A., and Davis, A. (1960). Meteorites as space probes for testing the spatial constancy of cosmic radiation, *J. Geophys. Res.* **65**, 3025–3034.

Suess, H. E. (1965). Secular variations of the cosmic-ray-prodoced carbon-14 in the atmosphere and their interpretations, *J. Geophys. Res.* **70**, 5937–5952.

Suess, H. E., and Wänke, H. (1962). Radiocarbon content and terrestrial age of twelve stone meteorites and one iron meteorite, *Geochim. Cosmochim. Acta* **26**, 475–480.

Tanenbaum, A. S. (1967). Clustering of the cosmic ray ages of stone meteorites, *Earth Planet. Sci. Lett.* **2**, 33–35.

Tilles, D. (1966). Implantation in interplanetary dust of rare gas ions from solar flares, *Science* **153**, 981–984.

Tilles, D. (1967). Extraterrestrial excess ^{36}Ar and ^{38}Ar concentrations as possible accumulation-rate indicators for sea sediments, *Icarus* **7**, 94–99.

Tilles, D., DeFelice, J., and Fireman, E. L. (1961). Argon-37 in recovered satellite material, *Bull. Amer. Phys. Soc.* **6**, 277.

Turkevich, A. L., Franzgrote, E. J., and Patterson, J. H. (1967). Chemical analysis of the Moon at the Surveyor V landing site, *Science* **158**, 633–637.

Turkevich, A., Patterson, J. H., and Franzgrote, E. (1968). Chemical analysis of the moon at the Surveyor VI landing site: Preliminary results, *Science* **160**, 1108ff.

Umemoto, S. (1962). Isotopic composition of barium and cerium in stone meteorites, *J. Geophys. Res.* **67**, 375–379.

Urey, H. C. (1959). Primary and secondary objects, *J. Geophys. Res.* **64**, 1721–1737.

Urey, H. C. (1965). Meteorites and the Moon, *Science* **147**, 1262–1265.

Urey, H. C., and Craig, H. (1963). The composition of the stone meteorites and the origin of the meteorites, *Geochim. Cosmochim. Acta* **4**, 36–82.

Valentin, L. (1965). Réactions (p, n) et (p, pn) induites à moyenne energie sur des noyaux legers, *Nucl. Phys.* **62**, 81–102.

Valentin, L., Albouy, G., and Cohen, J. P. (1963). Réactions induites par des protons de 155 MeV sur de noyaux legers, *Phys. Lett.* **7**, 163ff.

Valentin, L., Albouy, G., Cohen, J. P., and Gusakow, M. (1964). Excitation functions of (p, pn) and (p, 2p2n) reactions in light nuclei between 15 and 155 MeV, *J. Phys.* (*Paris*) **25**, 704ff.

Van Allen, J. A., and Frank, L. A. (1959). Radiation measurements to 658, 300 kilometers with Pioneer 4, *Nature* **184**, 219–224.

Van de Hulst, H. C. (1947). Zodiacal light in ten solar corona, *Astrophys. J.* **105**, 471–488.

Van Dilla, M. A., Arnold, J. R., and Anderson, E. C. (1960). Spectrometric measurement of natural and cosmic-ray induced radioactivity in meteorites, *Geochim. Cosmochim. Acta* **20**, 115–121.

Van Schmus, W. R., and Wood, J. A. (1967). A chemical–petrological classification for the chondritic meteorites, *Geochim. Cosmochim. Acta* **31**, 747–766.

Vernov, S. N., and Chudakov, A. E. (1959). The study of the terrestrial corpuscular radiation and cosmic rays during the flight of a cosmic rocket, *Sov. Phys. Dokl.* **4**, 338–342.

Vilcsek, E., and Wänke, H. (1961). Das Strahlungsalter der Eisenmeteorite aus Chlor-36-Messungen, Z. Naturforsch. **16a,** 379–384.

Vilcsek, E., and Wänke, H. (1963). Cosmic-ray exposure ages and terrestrial ages of stone and iron meteorites derived from Cl³⁶ and Ar³⁹ measurements, *In* "Radioactive Dating," pp. 381–393. IAEA, Vienna.

Vinogradov, A. P., and Zadorozhnii, I. K. (1965). Cosmogenic, radiogenic and primordial inert gases in stone meteorites, *Meteoritika* **26,** 77–90.

Vinogradov, A. P., Zadorozhnii, I. K., and Florenskii, K. P. (1957). *Geokhimiya* **6,** 443.

Vinogradov, A. P., Zadorozhnii, I. K., and Florenskii, K. P. (1958). Content of the inert gases in the Sikhote-Alin iron meteorite, *Geokhimiya* **6,** 443–448; *Chem. Erde* **19,** 275–285.

Vinogradov, A. P., Surkov, P. A., Chernov, G. M., Kirnozov, F. F., and Nazarkina, G. B. (1966). Measurements of the γ-radiation of the lunar surface on the Luna-10 Cosmic Station, *Geokhimiya* **8,** 891–899.

Viste, E., and Anders, E. (1962). Cosmic-ray exposure history of tektites, *J. Geophys. Res.* **67,** 2913–2919.

Voshage, H. (1962). Eisenmeteorite als Raumsonden für die Untersuchung des Intensitätsverlaufes der Kosmischen Strahlung während der letzten Milliarden Jahre, *Z. Naturforsch.* **17a,** 422–432.

Voshage, H. (1967). Bestrahlungsalter und Herkunft der Eisenmeteorite, *Z. Naturforsch.* **22a,** 477–506; "Radioactive Dating and Methods of Low-Level Counting," pp. 281–291. IAEA, Vienna.

Voshage, H., and Hintenberger, H. (1959). Kalium als Reaktionsprodukt der Kosmischem Strahlung in Eisenmeteoriten, *Z. Naturforsch.* **14a,** 828–838.

Voshage, H., and Hintenberger, H. (1961). Massenspektrometrische Isotopenhäufigkeitsmessungen an Kalium aus Eisenmeteoriten und das Problem der Bestimmung der K⁴¹–K⁴⁰-Strahlungsalter, *Z. Naturforsch.* **16a,** 1042–1053.

Voshage, H., and Hintenberger, H. (1963). The cosmic-ray exposure ages of iron meteorites as derived from the isotopic composition of Potassium and the production rates of cosmogenic nuclides in the past, *In* "Radioactive Dating," pp. 367–379. IAEA, Vienna.

Wänke, H. (1958). Scandium-45 als Reaktionsprodukt der Höhenstrahlung in Eisenmeteoriten. I. *Z. Naturforsch.* **13a,** 645–649.

Wänke, H. (1960). Scandium-45 als Reaktionsprodukt der Höhenstrahlung in Eisenmeteoriten. II. *Z. Naturforsch.* **15a,** 953–964.

Wänke, H. (1961). Über den Kaliumgehalt der Chondrite, Achondrite und Siderite, *Z. Naturforsch.* **16a,** 127–130.

Wänke, H. (1966). Meteoritenalter und Verwandte Probleme der Kosmochemie, *Fortschr. Chem. Forsch.* **7,** 322–408.

Wagner, G. A. (1966). Altersbestimmungen an Tektiten und anderen natürlichen Gläsern mittels Spuren der Spontanen Spaltung des Uran²³⁸—Fission-Track-Method, *Z. Naturforsch.* **21a,** 733–745.

Wasson, J. T. (1961). Radioactivity produced in Discoverer XVII by November 12, 1960, solar protons, *J. Geophys. Res.* **66,** 2659–2663.

Wasson, J. T. (1962). Radioactive cobalt and manganese in Discoverer XVII stainless steel, *J. Geophys. Res.* **67,** 3513–3517.

Wasson, J. T. (1963). Radioactivity in interplanetary dust, *Icarus* **2,** 54–87.

Wasson, J. T. (1964). Radioactivity in Sputnik 4 fragment, *J. Geophys. Res.* **69,** 2223–2230.

Wasson, J. T., Alder, B., and Oeschger, H. (1967). Aluminum-26 in Pacific sediment: Implications, *Science* **155**, 446–448.

Webber, W. R. (1961). Time variations of low rigidity cosmic rays during the recent sunspot cycle, *Progr. Elementary Particle Cosmic Ray Phys.* **6**, 77ff.

Webber, W. R. (1964). The spectrum and charge composition of the primary cosmic radiation, "Handbuch der Physik," Vol. 46, pp. 181–264. Springer, Berlin.

Wetherill, G. W. (1967). Collisions in the Asteroid Belt, *J. Geophys. Res.* **72**, 2429–2444.

Whipple, F. L., and Fireman, E. L. (1959). Calculation of erosion in space from the cosmic-ray exposure ages of meteorites, *Nature* **183**, 1315.

Winckler, J. R., Bhavsar, P. D., and Peterson, L. (1961). The time variations of solar cosmic rays during July 1959 at Minneapolis, *J. Geophys. Res.* **66**, 995–1022.

Wright, F. W., and Hodge, P. W. (1964). Compositional studies for extraterrestrial particles, *Ann. N.Y. Acad. Sci.* **119**, 287.

Yokoyama, Y., and Mabuchi, H. (1967). Radioactive nuclides induced by solar cosmic radiation in chondrites, *C. R. Acad Sci. Paris. Ser. B.* **264**, 655–657.

Zähringer, J. (1963). K–Ar measurements of tektites, *In* "Radioactive Dating," pp. 289–305. IAEA, Vienna.

Zähringer, J. (1966). Die Chronologie der Chondriten aufgrund von Edelgasisotopen-Analysen, *Meteoritika* **27**, 25ff.

Zähringer, J. (1968). Rare gases in stony meteorites, *Geochim. Cosmochim. Acta* **32**, 209–237.

4 Interactions of Elementary Particle Research with Chemistry*

Sheldon Kaufman   *Argonne National Laboratory*
Argonne, Illinois

I. NUCLEAR CHEMISTRY WITH HIGH-ENERGY ACCELERATORS

A. Introduction

Nuclear chemistry may be defined as the study of the structure and reactions of the atomic nucleus by chemical techniques. This field is to be distinguished from radiochemistry, in which radioactive tracers are used to investigate chemical phenomena, and radiation chemistry, in which the

---

* Based on work performed under the auspices of the U.S. Atomic Energy Commission.

chemical reactions caused by radiation are investigated. The discovery of nuclear fission and of the naturally occurring radioactive elements and the production and isolation of synthetic elements are classical examples of the work of nuclear chemists. Nuclear chemists have contributed substantially to studies in nuclear spectroscopy and nuclear structure, the determination of nuclear properties (half-lives, decay energies, isomerism, mass assignments, cross sections), studies of the angular distributions, ranges, and energies of nuclear reaction products, and the determination of excitation functions for nuclear reactions (i.e., the variation of reaction probability with the energy of the bombarding particle).

Many of these investigations involve the chemical isolation and identification of nuclear reaction products, often from complex mixtures—steps which are essential to an understanding of the phenomena involved. The work of the nuclear chemist supplements and complements that of the nuclear physicist. The border line between nuclear chemistry and nuclear physics has, in fact, practically disappeared in many areas of investigation, and the "nuclear chemist" carries out many experiments of a purely physical nature, observing the products of nuclear reactions with counters, ionization chambers, nuclear emulsions, etc. Although the distinction between nuclear chemistry and nuclear physics is, at best, rather hazy and probably unnecessary to draw, one might say that the nuclear physicist tends to concentrate on the nucleons and light nuclei emitted in nuclear reactions (i.e., neutrons, protons, deuterons, tritons, α particles) which can be identified by physical means, while the nuclear chemist examines the heavier fragments.

In the relatively new field of high-energy nuclear phenomena the roles of "chemist" and "physicist" are somewhat more clearly defined. The high-energy physicist is concerned with the production and properties of the "elementary" and "strange" particles and their role in the structure of matter. Extremely high-energy (multigiga electron volt) projectiles are required to produce the mass equivalent of these strange particles and to provide probes of short enough wavelength to "see" any internal structure. Experiments in elementary particle physics involve studies of the interactions of nucleons, mesons, and strange particles, employing nuclear emulsions, spark-discharge chambers, bubble chambers, cloud chambers, and counters for the detection of individual events.

The residual nuclei left after the bombardment of complex nuclei with high-energy nuclear probes represent the accumulation of products of many individual events, and these, as well as the nucleons and particles emitted, are the chief concern of the high-energy nuclear chemist. A knowledge of the formation cross sections, the kinetic energies, and the angular distributions of these product nuclei is the raw material from which an understanding of these nuclear reactions may be gained. Since elementary particle production may influence the properties of the residual nuclei and the emitted nucleons and particles, there is obviously (and desirably) much overlap of interest between the high-energy physicist and chemist. In studying a nuclear reaction, the goal is an understanding of the mechanism, i.e., how it occurs. An additional goal has been to obtain systematic knowledge of reaction cross sections and their dependence on bombarding particle, energy, and target nucleus.

The complete specification of a given nuclear reaction would mean specification, for reactants and products, of momentum (including direction), and internal state (excitation energy, spin, and parity). This sort of detailed experiment is at present only feasible for the simplest of reactions at low energy, where the number of possible final states is small. The number of states increases rapidly with available energy, and the experimental measurements quickly become averages over a large number of final states. At sufficiently high energy, nuclear reactions become so complex that it is a matter of arbitrary definition (or experimental convenience) just what are chosen as final products.

For example, a 400-MeV proton interacting with a heavy nucleus may give rise to the following typical sequence of events. In approximately 10^{-22} sec several neutrons and protons are emitted with energies of 10–100 MeV. At a later time, perhaps about 10^{-16} sec afterwards, more neutrons are emitted with lower energies. The nucleus then splits into two fragments of approximately the same mass, which start to move away from each other. Each fragment emits several more neutrons and some γ rays during the next 10^{-9} sec or so. About 10^{-3} sec after the initial event the fragments undergo a β decay. Further β decays occur, with half-lives increasing to months or years, before the final stable products are reached. The time gap after the last γ ray and before the first β particle is emitted is perhaps the logical place to call the products "final," but the previous stages, although difficult to isolate, are also of interest.

In the following sections we will describe the theoretical framework within which high-energy nuclear reactions are studied, and the types of processes which are thought to occur. The experimental techniques that are used will be described next, with some of the results and their significance. To a certain extent the stage of a nuclear reaction (as described in the previous paragraph) which one desires to investigate governs the choice of technique, and vice versa. Thus, cross-section measurements are done normally on a time scale of minutes or greater and the final products represent sums over all of the intermediate stages, including the different possible ways to form a given nucleus. On the other hand, track detectors reveal the charged particles emitted during the entire reaction process but are not able to identify the final nucleus with the resolution permitted by chemical techniques.

Following these sections, some of the numerical calculations which have been done to help interpret the results will be described. Two applications of reaction studies which are important in the operation of high-energy accelerators are discussed next, beam monitoring by activation and the problem of radioactivity induced in the accelerator and its shielding material. In the last section some other fields of research which have been stimulated by high-energy reaction work are described briefly.

B. Theoretical Framework

The energy at which a nuclear reaction becomes a *high*-energy reaction is probably best defined in terms set forth by Serber [1] in 1947. If the energy of a particle is much greater than the interaction energy between nucleons in the nucleus and if its de Broglie wavelength is smaller than the mean distance between nucleons, the impulse approximation is valid. This occurs at energies above about 100 MeV for incident protons. The impulse approximation consists of treating the reaction as the sum of the individual interactions between the incident particle and the nucleons, calculated as two-body interactions of free particles. The short wavelength of high-energy particles allows the approximation of a classical trajectory between collisions. The incident nucleon is assumed to make collisions with the individual protons and neutrons inside the nucleus. Each such collision may transfer enough energy to the struck nucleon so that it itself is capable of making further collisions. The resulting process, called the *intranuclear cascade,* is a stochastic one which suggests use of a Monte Carlo calculation to describe it. Each particle in the cascade is traced through the nucleus until it either escapes or loses so much energy that it can be considered to be absorbed into a kind of compound nucleus. After all particles have escaped or been absorbed, the final nucleus is characterized by momentum and excitation energy both computed by application of the conservation laws. The postcascade fate of the nucleus is then assumed to be deexcitation by particle evaporation, fission, or another relatively slow process, exactly as if it were a compound nucleus.

When the center-of-mass energy in a nucleon–nucleon collision exceeds 140 MeV, the rest mass of a π meson, it is energetically possible to create a meson in the collision. The probability of such an inelastic collision increases with energy, so that the process is significant at incident proton energies above about 400 MeV, and must be included in the development of the cascade. The π mesons inside the nucleus may be reabsorbed in addition to having elastic collisions; the absorption further spreads the available energy throughout the nucleus.

The Serber model has been used extensively to account for the characteristic features of high-energy nuclear reactions, and in general it succeeds. As the energy of the bombarding particle increases, there is a decrease in the importance of compound-nucleus formation. The latter is characterized by cross sections for producing individual nuclides which rise to a maximum from a threshold energy and then decrease as competing processes become energetically possible. In the high-energy realm many different products are formed with about the same cross section, and the energy dependence of the cross section is weak. The Serber model accounts for this in terms of the statistical nature of the fast cascade, which may result in many nucleons being ejected from the nucleus, or none, and residual excitation energies varying from nearly zero to the maximum possible (the latter corresponding to compound-nucleus formation).

The complexity of the process has led to the use of general descriptive terms to classify these phenomena. Such terms as simple reactions, spallation, fission, and fragmentation are used to describe classes of reactions, although there is often no clear-cut distinction between them. A simple reaction is one in which the final nucleus differs from the initial one by only a few mass numbers. The reason for this classification is that a simple or limited cascade may allow us to examine the details of the model which are obscured by more extensive cascades. Some examples of these reactions are (p, n) and (p, $p\pi^+$), in which there is no change in mass number, and (p, pn) and (p, 2n) in which the mass number decreases by one.

The (p, n) reaction at high energy is primarily a charge-exchange process in which the incident proton exchanges a positive meson with a neutron in the nucleus, but with only a small transfer of momentum. The incident particle, now a neutron, retains almost all of its original momentum and energy, and the residual nucleus has little or no excitation energy. If somewhat more energy is transferred, the residual nucleus may be able to evaporate a neutron, making the net reaction (p, 2n). It is much less probable that a (p, 2n) reaction could result from ejection of two energetic neutrons in the cascade, because such cascades must occur in at least two steps and are likely to leave the nucleus excited enough to evaporate additional neutrons. Above the meson threshold one expects inelastic proton–neutron collisions, so that the (p, n) product may also be formed by (p, $n\pi^0$) and (p, $p\pi^-$) reactions, or even multiple pion reactions.

The reaction (p, $p\pi^+$) results in a different nuclide, one with the same mass number but with one lower atomic number, while the charge exchange reactions yield the nuclide with one higher atomic number:

$$^{65}_{29}Cu + p \rightarrow {}^{65}_{28}Ni + p + \pi^+, \tag{1}$$

$$^{65}_{29}Cu + p \rightarrow {}^{65}_{30}Zn + n\,(+ \pi^0)\,(\text{or } p + \pi^-). \tag{2}$$

Reaction (1) cannot occur by nucleon emission alone; one or more mesons must be formed and escape in order to remove two units of charge and one unit of mass. Observation of the radioactive nuclide ^{65}Ni from a copper target bombarded with protons is thus equivalent to observing a proton and positive pion in coincidence.

The (p, pn) and (p, 2p) reactions, unlike the (p, 2n) reaction, may take place with high probability by ejection of two fast nucleons, because a single collision suffices, and the final nucleus may be formed with insufficient excitation energy to evaporate further particles. This mechanism has the descriptive label *clean knockout*. A two-step process may also contribute, namely, a collision with small momentum transfer, the incident proton escaping, followed by evaporation of a single nucleon. Experimental evidence, discussed later, indicates that about 70% of the ^{65}Cu(p, pn)^{64}Cu reaction takes place via the clean knockout mechanism. The probability of a clean knockout occurring is greatest in the surface region of the nucleus, where the nucleon density is low enough to give a high probability that both nucleons escape without further interactions.

The second descriptive term named above, *spallation,* refers to the typical high-energy process of fast cascade followed by evaporation of nucleons, α particles, etc. from the excited cascade product. Spallation products tend to be neutron deficient, undergoing β decay by emitting a positron or capturing an orbital electron, because the Coulomb barrier suppresses charged-particle evaporation, except at the highest excitation energies. The term spallation has also been applied to the evaporation of neutrons from a compound nucleus formed by low-energy bombardment of a heavy nucleus.

Fission at high energy is probably best defined as the breakup of an excited nucleus into two or more fragments of roughly comparable masses. It is not restricted to the heaviest elements, but occurs in elements as light as silver and even copper. In order to retain some similarity to the fission process occurring at low energies, high-energy fission is usually considered to be a relatively slow process in comparison with the intranuclear cascade. That is, it occurs after the cascade, in competition with evaporation, and the only angular anisotropy of fission is due to the intrinsic angular momentum of the fissioning nucleus. The angular distribution of fission fragments must therefore be symmetric about a plane perpendicular to the direction of motion of the fissioning nucleus, with the assumption that fission occurs after "memory" of the cascade is lost. A fission product can be distinguished from a spallation product by the lack of a suitable heavy partner for the latter, and the question of how heavy is suitable is somewhat arbitrary. As an example, we might say that a nucleus of mass 60 *evaporates* an α particle or *fissions* into a fragment of mass 20 and one of mass 40, but what do we call it when it breaks up into two particles of masses 10 and 50?

What we call it is not very important, but the question really involves whether there is a specific high-energy process, qualitatively different from fission and spallation, which contributes to this region between very asymmetric fission and evaporation of heavy particles. The name *fragmentation* is sometimes used to indicate such a process, and it will be used here to signify a fast process which occurs simultaneously with (or instead of) the intranuclear cascade.

C. Experimental Techniques and Reaction Mechanisms

In this section we will discuss in turn the various experimental techniques used to study high-energy nuclear reactions. In addition we will discuss in more detail the reactions and their mechanisms. This arrangement of discussing techniques and mechanisms together was chosen because of the relation between a technique and the stage of the reaction to which it is sensitive.

The classical technique used by nuclear chemists to investigate nuclear reactions is the chemical separation and measurement of the radioactive nuclides which result. In this way a cross section can be measured for the formation of an individual nuclide. If the measurement is performed after all shorter-lived isobars have undergone β decay into the nuclide being measured, the cross section is termed a *cumulative* one. The term *independent cross section* refers to the amount formed directly in the reaction before any contribution from radioactive decay of precursors takes place. There is no necessary restriction to radioactive nuclides, since a sensitive mass spectrometer can detect the small quantities formed in a nuclear reaction, and many cross sections for stable and long-lived nuclides have been measured in this way. The average kinetic energy and the angular distribution of a product can be measured by collection of the nuclei which recoil from the target. Detailed information about individual events can be obtained by measuring the tracks left by the recoiling nuclei in photographic emulsions or in the new solid-state track detectors such as mica. These techniques reveal directly the number and angular distribution of the charged fragments from individual events. Semiconductor detectors are being used to measure energy, mass, and angle of individual fragments from very thin targets. While the latter two techniques can reveal details of the kinematics of the process, they lack the resolution in mass and atomic number of the radiochemical and mass spectrometric techniques. Clearly the knowledge obtained by the different techniques is complementary.

1. Cross-Section Measurements

The cross section of a nuclear reaction is proportional to the probability of that reaction occurring, and may be defined for a beam of particles incident on a target as follows:

$$dN_i/dt = I\sigma_i nx. \tag{3}$$

In Eq. (3) σ_i is the cross section for production of nucleus i, I is the beam intensity, n and x are the target atom density and target thickness, respectively, and dN_i/dt is the rate of production of nucleus i. The usual cross-section measurement uses the integrated form of Eq. (3), and the total number of nuclei, N, formed during a finite bombardment interval is measured after the bombardment by means of their radioactivity:

$$R = \lambda N, \tag{4}$$

where R is the disintegration rate and λ is the decay constant. Alternatively, N may be measured directly with a mass spectrometer.

Determination of R involves the following steps. A chemical separation and purification of the desired element from the target material and all other radioactive elements is done. Isotopic separation may also be done to improve the resolution of different isotopes. The radioactivities present are measured with any of a variety of detectors (proportional, scintillation, semiconductor, etc.), with pulse-height analysis if necessary. A given nuclide is identified by half-life and characteristic radiations, and if the efficiency of the detector is known, R may be obtained from the observed counting rate. The decay scheme of the nuclide and the abundance of the radiation being detected must be known unless only relative cross sections are desired (e.g., comparisons between different targets, variation with energy, and angular distribution of recoils).

The radiochemical method is often laborious and time consuming, and is limited to radioactive nuclides of "reasonable" half-life. Mass spectrometers have been used successfully to measure cross sections for stable and long-lived isotopes of some of the inert gas and alkali metal elements. Recently this technique has been extended to short-lived isotopes by putting a mass spectrometer "on line" in an accelerator beam, with transit times of less than 1 sec. Track detectors (Section I,C,3) and counter (Section I,C,4) are of use in cross-section measurements, within the limits of their resolution. The fundamental advantages of the radiochemical method, namely, the ideal resolution (exact determination of charge and mass) and sensitivity (single atoms can be detected), insure its continuing use in nuclear reaction studies.

a. Excitation Functions. The variation of cross section with bombarding energy, called the *excitation function,* can be used to classify reactions by mechanism. In Fig. 1 we show some examples of excitation function for

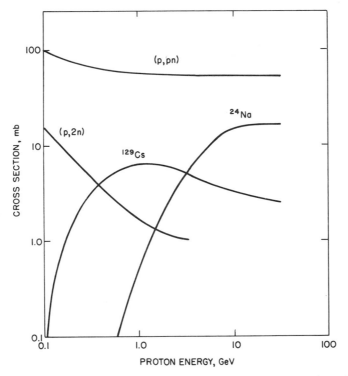

FIG. 1. Excitation functions for some typical products of the reaction of protons with a heavy element. The curves are meant to show only the general behavior of different kinds of reactions and do not refer to any one target element.

reactions of protons with a heavy element. These curves are not all for the same target, but are representative of the behavior expected, for example, for ^{238}U. The (p, pn) reaction has a cross section which decreases from its peak value at compound nucleus energies (i.e., 20–30 MeV) and then becomes constant above an energy of about 1 GeV. This constancy is related to the proposed clean knockout mechanism for the reaction, in which a proton–neutron collision occurs in the low-density surface region of the nucleus so that both nucleons escape with high probability, leaving the nucleus with insufficient excitation energy to evaporate any additional nucleons. The excitation function has a similar energy dependence as the total cross section for proton–neutron scattering.

When one or more mesons are formed in the collision, which occurs with high probability above 1 GeV, one would expect the cross section to be lowered as a result of the additional opportunities for further collisions, which would lead to more complex reactions. However, the cross section is constant above 1 GeV, indicating that the inelastic (meson-forming) collisions are as effective in leading to a simple reaction as the elastic ones. A possible interpretation of this is that the outgoing particles do not interact independently with their cross sections adding, but there is interference between them, or else they act as a unit, analogous to the nucleon "isobar." Then the total probability of leaving the nucleus without undergoing further collisions would not decrease with the number of mesons formed.

The excitation function for a (p, 2n) reaction, as shown in Fig. 1, falls off approximately inversely with proton energy up to about 0.5 GeV, after which it decreases more slowly, and is much smaller than that for the (p, pn) reaction. The two reactions must have different mechanisms, although they differ only in the change of a proton to a neutron in the products. The difference, as discussed in the previous section, is that a (p, 2n) reaction at high energy requires a charge-exchange collision with low-momentum transfer, so that the struck neutron is changed to a proton and captured by the nucleus, and the incident proton leaves as a neutron. If the residual nucleus has insufficient excitation energy to evaporate any nucleons, a (p, n) reaction results. If one neutron is evaporated, we have a (p, 2n) reaction, and so forth. These are rather restrictive conditions, so the cross section is small. The inverse energy dependence follows from this mechanism, if one assumes that the distribution of residual excitation energies is uniform, extending up to a maximum value equal to the bombarding energy. The (p, 2n) reaction samples a fixed energy range (the energy range in which one and only one neutron evaporates) ΔE, so the cross section is proportional to $\Delta E/E$, where E is the bombarding energy. In contrast, the (p, pn) reaction has much less restrictive conditions, since it is only necessary that all products of the first collision escape the nucleus.

The excitation function for ^{129}Cs shown in Fig. 1 is representative of that for a large number of medium-mass ($A = 80$–160, approximately) nuclides which are neutron deficient, formed from heavy targets. They have an energetic threshold below which they are not formed, and the cross section rises to a peak and then decreases. The excitation function for ^{24}Na is typical of that for light products from heavy nuclides; it is similar to that for ^{129}Cs but shifted to higher energies. The fragmentation hypothesis suggests that these two types of products are formed in similar ways; the medium-mass neutron-deficient product is the residue (after an evaporation sequence) from a cascade in which a light fragment is emitted. (It is clear from their different energetic thresholds, however, that ^{129}Cs and ^{24}Na are not such a complementary pair.) This is a question which is still open and remains a subject of some controversy.

The influence of the free-particle cross section on the nuclear reaction cross section is most apparent when there is pronounced structure in the former. This is the case for pion–nucleon cross sections where several prominent resonances occur. The pion–nucleon total scattering cross sections as a function of energy are shown in Fig. 2. Because of the charge independence of nuclear forces the $\pi^- + p$ and $\pi^+ + n$ cross sections are identical, as are the $\pi^+ + p$ and $\pi^- + n$ cross sections. The nuclear reactions of the type $(\pi, \pi N)$, where N is either a proton or a neutron, should be most sensitive to the resonance structure because of the predominant clean knockout mechanism. Unfortunately, the number of such reactions that have been studied is small, because of the low intensity of the available pion beams. In Fig. 3 excitation functions are shown for two of these reactions, ^{12}C$(\pi^-, \pi^- n)^{11}$C, and ^{40}Ar$(\pi^-, \pi^- p)^{39}$Cl. In both reactions we see peaks corresponding to the free-particle resonances, but they are broadened by the effects of the nucleon being inside the nucleus. The appearance of the free-particle structure is evidence that the clean knockout mechanism does indeed dominate the reaction. However, recent experiments at CERN [3] have shown that the cross sections for the $(\pi^-, \pi^- n)$ and $(\pi^+, \pi^+ n)$ reactions on the light targets ^{12}C, ^{14}N, and ^{16}O are equal at 180 MeV, the peak of the resonance, while the $\pi^- + n$ scattering cross section is three times as large as the $\pi^+ + n$ cross section (Fig. 2). It is difficult to reconcile this discrepancy with the clean knockout mechanism, while it is quite consistent with the alternate mechanism of scattering followed by neutron evaporation. The initial excitation process would be identical for either charge pion on a nucleus with $Z = N$, such as the three cases studied. The difficulty here is that the cross section for the $(\pi, \pi n)$ reactions is an order of magnitude larger than the measured nuclear excitation cross sections to levels in the nucleus which would evaporate one neutron. Much more experimental work on meson-induced reactions is needed, and the new "meson factory" accelerators will make the experiments easier and more detailed.

b. Mass and Charge Distributions. When the kinetic energy of the bombarding particle is greater than the total binding energy of the target nucleus, it is not surprising that the products of the reaction should include every nuclide lighter than the target since there are obviously no energetic restrictions on the products. What is surprising is that there is some

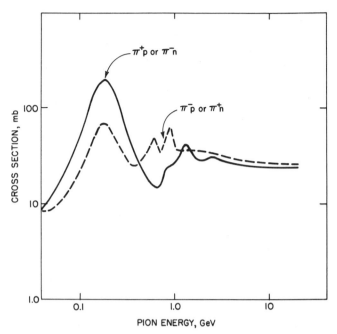

Fig. 2. Pion-nucleon total scattering cross sections as a function of pion kinetic energy.

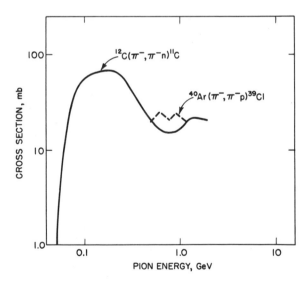

Fig. 3. Excitation functions for two simple pion-induced reactions: $^{12}C(\pi^-, \pi^-n)^{11}C$ and $^{40}Ar(\pi^-, \pi^-p)$. Data for these curves were obtained from sources in the literature [2a, b].

regularity in the yield of products rather than the complete breakup of the target into its constituent protons and neutrons. In fact, the Serber mechanism of high-energy interactions predicts a significant probability of small energy transfers and the simple reactions discussed above often have relatively large cross sections. The cross section of each product of a given combination of target and projectile is a function of the atomic number and mass of that product, and one may visualize the complete set of cross sections as forming a surface in the coordinates A and Z. Since β decay is the usual decay mode of the final products, it is useful to separate the cross section surface into the *mass–yield* curve, which is the cross section for all isobars of a given mass as a function of mass and the *charge dispersion* curves, which show the variation of cross section along a given isobaric chain (β-decay chain).

Charge–dispersion curves in most cases show a fairly sharp peak, so that most of the isobaric cross section goes into two or three nuclides. This peak is usually on the neutron-deficient side of the β stability line, but for light fragments the peak tends to be close to that line, while for a highly neutron excessive target such as ^{238}U, it may be on the neutron-excess side. In fact, the charge dispersion curve appears to be double peaked in the vicinity of mass number $A = 131$ for ^{238}U bombarded with giga electron volt-energy protons [4]. The neutron-excess peak is attributed to fission at rather low excitation energies, while the neutron-deficient peak is caused by spallation or some other high-excitation energy process.

In order to construct a mass–yield curve, one must measure or estimate enough isobaric cross sections to sum over the charge dispersion curve for each mass. In practice, the latter curve is assumed to vary only slowly with mass, and one curve is used for a range of masses. Stable product cross sections are found by interpolation unless mass spectrometric measurements are made. The resulting mass–yield curve shows sharp changes with energy for heavy targets and more gradual changes for lighter targets. Examples of some mass–yield curves at different energies are given in Figs. 4 and 5, for the elements Cu and Ta.

The curves for Cu (Fig. 4) show that the effect of increasing the bombarding energy is to spread the total inelastic cross section over more nuclides. There is little change in the curve above an energy of several giga electron volts, showing that the spectrum of deposition energy is constant. The remainder of the total energy must go toward the creation of new particles, which escape the nucleus without adding to its excitation. The approximate constancy of cross sections above several giga electron volts appears also to be true for targets lighter than copper.

The curves for Ta (Fig. 5) show the behavior typical of heavy elements which have low fission probability at low energy. At energies in the range

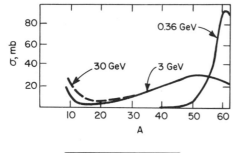

FIG. 4. Mass–yield curves for copper bombarded by protons of energy 0.36, 3, and 30 GeV. Data for these curves were obtained from sources in the literature [5].

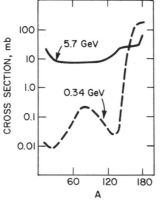

FIG. 5. Mass–yield curves for tantalum bombarded by protons of energy 0.34 and 5.7 GeV (adapted from Grover [6]).

0.1–0.5 GeV there is a distinct separation between the spallation region and the fission region. The former is characterized by a rapid decrease of cross section with increasing distance from the target, which is expected for the evaporation mechanism. The peak near one-half the target mass is due to fission, but it is not established whether the main process is fission of a highly excited nucleus before much evaporation has occurred, or whether the fission occurs near the end of the evaporation chain because the fission-ability parameter, Z^2/A, has increased sufficiently to permit effective competition with neutron evaporation.

At 5.7 GeV, however, there is no fission peak, only a flat curve, and there seems to be no clear distinction between the regions which were apparent at the lower energy. It is known from other techniques (track detectors and counters) that fission occurs (i.e., breakup into two large fragments), and therefore the mass–yield curve shows that there is considerable overlap between the various mechanisms.

Mass–yield curves for fissionable elements, of which ^{238}U is the most studied, retain a fairly distinct fission peak centered at about $A = 110$, even at 28 GeV, although the recoil studies discussed below and the double-peaked charge dispersion curves indicate that the neutron-deficient nuclides are formed by a different process than the neutron-excess ones.

2. Radiochemical Recoil Studies

The kinetic energy and direction of flight of a product nucleus are the result of the recoil momenta due to emission of all particles accompanying that product. Measurement of the distributions in these quantities for only one product, the radioactive nucleus "left over," is one step toward a complete description of the reaction. But it is probably as far as one needs to go, since the reaction itself is only defined in terms of this product nucleus and not the different combinations of particles which could have accompanied it. In fact, for certain simple reactions such as the (p, 2p) reaction the important information (the momentum distribution of the nucleon knocked out) is contained in the nuclear recoil properties.

Radiochemical recoil studies actually measure the range or distribution of ranges in some absorbing material. This method has certain advantages over a direct measurement of kinetic energy by a counter. It has the high resolution in mass and charge of the radiochemical measurement, it is suitable for kinetic energies down to 1 MeV or less, and it has enough sensitivity to study low-yield processes. The relation between kinetic energy and range must be known for the particular combination of recoil nucleus and absorbing material which is used. This need has led to a considerable amount of experimental data and theoretical treatments of the slowing and stopping of heavy ions in matter.

An "ideal" recoil experiment would use a target which is thin in comparison to the mean range of the recoils and a set of catcher foils, also thin, in order to measure the distribution of ranges. A gas may be used in place of thin foils. The recoils should be well collimated, and the range distribution should be measured at a number of angles. This kind of detailed differential study requires a high beam intensity and is time consuming; only a few reactions have been studied so completely. Fortunately, useful data can be obtained by integral experiments, which are much simpler. The most common of these is to use a target and recoil catchers which are thick compared to the range, and measure the fraction of nuclei recoiling out of the target in the forward and backward directions with respect to the beam (integrated over angle).

a. Thin-Target Experiments. In analyzing the results of high-energy recoil experiments, the fast cascade–slow evaporation model is usually used. The observed velocity in the lab of the recoil is resolved into components corresponding to the two steps (Fig. 6). The fast-cascade step forms an excited nucleus moving with velocity v, generally not in the beam direction. This is a contrast with low-energy compound nucleus reactions, where this velocity is parallel to the beam and is known from the incident energy.

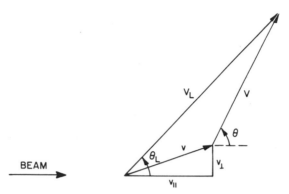

FIG. 6. Relationship between velocity vectors used in analysis of recoil experiments. Velocity v is due to the fast cascade and has components $v_\parallel$ and $v_\perp$ parallel and perpendicular to the beam direction, respectively. Velocity V is due to the evaporation stage and is at an angle θ to the beam. Velocity V_L is the resultant velocity in the laboratory system.

The components of v are $v_\parallel$ and $v_\perp$, respectively parallel and perpendicular to the beam. The evaporation of a number of particles from the moving nucleus results in a velocity V, which is assumed on statistical grounds to have a symmetrical distribution about a plane perpendicular to the beam. This will be true if the evaporation steps occur long enough after the cascade so that the nucleus has no "memory" of it.

The experimental data are then analyzed by assuming a simple form for the angular distribution of V in the system of the moving nucleus, usually $a + b \cos^2 \theta$, which is symmetric about 90°. The ratio $v_\parallel/V$ is estimated from the forward–backward ratio of recoils in the laboratory system, a quantity which is insensitive to $v_\perp$. The differential range curves are transformed to velocity curves, and the mean value of V is obtained. Consistency requires that the value of $v_\parallel$ derived from the forward–backward ratio be such as to lead to the proper velocity spectra in the laboratory system.

Consistent results were obtained for a fairly simple spallation reaction, $^{27}\text{Al}(p, 3pn)^{24}\text{Na}$ studied at 0.38 and 2.9 GeV by workers at Brookhaven National Laboratory [7]. The mean velocity V corresponded to a kinetic energy of the ^{24}Na equal to 1.25 MeV, which indicates the low energies that can be measured by this method. The value of $v_\parallel/V$ found was 0.25, so that most of the kinetic energy was due to evaporation recoil rather than the cascade. Fairly good agreement between the experimental results and the predictions of the cascade-evaporation model was found.

In contrast, a similar study of ^{24}Na formed from a heavy target (^{209}Bi + 2.9-GeV protons) could not be analyzed in a consistent way [8]. That is, there was no moving system in which the angular distribution would transform into one symmetric about 90° and in which the velocity distributions at different angles would transform into the same distribution. The angular distribution was peaked forward, if the velocity criterion was used to choose the moving system. This indicates that however ^{24}Na is formed (fission or fragmentation), the nucleus which breaks up still has some memory of the cascade. Therefore, the process occurs in a time shorter than nuclear rotation times and must be a fast nonequilibrium one, unlike low excitation energy evaporation or fission.

An interesting property of the kinematics of simple reactions has been used to separate and identify reaction mechanisms. When the mass numbers of the initial and final nuclides are the same, kinematics restricts the recoil angle of the final nuclide in the laboratory system to values less than 90°. If the reaction does not form new particles, i.e., is elastic in the elementary particle sense, the recoil nuclei are restricted to a narrow range of angles just below 90°. If part of the kinetic energy of the incident particle is converted to rest mass of mesons, the maximum angle is less than 90°, and there is no tendency toward peaking in the angular distribution. Figure 7 illustrates this for some simple reactions of 1-GeV protons with copper isotopes [9]. The reaction ^{65}Cu(p, pπ^+)^{65}Ni is an example of an inelastic reaction (meson creation) with no change in mass number. The kinematic maximum angle is 72° at 1 GeV, and we see that cutoff clearly. The case of an elastic process with no change in mass number is represented by the reaction ^{63}Cu(p, n)^{63}Zn, and we see a sharp peak in the angular distribution just below 90°. The small tailing to larger angles is due to scattering of the slow recoils and to the finite angular resolution. There is a broad distribution of recoils extending to 0°, similar to that of the (p, pπ^+) reaction. Since ^{63}Zn may also be formed by the inelastic reactions ^{63}Cu(p, pπ^-)^{63}Zn and ^{63}Cu(p, nπ^0)^{63}Zn, we attribute the forward recoils to these reactions. An estimate of the angular distribution of these reactions is shown as a dashed line. The different mechanisms are well separated and by integrating the two distributions it is found that the elastic charge-exchange (p, n) reaction contributes about 70% of the cross section at 1 GeV.

The other two cases shown in Fig. 7 are reactions in which the mass number of the nucleus decreases by one. Here there is no kinematic restriction on the recoil angle, but one can still distinguish between different mechanisms. The ^{63}Cu(p, 2n)^{62}Zn reaction is probably a two-step process, the first step being the formation of ^{63}Zn in an excited state by one of the above reactions, and the second step the evaporation of one neutron. The

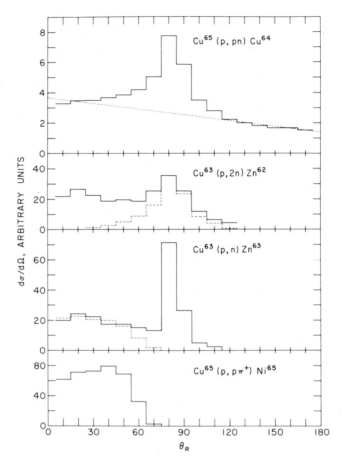

FIG. 7. Recoil angular distributions for products of simple reactions of 1-GeV
protons with copper isotopes (from Remsberg [9]).

neutron evaporation will broaden the distribution from the first step, and an
estimate of this broadened distribution for the elastic (p, n) peak is shown
as a dashed line. In this case, integration of the separated angular distributions gives a value of 65% of the cross section due to the elastic charge
exchange, in agreement with the value found for the (p, n) reaction.

Finally, one sees the 90° peak in the reaction $^{65}Cu(p, pn)^{64}Cu$ above a
continuous background. The peak is attributed to the two-step mechanism
of scattering, followed by neutron evaporation, since the first step of proton
scattering is kinematically identical to the (p, n) elastic scattering. The
smooth background comes from the clean knockout mechanism and is estimated by the dashed line. Integration of the separated distributions gives a

value of 30% for the contribution of the two-step mechanism and 70% for the clean knockout.

This very pretty series of measurements illustrates the power of the recoil technique in studying simple nuclear reactions. A large part of the analysis requires no models, only the use of kinematic arguments, making it particularly attractive.

b. Thick-Target Experiments. While the thin-target experiments described above can serve to test some of the assumptions of a model, thick-target experiments must make full use of the model to derive the quantities of interest, and thus cannot test it. The typical thick-target, thick-catcher recoil experiment measures the fraction of the total activity recoiling out of a target of thickness greater than the range in the forward, backward, and perpendicular directions to the beam. These fractions are measured with a sandwich of target and catchers, and are integrals over 2π steradians. The perpendicular fraction is measured with the target sandwich aligned nearly parallel to the beam. The recoil fractions can be related to the quantities defined in Fig. 6, assuming symmetry of V about 90° to the beam and an angular distribution of the form $a + b \cos^2 \theta$. The three quantities measured allow the determination of average values of V, b/a, and $v_{||}$, while $v_\perp$ must be estimated from model-dependent calculations which relate $v_\perp$ and $v_{||}$ to the fast-cascade process. The resulting numbers may be tested for internal consistency but are dependent on the assumed model. From the thin-target results, it is known that the model probably breaks down for light nuclei formed from heavy targets. However, even then, the above analysis is useful in survey experiments designed to reveal broad trends. It is in these experiments that the relative ease of doing thick-target measurements is so important because of the large number of nuclides to be measured.

The average recoil ranges of a number of isotopes of the element iodine formed in the 18-GeV proton bombardment of ^{238}U were measured by the thick-target technique [10]. The ranges (hence, kinetic energies) of the neutron excess iodine isotopes were about the same as in low-energy fission, while the ranges of the neutron-deficient isotopes were about one-half as large. The transition between the two values of the range was remarkably sharp, occurring within two mass numbers at about $A = 123$. Subsequent experiments have shown that this is a general phenomenon in the mass range $A = 80–160$ and that the decrease in range of the neutron-deficient products occurs rather sharply at a bombarding energy of 1–3 GeV.

The implication of these data is that the neutron-excess nuclides in the "fission" peak of ^{238}U at energies above 3 GeV are true fission products, their kinetic energies coming from the mutual Coulomb repulsion of two large fragments. In contrast, the neutron-deficient nuclides in the same region have no single massive partner; their low kinetic energies could be due to a variety of mechanisms. Some possibilities are: (1) multibody breakup of a heavy excited nucleus (ternary fission, for example); (2) fission of a much lighter nucleus after an extensive loss of nucleons; (3) a very long evaporation chain ending in that nuclide (the low kinetic energy being the resultant of many random recoils); and (4) the same type of fast nonequilibrium process responsible for the formation of light fragments like ^{24}Na (fragmentation). The sudden onset of these low kinetic energies both with respect to mass number of the isotope and the bombarding energy remains a puzzling feature.

3. Track Measurements

By track measurements we mean the observation of charged-particle tracks in cloud chambers, bubble chambers, photographic emulsions, and the so-called solid-state track detectors (glass, mica, and plastics, for example). The latter two have been the most used to study nuclear reactions. They are most useful in measuring energy and angular distributions of charged particles and in seeing individual events, but they have poor resolution for the charge and mass of heavy ions. Thus, emulsions can discriminate between ions of different charge only for atomic number less than about 10, and the mass resolution is poor even in that range. The useful feature of mica track detectors and similar materials is their threshold, which makes them sensitive only to ions above a certain charge. These techniques thus are used for special studies where the visualization of individual events is valuable. These include fission, identified by two or more heavily ionizing tracks in the same event, and emission of light ($Z \leq 10$) fragments.

The photographic emulsion is often used as both target and detector, and events ("stars") are classified as originating in light nuclei (C, N, O) or heavy nuclei (Br, Ag) on the basis of the number of prongs (evaporation particles). Alternatively, the emulsion is saturated with a heavy-element compound whose interactions are picked out because of that element's much greater fissionability than emulsion nuclei. Finally, the emulsion may be used as a detector for recoils from an external target, for example, ^{8}Li recoils, detected by their characteristic "hammer tracks" due to the decay, ^{8}Li$(\beta^-)^8$Be $\rightarrow 2\ ^4$He.

The general features found for high-energy interactions in emulsion nuclei are the following: α-particle energy spectra are consistent with an evaporation mechanism, provided that the velocity of the evaporating nucleus (estimated from the recoil track associated with the α track) is taken into account. The frequency of tracks which can be assigned to atomic numbers between about 4 and 10 (light fragments) increases rapidly above 1-GeV bombarding energy, as does the frequency of stars with two or more light fragment tracks. Stars with two or more tracks of fragments with Z apparently greater than 10 are seen at giga electron volt energies, and probably are due to fission of Ag and Br nuclei.

These fission events are usually associated with a large number of evaporation tracks, indicating a high excitation energy for the nucleus. Moreover, the probability of a star having two or more high-Z tracks increases rapidly with the total number of evaporation-type tracks. These observations show that it is a specifically high excitation-energy fission.

The new solid-state track detectors for heavy charged particles offer great promise in nuclear physics and chemistry. The tracks of atomic dimensions left by the ionizing particle in an insulating solid can be developed by chemical etching to a size suitable for optical viewing. Materials as varied as crystals, inorganic glasses, and plastics have been used. For a given material there is a threshold dE/dx of the charged particle below which no track will be developed. The threshold for mica is in the region corresponding to nuclei of charge $Z \approx 15$ with energy of ~ 1 MeV/amu, while plastics such as Mylar and Lexan are sensitive to nuclei above $Z \approx 8$, and cellulose nitrate registers α particles of several mega electron volts.

The useful features of these detectors include the following:

1. Ability to withstand large doses of particles below the threshold for development without fading or fogging. Thus, one can expose these detectors to a high-intensity primary flux and detect extremely small fission cross sections.

2. One can use various target elements incorporated into the detector or placed adjacent to it, without the detector itself causing events, by the use of a low atomic number material.

3. Detectors with different thresholds can be used as crude dE/dx spectrometers.

In high-energy nuclear reactions, these detectors are undoubtedly most useful in studying fission. Total fission cross sections are being measured at giga electron volt energies by irradiating a sandwich of two mica sheets and a thin target of the element being measured. After development of the mica, the two sheets are placed together again and pairs of tracks become apparent upon proper alignment. In this way one can reject single tracks due to spallation. Some three-track events are also seen with uranium targets bombarded with 18-GeV protons, and the evidence indicates that these are probably due to ternary fission into three roughly equal masses. Much higher frequencies of ternary fission were observed in the bombardment of uranium and thorium with 400-MeV argon ions.

4. Counter Experiments

a. Light Charged Particles. A well-known technique to measure the mass and charge of light particles is to use a two counter telescope composed of a thin (ΔE) counter and a thick (E) counter. For nonrelativistic velocities, the energy loss in the ΔE counter is approximately proportional to (MZ^2/E), so that a plot of ΔE against E yields a series of hyperbolas for different ions. The main separation is by charge, because of the Z^2 term, and it becomes difficult to resolve neighboring masses when Z is larger than about 9. Further difficulties for the identification of the heavier particles are caused by the fact that they have such a large energy loss that an extremely thin ΔE counter is necessary, and by a breakdown of the (MZ^2/E) proportionality. Various methods have been used to combine the two signals from the counters in order to obtain a "particle identifier" signal which is different for each particle and independent of energy. A recent technique uses an empirical range–energy relationship instead of the energy loss equation to obtain a particle identifier signal. The use of a second ΔE counter improves the resolution by eliminating events with abnormal energy loss.

Alpha particles are relatively abundant and are easy to separate from nearby masses. The group at Orsay [11] have measured energy spectra and angular distributions of α particles, as well as deuterons and tritons, emitted from targets of heavy elements bombarded with 157-MeV protons. These data were resolved into an isotropic evaporation part and a forward-peaked part, attributed to direct knockout. When the target was silver, all of the α particles seemed to be evaporation in agreement with the emulsion data at giga electron volt energies. It seems necessary to consider the existence of α particle substructure in the outer regions of heavy nuclei to account for the direct knockouts.

Isotopes of Li, Be, B, and C have been observed in the bombardment of uranium by 5-GeV protons [12]. The range–energy particle identification technique was used, and neighboring masses were well resolved. Figure 8 shows the particle-identifier spectra, with peaks corresponding to three previously unknown nuclides, ^{11}Li, ^{14}B, and ^{15}B. Energy spectra as a function of angle were measured and analyzed in a similar way to the thin-target recoil spectra, and the same conclusion is found as in the case of ^{24}Na recoils from Bi, namely, there is no unique moving system which makes the angular distribution symmetric and the velocity spectra the same. This again implies that the emission of light fragments is at least partially due to a fast process, rather than equilibrium evaporation from a moving nucleus.

b. Fission. The complexity of fission at high energy means that counter experiments must be designed to measure as many physical quantities as is

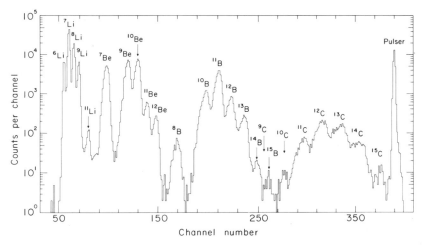

FIG. 8. Particle identifier spectrum showing light nuclides observed in the bombardment of uranium by 5.7-GeV protons (from Poskanzer *et al.* [12]).

practical. As an example, it suffices at low energy (spontaneous fission and thermal neutron fission) to measure the kinetic energy of the two fission fragments, because conservation of momentum relates masses and energies:

$$M_1 E_1 = M_2 E_2. \tag{5}$$

After a small correction is applied for neutron emission, Eq. (5) and a knowledge of the mass of the fissioning nucleus allow calculation of the fragment masses from coincident kinetic energy measurement. The 180° angular correlation between fragments simplifies the experiment. In contrast, high-energy fission has a broad angular distribution, takes place with a moving, excited nucleus whose mass and charge are unknown, and many neutrons and other particles may be emitted. In this case Eq. (5) is at best a rough approximation to the *average* fission event.

The general features of high-energy fission which have been observed by counter studies are well accounted for by the cascade-evaporation model. The angular correlation between coincident fragments shows that the most probable laboratory angle is a few degrees less than 180° (forward with respect to the beam), and that it decreases with decreasing target mass (targets of uranium, bismuth, and gold). The forward peaking in the laboratory system indicates that fission takes place in a nucleus moving in the beam direction. The less fissionable elements need more excitation energy in order to undergo fission, and the increase in excitation energy is correlated with an increase in momentum transferred by the incident particle. The kinetic energy of single fragments has a symmetrical distribution, unlike the asymmetric double-peaked distribution seen at low energy. This is directly related to the symmetrical distribution of masses seen in the mass–yield curves (Fig. 5).

The average total kinetic energy of the two fragments in uranium fission has a weak dependence on bombarding energy. With 156-MeV incident protons it is about the same as that for low-energy fission of uranium, and it decreases slowly with increasing bombarding energy. This effect indicates that the kinetic energy results from the Coulomb repulsion of the fragments; thus, as the cascade becomes more extensive with increasing bombarding energy, the average charge of the fissioning nucleus decreases due to ejection of charged particles, and the maximum Coulomb repulsion of the fission fragments decreases. In addition, three other effects can reduce the observed kinetic energies. If the fragments are formed with enough excitation energy to evaporate neutrons, this will tend on the average to lower their kinetic energy. The highly asymmetric events which form the extreme wings of the mass–yield curve will have less Coulomb repulsion than the more probable symmetric events, and will tend to lower the average. Finally, the average distance between the two fragments at the instant of separation may increase with increasing energy (i.e., the fissioning nucleus is more distorted), leading to lower Coulomb repulsion. In more detailed experiments the time of flight of a fragment over a known distance is measured, so that its velocity as well as its kinetic energy is known. This allows its mass to be calculated, without assuming the validity of Eq. (5) at high energy.

D. Calculations

The experiments described above, with few exceptions, measure properties which are integrated or averaged over many nuclear states, in general, all states which can decay to the ground state of the nucleus which is measured. Since most of the states involved have not been characterized individually, any calculations, of necessity, must be statistical in nature. For the simple nuclear reactions, typified by the (p, pn) reaction, fairly detailed calculations have been done, based on the impulse approximation. For other reactions recourse to a Monte Carlo technique is necessary. Following the initial intranuclear cascade calculation, the fate of the resulting excited nuclei is calculated by statistical evaporation theory, again using a Monte Carlo technique. Fission competition is included where appropriate.

1. Simple Nuclear Reactions

Calculations of differential and total cross sections have been done, usually for the (p, pn) reaction, for which the most experimental results are available. The mechanism is assumed to be clean knockout.

In the most detailed of such calculations [13], a distorted-wave impulse approximation, recoil angular distributions for the reactions $^{12}C(p, pn)^{11}C$ and $^{58}Ni(p, pn)^{57}Ni$ were calculated. Several different nuclear wave functions were used, as well as an optical model nuclear potential. The results indicated that the outer shells (1p in ^{12}C and 2p in ^{58}Ni) of neutrons lead to a sideways peaking (near 90° in the laboratory), which is due to the equatorial localization of the reaction sites for these outer-shell nucleons. There does not seem to be enough of this peaking to account for the very prominent peak seen in the $^{65}Cu(p, pn)^{64}Cu$ reaction (Fig. 7), which was interpreted as being due to the two-step inelastic scattering–evaporation mechanism.

A different approach [14] has been used to calculate the excitation function and recoil properties for reactions of the type $^{65}Cu(p, p\pi^+)^{65}Ni$. The reaction is assumed to proceed according to Eq. (1) and the cross section is calculated using the one-pion exchange theory in the pole approximation. This calculation is model dependent, unlike the purely kinematic arguments used in the analysis of the angular distributions of Fig. 7 previously discussed. The most striking feature of the calculation is the importance of the Δ (1236-MeV) nucleon isobar in the reaction, which it dominates up to about 3 GeV. The excitation function has a peak near 1 GeV and the calculation shows that the peak is due to the Δ isobar, which is also formed in the 180-MeV peak in pion–nucleon scattering (Fig. 2).

2. Intranuclear Cascade Calculations

In the spirit of the Serber model, the nucleus is assumed to be an ideal Fermi gas of neutrons and protons, whose interactions with energetic particles are the same as the corresponding interactions in free space. The fact that these interactions actually occur in nuclear matter is manifested through certain modifications. For example, the Pauli exclusion principle is taken into account by forbidding all collisions in which the final nucleons would be in an occupied state (i.e., below the Fermi level). The target nucleons have a momentum distribution, rather than being stationary, and the free-particle cross sections must be modified to correct for this. The existence of the nuclear potential means that the energy of a particle is a function of its position; the classical trajectories which are assumed may be modified to include the effect of refraction or reflection when the particle moves through a nonuniform potential.

The spatial and momentum distributions which are assumed for the nucleons depend on the nuclear model: the simplest model is a spherical, uniform-density nucleus. A more realistic model would include a radial variation of both density and potential. Some calculations have also been done using a model nucleus which contained α-particle subunits, in addition to individual nucleons.

At energies above approximately 400 MeV, pion production has a large enough probability that it cannot be ignored. This introduces uncertainties as well as complications into the calculations. The uncertainties involve such questions as the magnitude of the nuclear potential of pions, the formation, interactions, and decay of nucleon isobars and meson resonances, and the probable existence of multibody processes, such as meson absorption which cannot be included in strictly two-body interactions. The complications refer to the large number of inelastic (meson-formation) interactions whose cross sections must be included in the calculation.

The results of the calculation may be directly compared with experiment for the energy and angular distributions of the cascade nucleons. The general features of the experimental data are fairly well reproduced by the calculation; this shows that the Serber concept of essentially free two-body interactions is valid. However, the experiments revealing forward-peaked, high-energy spectra of light fragments (^{4}He, ^{8}Li) emitted from heavy nuclei suggest the existence of subunits or clusters which can be knocked out in the cascade.

This is in contradiction to a strict interpretation of the model, which would deny the stability of even an α particle in an interaction whose total energy is several orders of magnitude greater than its binding energy. Calculations which include the existence of clusters have been done but the *ad hoc* nature of the assumption is not very satisfactory.

Comparisons of the results of cascade calculations with radiochemical cross-section data make it clear that the details of the nuclear model used are important. For example, the cross section for a simple reaction of the (p, pn) type is underestimated by a factor of 2–3 by the calculation when a uniform-density nucleus is used. The reason for this discrepancy is that these reactions occur mainly in the nuclear surface, where the low density favors escape of both particles. Introduction of a nuclear model with a radially decreasing density into the calculation increased the cross sections for these reactions to the correct fraction of the total inelastic cross section.

The results of the calculation include the excitation energy, linear momentum, and angular momentum of each residual nucleus. The distributions of these quantities may be used as the initial configuration for statistical calculations of the deexcitation process. The mass–yield curve of experiment is closely related to the excitation energy distributions after the cascade, in the sense that a given excitation energy will result in a given average mass loss, independent of exactly what combination of particles are evaporated. The agreement with calculations is good here, except for heavy-target elements and the giga electron volt-energy region, where the uncertainties due to the assumed meson processes are magnified by the long path length of the cascade. The calculated linear momentum of the residual nucleus has been found to bear a simple relationship to its excitation energy: the average momentum parallel to the beam is nearly proportional to the average excitation energy, independent of target and bombarding energy. This relationship has been extensively used in interpreting the results of recoil experiments, and a self-consistency argument supports the validity of the calculation here.

3. Evaporation Calculations

An evaporation calculation is usually necessary in order to connect the results of the cascade calculations to experimental data. The treatment is based on the statistical model of the nucleus. Although the calculation may be done by numerical integration, the number and multiplicity of the integrals are excessive if more than two or three successive particles can be evaporated. Therefore, a Monte Carlo technique is used, analogous to the intranuclear cascade, except that the evaporation cascade consists of successive evaporation steps, and is concluded when the excitation energy falls below the separation energy of the least-bound particle. In this way, the results of the intranuclear cascade calculation, in the form of a distribution in A, Z, and E^* (mass, charge, and excitation energy), are used as starting points for the evaporation cascade, and the final distribution of products in A, Z are calculated.

Fortunately, it is possible to establish the range of validity of evaporation calculations by using experimental data on compound nucleus reactions. In this way, the best parameters to reproduce the low-energy data may be found for use in the high-energy applications. It is therefore expected that the intranuclear cascade calculation is being tested, rather than the combination.

Calculations of the probability of evaporating heavy particles (up to ^{24}Na) from highly excited nuclei have been done in order to find out if the products in the mass range $A = 10$–30 can be accounted for in this way. It was found that the dependence of cross section on the neutron–proton ratio of the target was the same for the evaporation calculation and the experimental results for a number of light nuclei. Although this does not prove the case, it shows that the mechanism for producing the light fragments must be one which is sensitive to the details of the mass–energy surface, as is evaporation.

E. Applications of Reaction Studies

1. *Beam Monitoring*

Most cross-section measurements are indirect, using a monitor reaction as a comparison standard, rather than measuring the beam intensity directly. This is especially the case when internal beam irradiations are done, because of the difficulty of a direct beam measurement. A sandwich of target and monitor foils is irradiated so that the same beam passes through both, and the unknown cross section is determined relative to the monitor cross section. Foil monitors are extensively used at high-energy accelerators because of their simplicity and convenience, requiring no electronics on the spot, and their freedom from interference by magnetic fields. They are not only used in nuclear chemical experiments but also in many elementary particle experiments to measure absolute cross sections. Only a few reactions are suitable for use as a monitor, and the important characteristics of a good monitor are discussed here.

The most used monitor reactions are $^{12}C(p, pn)^{11}C$ and $^{27}Al(p, 3pn)^{24}Na$. Others which are useful for special purposes are $^{12}C \rightarrow ^{7}Be$, $^{27}Al \rightarrow ^{18}F$, and $^{197}Au \rightarrow ^{149}Tb$. The first requirement of a monitor reaction is that the product radioactivity should be measurable without chemical separation. For light-target elements this results because of the small number of possible products; a half-life determination is sufficient to resolve the activity of interest. The production of ^{149}Tb from heavy elements is measurable without chemical separation by α-particle counting, effectively discriminating against the vastly greater β and γ activity present. In this case, however, the monitor element must be light enough not to form any of the heavy-element α emitters; therefore, gold is the most common monitor.

The monitor must be available as high-purity foils of uniform thickness. Aluminum and gold are ideal, and carbon is used in the form of various plastics. However, the latter may give problems in high-intensity beams due to melting or decomposition.

The half-life should be longer than the irradiation time, but not so long that small activity levels result. The 20-min half-life of ^{11}C makes it suitable only for short irradiations, which has led to the use of 15-hr ^{24}Na in preference. The isotope ^{7}Be, with a 54-day half-life, is suitable for very long irradiations.

Cross sections for producing the product of interest by low-energy particles should be small to prevent interference from secondary particles (especially evaporation neutrons) produced in the target or its surroundings. The production of ^{24}Na from aluminum is sensitive to these particles because of the fairly large cross section for the reaction ^{27}Al(n, α)^{24}Na at energies of 10–20 MeV. The reaction ^{197}Au $\rightarrow$ ^{149}Tb has a threshold energy of about 0.5 GeV, and thus is useful when such secondary reactions might be troublesome, as when a thick target used for meson production must be monitored.

In practice, the ^{12}C(p, pn)^{11}C reaction has been used for most of the determinations of absolute monitor cross section as a function of energy. The other reactions mentioned are then determined relative to it, in stacked-foil bombardments. Very little work has been done for other particles than protons. The ^{27}Al $\rightarrow$ ^{24}Na cross section for α particles is known only up to 0.38 GeV, and the ^{12}C $\rightarrow$ ^{11}C reaction has been measured for negative pions up to 2 GeV. It must be remembered that the activation technique cannot distinguish between different particles, except between those that do and do not (electrons and muons) have large nuclear cross sections.

2. Induced Activities in Accelerator Materials

The same nuclear reactions we have been discussing occur in the materials of a high-energy accelerator during its operation, as the beam strikes vacuum chamber walls, magnets, shielding blocks, etc. The radioactive products so formed create a long-term radiation hazard which may persist long after the accelerator is turned off. This problem becomes more serious as the energy of the beam is increased, because the main mechanism of absorption in matter of high-energy protons is energy loss by nuclear reactions and not by ionization. Thus, absorption of a primary proton results in formation of a number of secondary particles, each of which can cause more reactions and form more particles. The higher the energy of the primary beam, the more secondaries are produced.

A knowledge of reaction cross sections and excitation functions is necessary to predict the radiation hazard of any given material. However, one must know these cross sections for all the major components of the secondary beam, such as neutrons, mesons, and γ rays, and must estimate the energy spectrum and intensity distribution of each of these. Most of the required numbers are not known experimentally, but calculations have been made using various approximations. Experimental studies in which samples of various materials were left inside an accelerator for long periods, after which their activity was measured, have supplemented the calculations. Some general conclusions from both types of study are as follows: the best materials from the radiation hazard point of view are those of low atomic number, because of the small number of radioactive products formed. Their density is generally also low, which makes them too bulky to be used for shielding. Most of the metals near iron or copper result in similar radiation levels except for nickel, which is significantly worse due to the large cross sections for formation of ^{56}Co and ^{58}Co, which are relatively long-lived nuclides and emit energetic γ rays in their decay. Aluminum is nearly as good as lower-atomic-number elements for short exposures, but when it is exposed for several years, the buildup of 2.3-year ^{22}Na results in a long-lived radiation hazard.

F. Summary and Future Prospects

Chemists have participated in the study of nuclear properties and reactions since the discovery of radioactivity. The discovery of the new elements which were formed by radioactive decay of uranium and thorium was the first achievement of the new field. The development of accelerators permitted the study of nuclear reactions induced by energetic particles, and revealed a host of new radioactive isotopes of the known elements. The discovery of nuclear fission was a result of chemical identification of the radioactive products as isotopes of elements with about one-half the atomic number of the fissioning nucleus. Chemical separations of the mixture of isotopes resulting from fission and other nuclear reactions were required, and many ingenious techniques were developed for rapid and highly specific separations.

As the energy of the particle accelerators increased, so did the complexity of the nuclear reactions caused by their beams. A wide variety of techniques, both physical and chemical, were used to study these reactions. It turned out, unexpectedly, that even when the available energy exceeded the total binding energy of the target nucleus, simple reactions which used only a small fraction of this energy were still quite probable. This comes about because the energy must be transferred to the nucleus in a series of individual-particle collisions, and there is a fair chance that the struck particles will all leave the nucleus before much damage has been done, especially if the collisions occur in the low-density surface region.

The more complex reactions occur when more of the incident energy is transferred to the nucleus. Spallation, which is the formation of products by evaporation of many particles from the excited nucleus, and fission, which may occur at any step of this deexcitation in competition with evaporation, are both considered to be equilibrium processes, in the sense of statistical mechanics. There is a variety of evidence which suggests that a fast nonequilibrium process called fragmentation occurs at energies greater than about 1 GeV, which makes itself known mainly in the formation of very light fragments from heavy nuclei. At present, much of the experimental work is aimed at elucidating the nature of such a process.

Although we understand these phenomena in a general way and can predict qualitatively what products will be formed by a high-energy reaction and even estimate their cross sections, there remain many problems. For example, the exact nature of the fragmentation process is not clear. In fact it is not clear if such a new process must be postulated to account for the experimental data, or whether an extension of the cascade-evaporation model will suffice. The data indicate a fast nonequilibrium process like the cascade, but involving clusters or fragments. Between the cascade and the stage of the reaction where equilibrium evaporation is a good approximation there must be a period in which the nucleus approaches equilibrium. During this period it is possible that the phenomena attributed to fragmentation occur. The nucleus would be highly excited, and the excitation energy and internal motions of its constituents would not yet be uniform. The emission of fragments with high kinetic energies in the forward direction is reasonable from such a system, and it could probably be treated as a nonequilibrium evaporation.

Another problem is why cross sections for most reactions remain nearly constant as the bombarding energy increases from 3 to 30 GeV. The number of mesons created (multiplicity) rises by about a factor of 7 between these energies, and one might have supposed that the average amount of energy deposited in the nucleus would go up by about the same factor, not only because the number of interacting particles increases but because pions have larger interaction cross sections with nucleons than other nucleons (the resonances) and because of the additional process of pion absorption, which deposits the pion rest energy of 140 MeV. The average deposition energy spectrum, however, must remain nearly unchanged in order that the pattern of cross sections remain constant.

This leads to the question of how pions interact inside the nucleus—are they free particles in the spirit of the impulse approximation or do they form nucleon isobars with the nucleons? An isobar is just a pion–nucleon resonance treated as a particle. If such a viewpoint is valid, it raises the possibility of studying interactions between such a highly unstable particle and the nucleons inside the nucleus. The use of pions as bombarding particles may help to answer this question.

Much more work needs to be done with bombarding particles other than protons, not only pions but neutrons, α particles, and heavier nuclei. Accelerators to obtain heavy ions in the giga electron volt energy range are being planned, with one objective the production of new superheavy elements. This will open up a new field of nuclear reactions, combining heavy ion and high-energy phenomena. The use of K mesons, antiprotons, and other unstable particles to study nuclear reactions may become possible as the intensity and energy of accelerators increase.

What do we expect to happen to nuclear reactions as the bombarding energy rises to 200 GeV or more? One can extrapolate the presently observed trend from 30 GeV, the highest energy at which they have been studied. If that trend continues, cross sections will change slowly with energy, but the overall pattern of mass and charge distributions will remain constant. However, past history has shown that it is dangerous to make such extrapolations. At each previous increase of an order of magnitude in energy there were surprises. When the Brookhaven Cosmotron provided 2–3-GeV protons, some of the surprises were the large cross sections for forming middle-mass nuclides from heavy targets such as tantalum and lead, and the greatly increased cross sections for forming light fragments. The biggest surprise at energies of 20–30 GeV was probably the lack of surprises. Who knows what will happen at 200–300 GeV? There is always a feeling of excitement when new territory is explored because of the habit nature has of surprising us.

II. OTHER RESEARCH STIMULATED BY HIGH-ENERGY REACTION STUDIES

A. Nuclear Structure and Spectroscopy

The interaction between this field and that of nuclear reactions is particularly strong, since they share the same subject matter and experimental techniques. The production of a radioactive nuclide, measurement of its cross section, and determination of its half-life and decay scheme are all interrelated. Probably the greatest influence the study of reactions has on that of structure is in the production process. A high-energy accelerator is a good production source of isotopes because of the rather nonspecific nature of high-energy reactions. Many isotopes can be made in adequate amounts by a suitable choice of target. For example, medium-to-heavy neutron-deficient isotopes are formed in abundance by bombarding a target several elements heavier than those desired. Medium-mass neutron-excess isotopes are formed from uranium in high-energy bombardments, but slow-neutron fission is usually a better method. Although high-energy accelerators do not have beam currents comparable to cyclotrons and linear accelerators, the high energy allows one to increase the quantity of radioactivity produced. This can be done by making the beam traverse a thin target many times, each time losing only a small fraction of its energy (multiple traversals), or by aligning a target foil edge on to the beam so that one uses the long range of high-energy particles. Since the electronic stopping is a minimum, most of the beam particles are lost by nuclear interactions, which means a high efficiency. High production efficiency is particularly important when an isotopic separation is used in addition to chemical separation, because of the low efficiency of the former. Such isotopic purification is usually necessary to obtain a pure source, because of the large number of isotopes of a given element formed at high energy.

High-energy accelerators have been useful in producing new nuclides. For example, the unusual nuclide ^{8}He was formed by spallation of carbon and oxygen by 2-GeV protons, and its half-life and decay radiations measured. The particle stability of the new nuclides ^{11}Li, ^{14}B, and ^{15}B was established by detecting them in products of the 5.7-GeV proton bombardment of uranium (Fig. 7). It seems that if enough energy is supplied, every possible product is formed, and may be found with sensitive enough detectors.

The use of on-line isotope separators at high-energy accelerators has recently begun with the ISOLDE project at CERN. Thick targets of various elements are bombarded by 600-MeV protons in an external beam and volatile radioactive products are extracted into the ion source of the isotope separator. Some of the elements separated and the target used are mercury from lead, xenon from cerium, and cadmium from tin. Nuclides with half-lives shorter than 10 sec are seen, and their decay schemes studied. The great interest in investigating the properties of nuclides far away from the stability line insures that this type of study will be vigorously pursued in the future.

B. Geochemistry, Cosmochemistry, and Astrophysics

Although these topics are the subject of other chapters in this volume, it is appropriate here to discuss briefly the specific applications of high-energy nuclear chemistry to them. These applications involve a knowledge of high-energy production cross sections to interpret the effects of cosmic ray bombardment of natural targets.

In the earth's atmosphere, cosmic rays interact with N and O nuclei to produce several radioactive nuclides of interest. The most important one is, of course, ^{14}C, because of its usefulness in radiocarbon dating. The major portion of ^{14}C results from the low-energy reaction ^{14}N$(n, p)^{14}$C, induced by secondary neutrons which are emitted when the primary particles interact. Tritium is formed both as a direct product of a high-energy reaction and in the low-energy reaction ^{14}N$(n, T)^{12}$C. These are the nuclides which have been most extensively studied, but the presence of $\sim 1\%$ of ^{40}Ar in the atmosphere results in the formation of heavier radioactive products, many of which have been detected in rain water.

From a knowledge of the intensity and energy spectrum of the cosmic rays, and excitation functions of the radioactive products, one can calculate the total production rate in the atmosphere. Measurement of the concentration of different products in air, rain, etc. may then allow one to estimate mixing rates between these geochemical reservoirs. The cosmic-ray-produced isotopes serve as radioactive tracers on a worldwide scale.

Meteorites which fall to earth also contain the trace products of high-energy reactions which took place while they were in interplanetary space. Both the radioactive and stable (inert gas) products have been measured, and knowledge of their production rates allows one to use radioactive dating methods to investigate the past history of such bodies. The importance of precise knowledge of the cross sections is pointed up by the practice of not relying on measurement on elemental targets, but of bombarding a sample of the actual meteorite in an accelerator beam.

The formation of the light elements, deuterium, lithium, beryllium, and boron, in the solar system represents a problem for any theory of stellar formation, because these elements are rapidly consumed by low-energy reactions in the stellar interior. A current hypothesis attributes their existence to formation in high-energy reactions in cold planetary matter by cosmic rays or solar protons. The importance of obtaining cross-section data for all isotopes of these elements in order to test this hypothesis is obvious. The mass spectrometric and dE/dx particle identifier techniques are best suited for these measurements.

REFERENCES

1. R. Serber, *Phys. Rev.* **72,** 1114 (1947).
2a. P. L. Reeder and S. S. Markowitz, *Phys. Rev. B* **133,** 639 (1964); A. M. Poskanzer and L. P. Remsberg, *ibid.* **134,** 779 (1964); S. Kaufman and C. O. Hower, *Phys. Rev.* **154,** 924 (1967).
2b. C. O. Hower and S. Kaufman, *Phys. Rev.* **144,** 917 (1966).
3. D. T. Chiuers, E. M. Rimmer, B. W. Allardyce, R. C. Witcomb, J. J. Domingo, and N. W. Tanner, *Nucl. Phys. A* **126,** 129 (1969).
4. G. Friedlander, L. Friedman, B. Gordon, and L. Yaffe, *Phys. Rev.* **129,** 1809 (1963).
5. R. E. Batzel, D. R. Miller, and G. T. Seaborg, *Phys. Rev.* **84,** 671 (1951); J. Hudis, I. Dostrovsky, G. Friedlander, J. R. Grover, N. T. Porile, L. P. Remsberg, R. W. Stoenner, and S. Tanaka, *ibid.* **129,** 434 (1963).
6. J. R. Grover, *Phys. Rev.* **126,** 1540 (1962).
7. A. M. Poskanzer, J. B. Cumming, and R. Wolfgang, *Phys. Rev.* **129,** 374 (1963).
8. J. B. Cumming, R. J. Cross, Jr., J. Hudis, and A. M. Poskanzer, *Phys. Rev. B* **134,** 167 (1964).

9. L. P. Remsberg, *Phys. Rev.* **188,** 1703 (1969).
10. R. Brandt, "Physics and Chemistry of Fission," Vol. 2, p. 329. IAEA, Vienna, 1965.
11. M. Lefort, J. P. Cohen, H. Dubost, and X. Tarrago, *Phys. Rev. B* **139,** 1500 (1965).
12. A. M. Poskanzer, S. W. Cosper, E. K. Hyde, and J. Cerny, *Phys. Rev. Lett.* **17,** 1271 (1966).
13. P. A. Benioff and L. W. Person, *Phys. Rev. B* **140,** 844 (1965).
14. L. P. Remsberg, *Phys. Rev. B* **138,** 572 (1965).

5   Accelerated Particles in Biological Research

C. A. Tobias*   *Donner Laboratory*
University of California
Berkeley, California

H. D. Maccabee   *Lawrence Livermore Laboratory*
University of California
Livermore, California

* The author's research is supported by the Atomic Energy Commission and the National Aeronautics and Space Administration.

I. INTRODUCTION

The internal energy of the stars comes from fusion and "burning" of heavy particles; protons interact to form helium, protons impinge on carbon to form oxygen, and so on. The solar corona contains nuclei of the most common elements; and high-energy protons, helium, and heavier nuclei all the way up to and perhaps beyond uranium impinge on the outer atmosphere of the planets. Thus, heavy particles along with electrons, electromagnetic radiation—x rays, γ rays, ultraviolet, visible, and infrared radiations—form the radiation milieu in which the planet earth has developed. These radiations no doubt had a major influence on the origin of organic compounds, the inception of life, and perhaps on current evolutionary trends.

During the past 20 years, many of the particles known to be present in cosmic space have been accelerated at ground level in cyclotrons, linear accelerators, and synchrotrons to increasingly higher energies. The heavy accelerated particles are quite useful in the study of some of the most basic biological problems—from the level of single cells to more complex organisms—and have furnished therapeutic tools for man [1].

II. NATURAL PROTECTION OF LIVING ORGANISMS

Learning about the delicate inner structure of cells and about the intricate control mechanisms of our body has progressed slowly throughout the centuries. As one reason for this, living organisms are constructed in such a way that natural, physical, or chemical protection is available against interference from a hostile external environment. If the external interference is too violent, organisms either are badly deranged or die from the assault. Accelerated heavy ions have supplied us with rather refined tools which, due to their special properties, allow delicate probing into the inner milieu of single cells and complex organisms, causing profound changes in local regions but affecting the whole organisms in only relatively minor ways. The reason for this is as follows: heavy particles, due to their interaction with atoms and molecules and their production of ionization and excitation, can deposit significant amounts of energy internally in tissue along the pathway of their travel. Figure 1 shows the energy transfer from various fast particles in tissue. Taking a fast helium ion as an example, we see that the highest energy loss is from a helium ion of about 1 MeV kinetic energy. Here about 8300 ion pairs form for each micron of tissue, thus, many of the atoms are

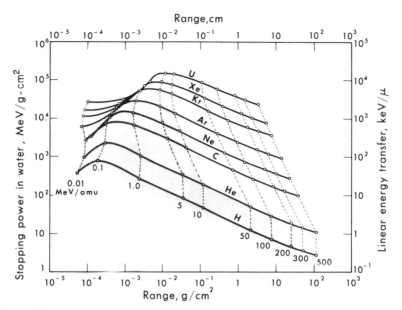

FIG. 1. Linear energy transfer and stopping power of energetic heavy particles as functions of their range penetration, plotted on logarithmic scale. The energy transfer increases approximately as the square of atomic charge, and a given particle transfers more and more energy as it slows down in tissue. The energies of the particles are indicated on the graph in units of mega electron volts per nucleon. It is clear from the graph that in order to have particles of sufficient penetration to be useful in radiation therapy or in brain study, they must be accelerated to several hundred mega electron volts per nucleon. (Based on studies by P. Steward [3].)

ionized, and we may expect strong local effects on molecular structure and function. Usually higher-energy particles are used with considerable range so that interaction between the particle and tissue increases as the particle penetrates inside and slows down. This is an advantage, since the skin can be protected while an internal organ gets a higher dose. An example may illustrate what is said above about energy disposition: If we wished to expose the "master gland" of the body (the hypophysis) of an individual to accelerated helium ions, we would use particles of about 400-MeV energy at the point of entry to the skin; where the particles cross the skin at the outside of the head, the density of some ionization events would be less than one hundredth of the density of many of these events at the pituitary gland. Because of statistical variation in the interaction of different particles, the ionization from a beam of particles is different from that of a single particle. A typical beam ionization curve for fast helium ions is shown in Fig. 2. The particles travel on an almost optically straight path so they can be aimed; when they have exhausted their kinetic energy, they stop with only negligibly

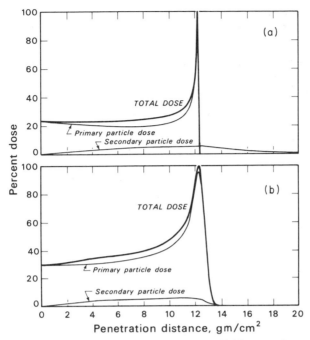

FIG. 2. Examples of Bragg ionization curves for parallel beams of neon ions (a) and helium atoms (b) in water, penetrating in tissue to about 10 cm depth. Contributions from direct energy transfer of the particles and from secondary heavy particles produced in tissue by inelastic collisions are plotted separately. Neon has a narrower ionization peak, an important property when the particles are used to produce small lesions in tissue. (Based on Litton [4].)

small secondary radiations continuing in their path. Thus, large doses of helium ions from the Berkeley 184-in. cyclotron have been used for more than 10 years for radiation *hypophysectomy,* a therapeutic procedure for a number of human diseases [1, 2]. As much as a 20,000-rad dose can be given to the pituitary gland by accelerated particles, and the dose on the surface of the skin could be kept so small that essentially no visible lesion occurs there. It is of interest to remark at this stage that deep, localized energy deposition can be achieved either at the local level within a single cell, with low-energy particles, or magnifying the scale perhaps 10,000-fold, at virtually any part of the human body with multi-MeV particles.

III. PHYSICAL PENETRATION OF ACCELERATED PARTICLES

Exact penetration properties and the exchange of energy between particles and atoms of tissue have been measured with considerable accuracy, and a very good theory exists which was initially developed by Niels Bohr and John Wheeler. In Fig. 1, one may note that there is a 10,000-fold domain of energy transfers, the highest energy exchange being due to very heavy particles with high atomic number and the lowest to extremely fast protons. Many of the quantitative biological studies on the effects of particles have concerned themselves with the specific effect of each particle, at each range of penetration. Data of this kind have been presented in particularly useful form for biological studies by Steward [3] and Litton [4]. Figure 1 also indicates that we still have to construct additional accelerators before we can use each particle with each energy loss at necessary depth penetrations for biomedical studies. The shaded area indicates domains where particles are available in accelerators such as the Berkeley 184-in. cyclotron or the heavy-ion linear accelerator (Hilac). The unshaded parts are domains where accelerated particles are not available as yet. As we shall see, it would be interesting and useful to medicine to have heavier accelerated ions available with deeper range penetration. Therapeutic applications in man generally require high energies up to several hundred mega electron volts per nuclei.

Individual particles all exhibit the properties elucidated in Fig. 1. However, there are small statistical differences between the behavior of each particle due to the fact that the particle undergoes many collisions in its passage through matter and that the energy exchange, which consists of individual ionization events and excitations, exhibits statistical fluctuations. When the entire beam is used as a parallel bundle of particles, the penetration and energy transfer are characterized by two relationships. The first one is the "number–distance" curve, which indicates the number of particles that penetrate to a given depth in tissue. From the rapid falloff of this curve it is obvious that almost all particles penetrate to approximately the same distance before stopping. The second relationship is the Bragg ionization curve. This shows the overall dependence of ionization on depth. Figure 2 shows the number–distance curve as well as the Bragg ionization curve currently available from experimental measurements for helium ions, and calculated for neon ions. For each depth penetration essential parameters are the relative height of the peak compared with the ionization at the entrance, the so-called Bragg ratio and the width of the peak, which determines the width of the lesion that can be made at a given depth.

Neon particles, because of less scatter in tissue, make more sharply defined lesions than protons or helium ions and the lesions are more sharply defined as the atomic number of the particles increases up to about $Z = 20$. Above $Z = 20$, inelastic nuclear collisions distort the depth ionization properties.

In order to understand the action of heavy-ion beams on biological tissue, we must also understand the physical interactions with precision. We know that the central portion of the ionizing track forms a "hot core," and that short-range electrons, δ rays, form a penumbra to the track. The track contains mostly secondary electrons but there are also heavy nuclear recoils. The distribution of such events was predicted by Landau and Vavilov and measured recently in detail by Maccabee *et al.* [5]. The radiological physics of the use of heavy beams has been described [6].

IV. RELATIONSHIP OF BRAIN ORGANIZATION TO BRAIN FUNCTION

It is well known that great similarities exist in the spatial organization of the gyri of the cerebrum from one individual to another in the same species. This has allowed the construction of very detailed brain maps that give descriptions of the structure down to each millimeter [7, 8]. Most of the observations to obtain brain atlases come from stained autopsy material.

There has been a classical idea that brain function is in rigid relationship to brain anatomy, and a certain amount of information substantiates this idea. In the cortex the microscopic organization of the various kinds of neurons, including that of nerve trunks, is remarkable. Functionally, we do know of a certain amount of localization of sensory and motor phenomena. Most of the information on the localization of function comes from individuals with brain injuries, *congenital malformations,* or diseases, that is, tumors. Comparison of the anatomical extent of injury and the nature and extent of lost function has yielded much valuable information. There remain many ambiguities about the exact anatomical localization of brain function. In the first place, it can be shown by electrical techniques that nerve impulses related to a specific action travel to, and are processed in, many areas of the brain nearly simultaneously—electrical activity does not remain only in the local region thought to be specifically responsible. Further, it is well known that considerable ability to think and to learn and remember remain even after loss of sizable parts of the cerebral hemispheres. It would appear that gross anatomy may not have all the answers. We should know more about the microscopic anatomy of the brain as well. The human brain has more than ten billion (10^{10}) neurons. According to some estimates there are as many as 1000 interconnections (synapses) for each neuron with other neurons. Are these interconnections established by mere chance? We do know that synapses are not completely random, since we do have knowledge of many definite nonrandom pathways, nerve tracts. At least we would like to know where, to what places, the main nerve trunk, the axon, reaches from each neuron. This turns out to be difficult to determine because of the small size of the axons (a few microns in diameter) and because they often traverse several inches of tortuous pathway. Here use can be made of histological techniques based on the fact that when a nerve trunk is cut, the distal portion (the part not linked up with the neurons) degenerates. Special staining techniques are available to indicate the pathway of such degeneration. If a lesion is made in the brain with a knife, often many nerve fibers are cut instead of a few. Since there can be ten million fibers crossing each square inch of area, it is not unusual to cut more than a million fibers. This, of course, may tremendously complicate the task of uniquely identifying just one pathway. The heavy particles, as we shall see, can be quite useful in the search for anatomical localization.

V. THE FIRST EXPERIMENTS WITH BEAMS LOCALIZED IN THE PITUITARY GLAND

The first experiments with localized beams were carried out in a study of the effects of deuteron particles on the pituitary gland of rats [9]. This gland is located at the base of the skull in a well-defined, but protected, location. Its removal in man is a major surgical task. In serious diseases it can be accomplished with precision—a triumph to modern neurosurgery. However, the surgery either involves a risk of injury to the brain, the frontal lobe of which must be moved aside, or it involves drilling a pathway through the bony floor of the *sella turcica,* housing the pituitary. This procedure opens up the possibility of causing cerebrospinal infections, and possibly meningitis, by admission of microorganisms from the intranasal or oral passages, which usually carry microbial contamination.

The size of the pituitary gland in the young rat is about 1×1.5 mm. It was found that with a passage of a single, fine beam of deuterons or of α particles, a partial or full hypophysectomy could be obtained. The pituitary gland is the "master" gland of the body. By outpourings of a number of hormones, the pituitary controls, either via the mediation of endocrine target organs or directly, much of cell proliferation in the body, as well as cellular metabolism, turnover rate of certain regulatory substances, the sex functions of the body, and others. To cause complete regression of the pituitary, rather high doses were necessary: 20,000 rads in a single dose for rats, 10,000 for dogs, and more than 5000 rads for young primates. It seemed necessary for the radiation to kill every cell without undue injury to neighboring tissues, hence, the high dose. There was a kind of inverse relationship between the dose to the pituitary and the time taken for complete regression of this organ. A high dose of 20,000 rads produced in rats a state resembling complete hypophysectomy in about 20 days; whereas after 5000 rads some functioning pituitary was left for as long as one year. This kind of radiation has demonstrated interesting properties of tissues. When surviving cells were found some time after irradiation, this usually occurred at the edge of the gland adjacent to intact blood vessels and circulation, even if that region received a high dose. It appeared that in the postirradiation period, a reorganization in tissue architecture must have taken place which may have resulted in migration of some of the healthier cells to the gland periphery. It was also demonstrated that if a few unirradiated cells are left in the gland, these can rapidly proliferate under the influence of hypothalamic secretions and reestablish the gland with relatively normal functions. Some measurements were made on the endocrine organs in the postirradiation period to observe possible differential effects (see Fig. 3). Studies like this are com-

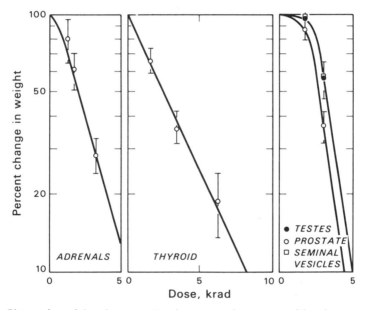

FIG. 3. Change in weight of some endocrine organs in a rat resulting from regression in size and function caused by local irradiation of the pituitary.

plex, because the effects depend on time as well as dose. We found that in the rat the secretion of somatotrophic growth hormones (SH) and thyrotropic hormones (TSH) are about equally sensitive to radiation. The SH and TSH are chiefly responsible for growth and for anabolism and catabolism of bone. The development of gross body weight depended on the effects of the somatotrophic hormones, except that even at low doses there was an increase in weight (not a loss) followed by a later decrease. The animals that received relatively low doses became obese (see below). The pituitary seemed more resistant when it came to the control of the adrenal hormone (ACTH) and the follicle-stimulating hormones responsible for development of the sex glands. It appears that the pituitary cells which produce the sex hormones were sensitive to radiation, but such overcapacity for hormone production is present in the rat (probably 100 times the amount needed) that effects on the sex hormone cycle are not noticeable until enough dose is given to cause destruction of the entire pituitary gland.

In later studies it became more and more apparent that different strains of animals and man have different hormone reservoirs and that radiation affects production of the various hormonal fractions in a differential manner. For example, adult humans appear to deplete their sex hormone reservoir under the influence of pituitary radiation much more readily than rats. Secretion of the growth-inducing somatotrophic hormone (SH) is quite sensitive to radiation in man, a fact that later turned out to be of value in connection with the treatment of *acromegaly*.

Acromegaly is a debilitating disease due to a tumor of the pituitary gland. The cells of the tumor sometimes manufacture great excess of the growth hormone (SH). The hands, face, feet, and ears of individuals suffering with this disease grow abnormally large and the features of the face take on characteristic appearance. The size of the pituitary and of the sella turcica may enlarge as much as tenfold (Fig. 4), and the resultant pressure on the brain and optic nerve from such growth may cause severe headaches and even loss of vision. Treatment is usually attempted only if it is believed that the tumor is still contained in the closed membrane (dura) in the sella turcica and that it has not metastasized to other areas of the brain. More than 100 patients have received helium ion treatment over the last ten years at Berkeley [10] and a sizable group has been treated with proton therapy at Harvard [11]. A dose of less than 10,000 rads (delivered in a two-week interval), or only about one-half that necessary for total human hypophysectomy, is sufficient for therapy. In the months following treatment, an exponentially decreasing titer of growth hormone was measured in the circulating blood of patients, eventually leveling off in the normal range (Fig. 5). There is also relief from subjective symptoms of acromegaly and attendant symptoms of diabetes. The rate of calcium turnover has also been measured by the addition of radioactive calcium in a whole-body counter. Whereas calcium turnover is quite slow due to abnormally high bone deposition in acromegaly, it is restored to normal after radiation. The heavy ions appear to be essential in this treatment because it is important to avoid the optic nerve, hypothalamus, and areas of the temporal lobe of the brain, conditions that usually cannot be met with other available radiations.

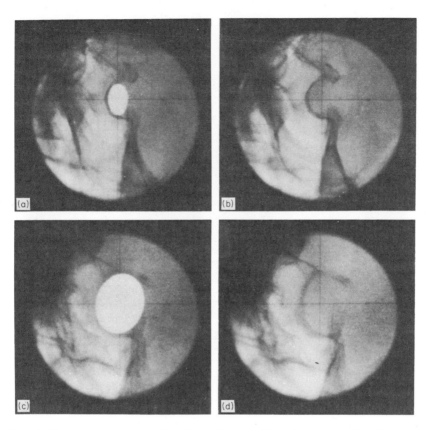

Fig. 4. Comparison of normal sella turcica (a, b) and acromegalic sella turcica (c, d). In acromegaly, the slowly growing tumor of the pituitary gland often causes marked enlargement of the sella turcica (the bony cavity housing the gland). The autoradiograph of the helium beam used in therapy is superimposed on two of the pictures.

In the early pituitary radiation series it was also noted that growth stunting of animals was always observed following radiation even from small doses, in spite of the fact that histologically the gland seemed to have recovered after small doses. It appeared as though the overall growth hormone synthetic capacity of the gland was limited, and radiation would always reduce this limit. The rats treated with moderate pituitary doses appeared quite normal, had fine shining hair, and appeared playful, perhaps somewhat less aggressive than those animals that grew to full size. As the dose was increased, the animals became more and more stunted. They eventually became very susceptible to external infections, when the dose became high

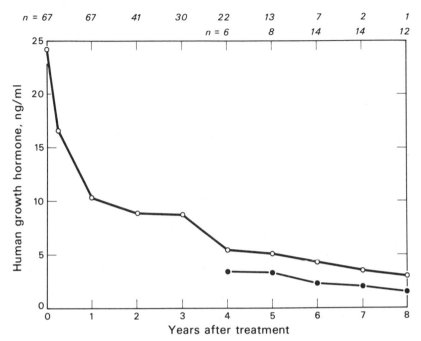

Fig. 5. Growth hormone titer following pituitary helium ion therapy. Tumor cells in patients with acromegaly produce abnormally large amounts of growth hormone (SH). Following therapy, the circulating hormone titer gradually decreases to normal levels. Since the concentrations of growth hormone are exceedingly small (measured in nanograms per gram tissue) special methods had to be developed for hormone assay. The assay used here, radioimmunoassay, became available only in the fourth year of our first therapy series (solid circles). Results from a later series are denoted by open circles [10].

enough to seriously limit the role of the adrenal glands in response to inflammation.

That a series of gentle changes in an animal takes place according to the function of a tiny gland that is 1/50,000 the size of its body seems rather remarkable. It has been suggested that eventual control of the limits of activity of the pituitary gland could help solve certain problems of mankind at some future time. If the somatotrophic function were limited in each gland, the apparent evolutionary trend toward greater and greater size in man, which seems to have been effective over the last 100 years, could perhaps be checked. We may remember that dinosaurs, the largest of all known animal species, have become extinct. Perhaps partly due to their gargantuan size they became inefficient and vulnerable. Americans have grown larger and larger in the last few generations; perhaps this was due in part to the availability of unlimited food resources. At the same time, some data point to the likelihood that smaller people have a longer life span and that very heavy people seldom live to a very old age. One may speculate that, as man learns how to control his own genetic development and hormonal control mechanisms, he will be able to control not only his own size, but his own fertility as well.

A smaller human with at least equal faculties for intellectual and emotional life would need less living volume and might consume fewer nutrients and produce less waste. The populations of the biggest human species (Russian, American, Swedish) appear to be growing at a slower rate than their smaller counterparts in China, Southern Asia, and India.

The tissues that surround the pituitary gland are more essential to the survival of the animal than the pituitary itself. The pituitary is connected to the *hypothalamus* via an innervated stalk. The hypothalamus controls the hormonal functions of the pituitary via direct innervation of the posterior pituitary and by means of neurosecretion via a portal circulation in the anterior pituitary [12]. The hypothalamus also has direct nervous connections to many of the organs of the body. The Berkeley group has placed single doses of deuterons as high as 40,000 rads in the rat pituitary without acute killing effects. In contrast, when the whole hypothalamus is radiated, then, following a time delay necessary for development of a radiation lesion, the animals exhibit an acute lethal hypothalamic syndrome even from a highly localized 5000-rad dose [13]. The syndrome includes complete cessation of eating and drinking on the part of the animal; it becomes less able to control its body temperature against external cold, and suffers a continuing loss of body water. Frequently death is preceded by a "hypothalamic coma." These are symptoms also seen in man with hypothalamic tumors and injuries.

VI. STUDIES ON TUMORS OF THE ENDOCRINE SYSTEM

It was known for many years that the origin of neoplasms (tumors) of the endocrine system are related to hormonal balance. It was shown more than 30 years ago by Lacassagne that castrated male mice, when treated with one of the female hormones (estrogen), develop mammary cancer. More recently, it has been demonstrated that hormonal contributions from the ovaries are essential to the development of radiation-induced mammary cancer in female rats [14]. A large percentage of normal female rats develop mammary neoplasms following whole-body exposure to γ radiation. The number of induced tumors declines sharply when the ovaries are removed, thus depriving the animals of most of their female sex hormones. Many experiments similar to the ones quoted have given rise to a "feedback model" of endocrine carcinogenesis (see Fig. 6). Certain endocrine organs, for example, the mammary glands, depend continually for their development and function on the availability of essential hormones. We know that the growth hormone and secondary female sex hormones as well as the pituitary lactogenic hormone are essential for mammary development. The homeostatic control systems of the body presumably acting with some feedback from various parts of the body continually monitor the status of each organ

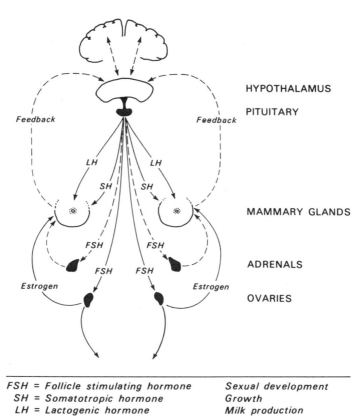

FSH = Follicle stimulating hormone	Sexual development
SH = Somatotropic hormone	Growth
LH = Lactogenic hormone	Milk production

FIG. 6. Schematic representation of homeostatic and hormonal control of the development of mammary tissue. Proliferation and secretions of the cells of this tissue are probably under the influence of the growth (somatotropic) hormone (SH), lactogenic hormone (LH), and secondary sex hormones, elaborated by the ovaries. Carcinogenic influences might originate from the cells of the mammary gland or from the secretions of the pituitary and hypothalamus. We do not understand completely the pathways of communication, neither do we know exactly the specific role of each hormone. Therapeutic studies with the helium ion beam may be helpful in unraveling the detailed mechanisms.

and supply the necessary milieu for its continued functioning. The hypothalamus, pituitary, and ovaries are involved in the control system of the mammary glands. We know more about the part of this system dealing with circulating hormones than the rest of the system. It is not well understood how information from the mammary glands is fed back to the hypothalamus and what the role of afferent and efferent nerves is in the control system. The dynamic balance of cells in the target tissue is very delicate. Just enough of its *stem cells* divide to make cells available for differentiation and specialized function. The differentiated cells do not seem to divide but may be extruded or eliminated with age, to be replaced by newly differentiating cells from the stem cell pool. It is believed that in the course of the carcinogenic transformation, cells that were initially under complete hormonal control with respect to cell division and differentiation become liberated from such controls. In the process, cells become undifferentiated and independent of the body control mechanisms and may acquire potential to grow anywhere in the body as *metastatic lesions*. The role of a carcinogen, whether it be a viral agent, chemical substance, or radiation, is to cause transformation of normal cells in the target into cells with abnormal characteristics. Such cells may have chromosome rearrangements and other genetic abnormalities. They probably do not respond well to the usual hormonal stimuli. The homeostatic feedback system apparently can recognize abnormal cellular function in the target organ and responds to it in an appropriate manner. We do not understand this response in detail; it is believed that under influence of the hypothalamus the pituitary sends out more stimulating hormone, thus causing cell proliferation in the target organ. Following the initial action of a carcinogen, a long time may elapse in experimental animals, sometimes extending for years. During this period, pretumorous cells persist and begin slow proliferation; the hormonal milieu is assumed to remain normal. Unfortunately, we do not know enough about the status of the endocrine system in the period between carcinogenic influence and the actual development of tumors. It is possible that the pituitary secretes an excess amount of normal and of abnormal hormones in an effort to maintain *homeostasis*. The cells of the mammary gland meanwhile undergo transformations, probably including chromosomal and genetic rearrangements. When malignant cancer cells appear, they may be in some measure independent of homeostatic control.

Carcinomas of the endocrine glands are among the most frequent cancers in the human population. Mammary cancer, for example, is the greatest killer in the age group of American women near 40, while its incidence peaks around menopause. This disease can appear in women of all ages; two percent of all cases of mammary cancer are in men. About 25,000 people die yearly in the United States of mammary cancer.

For about 30 years it has been known that oophorectomy (removal of the ovaries) caused temporary regressions of mammary carcinoma, lasting for about a year. The beneficial effect was attributed to the removal of estrogenic substances from the circulation. However, a number of months following oophorectomy the renewed proliferation of tumor cells occurred in most patients. It was shown by Huggins in 1945 that, in the absence of the ovaries, the adrenal glands took over synthesis of estrogenic substances [16]. He demonstrated that surgical removal of the adrenals resulted in additional striking remissions of mammary cancer. Unfortunately, relief from adrenalectomy also had limitations and did not last indefinitely. At this point (about 1950) it was reasoned that hypophysectomy might even more effectively remove the undesired hormones from the circulation than oophorectomy or adrenalectomy. Surgical hypophysectomy as a therapeutic procedure for metastatic mammary cancer was initiated by Luft and Olivercrona in 1952 [15].

When it became apparent that heavy-particle hypophysectomy could be useful in cancer, the Berkeley group sought ways to prove the effectiveness, safety, and applicability of the heavy accelerated particles. There were several roadblocks: heavy particles had never before been used in humans, particularly with the large doses contemplated for hypophysectomy. In 1954 Huggins of the University of Chicago made an appeal on a Chicago radio station for dogs suffering from cancer to be used as test subjects for therapeutic purposes. Our first experimental subject was one of these, a beautiful Doberman pinscher, "Gina," suffering with terminal stages of mammary cancer. Gina had ulcerated and painful metastatic lesions; her tumorous mammary glands were lactating continuously. Within 10 days after pituitary treatment with 190-MeV deuterons the lactation stopped entirely and regression of the tumor was under way. Gina showed rapid signs of recovery and her tumor regression lasted for two years, a significant period in a dog's life. This and other work with animals has convinced us that therapeutic trials in humans were warranted and could be done safely. Radiation hypophysectomy by protons was used initially in 1955, and later helium ions were also used [1, 2]. In this procedure the aim is to give a sufficiently high local dose so hormonal synthesis by the pituitary stops. Replacement therapy for some hormones must be instituted. Even though only the pituitary gland receives a heavy dose of radiation and the mammary glands and metastatic tumor lesions are not exposed to radiation, the response can be very dramatic, consisting of the gradual disappearance or "melting away" of extensive metastatic tumors. Good remission can be achieved when 21,000 rads are delivered to the center of the pituitary in a two-week irradiation schedule. Objective evidence of hypophysectomy, as measured by tests on gonadotropins, thyroid function, etc., was obtained. A number of remissions of several years' duration were obatined, the longest one lasting for eight years. Delivery of a hypophysectomizing proton or helium ion beam to the pituitary of a patient is a relatively easy procedure, free of trauma. The optic nerves responsible for vision and the hypothalamus are entirely protected from radiation (Fig. 7). However, when adequate dose is given to the pituitary for hypophysectomy, there is also a hazard that some of the cranial nerves could be injured. The third and sixth cranial nerves, in particular, lie in such close proximity to the temporal aspects of the pituitary that they also receive a sizable dose. The limitation in the cur-

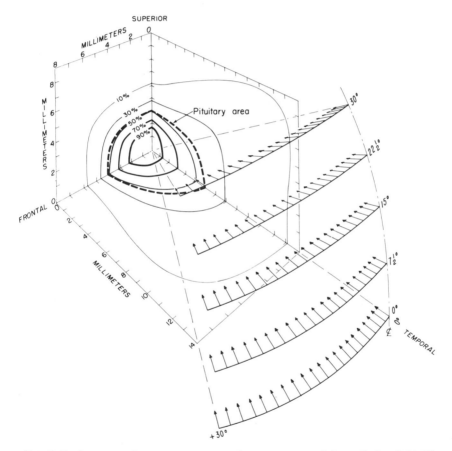

FIG. 7. Isodose curve in mammary cancer for one octant of the radiation field. The use of the high-energy particles allows intricate three-dimensional dose distributions, not achievable by other types of beams. The distribution shown is for high dose irradiation of the pituitary gland. The dose field is shaped to protect the optic nerves, hypothalamus, and brain stem and to fall off rapidly in three directions of the temporal lobes of the brain.

rent procedure, which severely curtails therapeutic work at present, appears in an insufficient sharpness of the Bragg ionization peak, and a somewhat excessive amount of scattering of the particles out of the beam. We have made calculations that indicate that, by use of the Bragg peak and mono-energetic carbon or neon particles, instead of helium ions, we should be able to perform hypophysectomy in cancer patients more effectively than at present. The heavier ions would allow directing relatively higher doses to the pituitary gland and relatively less to the surrounding brain. Thus, in all likelihood cancer research and therapy would benefit from such beams. Unfortunately, the heavier beams will not be available until a new acceierator has been constructed, or an existing one modified.

In spite of many successful hypophysectomies by radiation or by surgery only about one-half of the cancer cases treated have objectively demonstrable tumor regressions. It is not possible at present to predict which patients will respond favorably to radiation hypophysectomy. In any case, following a possibly prolonged period when the tumor is in regression, there is often another period of renewed metastatic proliferation. We do not know the exact reason for this. Some possible explanations are as follows:

1. Hypophysectomy is incomplete: the beam (or the surgeon) did not cover the pituitary gland completely.

2. There is some delayed recovery of pituitary function beginning with the regeneration of cells at the pituitary stalk.

3. There may be "accessory" pituitary glands lodged near the main gland. These become active when the main gland has regressed.

4. The hypothalamus itself may take over some of the functions of the pituitary.

5. The tumor tissue can make some estrogenic substances from precursors which may replace the hormones usually supplied by the endocrine glands.

6. By some curious process tumor cells may become entirely independent of hormones, and, as they acquire this property, they may begin to proliferate again, even in the absence of the pituitary.

These manyfold possibilities are listed here to give an idea of the great complexity of the cancer problem and of the relative state of ignorance scientists find themselves in with respect to detailed control processes. The responses of each patient to radiation hypophysectomy have potential clues that could serve us with explanations regarding cellular and hormonal responses. The cyclotron technique of exposure has become quite efficient. We are still hampered by the imperfections of other investigative techniques, particularly in the field of hormone assay. Most hormones have significant biological activity when their concentration in the circulation is 1 nanogram/cm^3 or less. Our knowledge of how the hormonal substances are being carried in the circulation is very incomplete and physical methods of hormone assay are at best very unreliable. For example, estrogenic substances are usually measured in the urine as they are being excreted. We do not know if such measurements truly reflect the concentrations in the body or merely signify the amounts of waste products. There are at least 16 different estrogenic substances that could be implicated in cancer; we can only measure routinely about three of these at a given time. Because of limitations of this kind, even with excellent new therapeutic tools, progress is limited in our rate of understanding of the detailed processes. Promising newer techniques are being developed—some of these use mammalian cells in culture for hormone assay. Cells can be taken from the patient by biopsy and their hormonal requirements analyzed. Some progress is also being made concerning the mechanism of hormonal action in controlling cellular processes.

VII. CANCER THERAPY VERSUS CANCER INDUCTION

Practically all cancer therapeutic agents are also cancer-inducing agents. This is a peculiar, paradoxical situation. These agents are effective carcinogens when they are administered to normal animals early in life and in repeated, small doses—not enough to kill many cells. The tumors appear after a long time delay. The same agents are effective therapeutically after the appearance of tumors. They are given in large doses, enough to kill all or most tumor cells and many normal cells. Their carcinogenic potency does not appear to be important at this time since it is the existing tumor that limits the subject's life and the carcinogenic effect of the therapeutic agent usually would not manifest itself for many years.

Although high doses of proton and helium ion particles were clearly useful for pituitary radiation in humans for control of tumor proliferation, locally or in metastatic sites, it remained important to investigate the action of these particles on various systems at all dose levels. Groups of four-week-old rats were given a single pituitary dose of deuterons; their subsequent development was studied during their entire life span [17]. Late in life these animals developed a spectrum of pituitary tumors. At the dose level of 950 rads there were 3.5 tumors per pituitary gland, of all cell types, many tumors not characteristic of the normally occurring "old-age" pituitary tumors in the particular strain studied. Pituitary tumors are not easy to produce experimentally in animals; only a few examples are reported in the literature. For example, Furth *et al.* [18] found hypophyseal tumors in animals exposed earlier to neutrons in atom bomb tests.

An explanation of the origin of pituitary tumors in animals exposed to heavy particles can be found in the feedback theory of carcinogenesis. We believe that many of the pituitary cells received a sublethal dose of radiation. The entire irradiated gland was unable to fulfill hormonal demands of the body. The pituitary then was subjected to increased stimulation from the hypothalamus for more efficient hormone production, and was unable to respond with hormone production. The gland first hypertrophied and later neoplasms formed in it. It was significant in this series of experiments that only animals that received less than 3000 rads developed pituitary tumors. Not a single animal that received an exposure of 3000 rads or more, akin to a pituitary ablation procedure, was shown to have developed a pituitary tumor later in life. No doubt this is due to the fact that most pituitary cells were killed.

The origin of neoplasms by the feedback mechanism is not an isolated incident. We know that thyroid glands, mammary glands, or pituitary glands develop neoplastic tissue in the period subsequent to local sublethal radiation exposure or equivalent chemical poisoning. Furth has taken pituitary tumor tissue from animals in which the tumor was induced by radiation, and transplanted bits of this into the body of other normal rats. Such transplanted tumor tissue could cause, in the new host, neoplasms of the thyroid, adrenals, or mammary gland [19]. It has also been possible recently to produce in rats tumors of the endocrine system which were due to lesions produced in the hypothalamus [51]. These are additional pieces of evidence for the intricate control relationships that seem to exist between neural tissues of the hypothalamus, the pituitary, and endocrine organs of the body.

VIII. LOCALIZED RADIATION STUDIES IN VARIOUS NERVOUS STRUCTURES

It seems remarkable that the application of a small beam of particles can cause specific, profound, and long-lasting effects influencing not only the physiology but also the behavior of animals. Since the nervous system has a very large number of autonomic controlling functions and the problems of higher nervous system function appear even more complex, it will probably take many years before all possibilities for research in this field are exhausted. Thus far, little more than pilot studies are available. Some of these will be briefly reported here. Solutions to most of the problems of the nervous system will obviously require technically more complex approaches in the future.

A. Hypothalamus

A low dose of deuterons in the region of the pituitary has resulted in development, over a period of a year, of groups of very obese rats, some of which have as much as 60% body fat. The thyroid glands of these animals were also found to be hypertrophied, but exhibited relatively normal iodine uptake. The skeletal growth of the animals remained slightly subnormal. When food intake was reduced by restricting the amount of food available, then the animals seemed to develop relatively normally. Thus, it was believed that the obesity was a result of radiation injury of the "appetite center," located in the ventromedial nucleus of the hypothalamus. However, attempts to obtain this syndrome by hypothalamic irradiation alone failed; irradiation of the hypothalamus by a large dose leads to loss of appetite and loss of weight instead.

In order to be fully effective in altering homeostatic control, hypothalamic lesions must be produced on both sides of the head, symmetrically, bilaterally. The usual method of producing lesions is by needles passed into the brain. This method, however, sometimes leads to trauma and bleeding. The surgical method produces lesions immediately; however, the size of the lesions may expand with time. Heavy-particle lesions can be accurately placed by the use of diagnostic x-ray landmarks. However, time must be given for development of the lesions before physiological changes are observed.

Small radiation lesions generally are studied in the laboratory rat. Although they appeared to be well localized histologically, they have usually caused time-dependent changes in several observable physiologic variables, rather than a single variable [13]. At this time we do not know whether it will be necessary to refine the size and location of the lesions to obtain simpler changes, or whether it is a property of the hypothalamic system that, due to many interconnections, any lesions will lead to complex changes with time.

Whenever lesions were produced in the hypothalamus of a young animal, some reduction occurred in the bone growth of the animal, indicating a deficiency of somatotropin production in the pituitary [20]. When the rat hypothalamus was divided into five separate regions, growth stunting effects, due to lesions in each, were observed.

When a lesion was produced in the pituitary stalk (a bundle of nerves connecting the hypothalamus and the posterior pituitary gland), the animal exhibited increased water turnover very markedly (diabetes insipidus). This effect was minimized, however, when both the anterior and posterior pituitary were included. The Swedish investigators Andersson *et al.* [22] made a detailed study, with accelerated protons at the Uppsala cyclotron, of hypothalamic lesions induced in goats which caused alterations in water metabolism.

Exposing the region of median eminence in rats to a lesioning dose of deuterons resulted in many animals that suffered from abnormally high water turnover, glycosuria, and a thyroid function abnormality. Many of these animals developed "rage." They exhibited extreme nervousness, responded with great hostility to noises, and would attack a man or other animals within range [13]. This syndrome has been previously described to have resulted from hypothalamic injury in the ventromedial nucleus.

In order to better understand homeostatic control mechanisms, it is necessary to know more than the localization of the regions of nervous control. We wish to know the identity of the molecules that participate in the control mechanisms and the nature of their action of the target cells. In the case of activities of the thyroid gland, it is already known that a control center in the anterior hypothalamus is involved and that the pituitary gland is stimulated to secrete thyrotropic hormone. Under the action of this hormone the cytological organization of the thyroid undergoes a profound change and the thyroid actively begins to take up iodine. The iodine is first incorporated into monoiodotyrosine, and later into thyroxine (tetraiodotyrosine, T_4) as well as triiodothyronine (T_3). These two substances circulate in blood and are responsible for metabolic and biosynthetic action at the tissue level. Blanquet [23] noticed that radiation of the hypothalamus of the rat can give rise to a new kind of thyroid disorder not previously described. Iodine uptake by the thyroid gland remains normal, and monoiodotyrosine is made in normal amounts. Studies with iodine label have shown, however, that T_4 and T_3 are made in subnormal amounts. These studies point to two steps in the mechanisms of thyroid control. One of the actions of thyroid-stimulating hormone (TSH) is the initiation of internal structural changes in thyroid cells. Cytomembranes form that are presumed to become sites of production of thyroxine. It was demonstrated by electron microscopy that the morphological appearance of cytomembranes becomes abnormal following certain types of particle lesions in the hypothalamus. It also was demonstrated that thyroxine can localize in hypothalamic and posterior pituitary tissue. The assumption has been made but is not completely proved that T_4 and T_3 may be the feedback agents conveying information to the hypothalamus of the needs of the body for TSH.

B. Neural Lesions and the Ecology of Cell Types

In some ways, the neurons in the nervous system may be regarded as the decision-making elements in a very large computer: they sort and analyze the electrical impulses arriving from various parts of the body at their many synapses and send electrical messages to other neurons in the form of action potentials which move along axons and dendritic potentials. These signals have up to about 80-mV amplitude. It is believed that learning, memory, and thinking processes are in some way related to them.

There is a very complex population of cells surrounding the neurons that somehow participate in, aid, and support the neurons' highly precise mission. These additional cell types are called *glia* and the degree of their organization is truly astounding. Their missions and interdependence are only beginning to be unraveled. For example, there are some glia (oligodendrocytes) that have an apparent function to electrically insulate the axons so that their electrical signals can travel faster and so that electrical "cross talk" between neighboring neurons can be minimized in neuronal messages. The insulation is accomplished by having the oligodendrocytes literally wind themselves many times around a given axon, providing it with a spiral multi-layered lipid insulation, "myelin." When neurons lose their myelin sheaths, this is a ⁀ign of serious illness in the individual. For example, the debilitating disease of multiple sclerosis has as one of its major manifestations the loss of myelinization. The absence of myelin appears to sometimes cause unchecked, uncontrolled oscillating discharges in neural nets. Other glia (the astroglia) have the apparent function of predigesting the nutrients, as they arrive in the capillary circulation, and feeding the more delicate neurons. Thus, they interpose themselves between blood vessels and neurons.

Radiation can disturb the ecological balance of cells in the brain. Klatzo *et al.* [24] have shown that following a dose of particle radiation the astrocytes developed "indigestion," as evidenced by the accumulation of undigested glycogen granules in their cytoplasm. Inside the radiated astrocytes dye substances and proteins can be found which are usually forbidden entry by a membranous structure, the "blood–brain barrier." Some cells (the microglia) act as policemen—they possess the mysterious ability of ingesting foreign cells and molecules and disposing of them. They appear to have an uncanny ability to diagnose if a neuron is abnormal, injured, or perhaps too old. Microglia move through the brain and eliminate such cells. When local injury occurs, thousands of microglia appear at the site of injury. The possibility of such movement across brain tissue in itself appears to be miraculous. One does not know how the microglia sense, across many cell layers, where they are needed and how they manage to move across the tissue, which appears to be completely filled with cells and neuronal processes with no apparent passageways between them. Some glial cells apparently have their own brand of communication. Cytoplasmic bridges appear to form between neighboring cells, as was recently shown by studying electrical impedance between such cells [25].

There are even cells (the ependymal cells) that possess a microscopic fan system. These are processes extending from the cells (cilia) which are in continuous, rapid, whiplike motion. Such motion causes effective stirring and flow of the cerebrospinal fluid which surrounds the neurons. Turbulence created by these cells allows the proper exchange of new nutrients and elimination of waste materials.

The neurons themselves are precious structures. Unlike most other cell types in our bodies, they do not divide. The same cells must perform, over our entire life span, the autonomous controlling functions as well as the higher neuronal processes, which culminate in intelligent thinking. The nervous system also appears to be the main site of our emotional life. To this author, trained initially as a physicist, it appears odd how careless many people are with their neurons, subjecting them to a long series of abuses due to intoxicants and drugs. Anoxia of brain tissue can result from excessive use of tobacco or alcohol or carbon monoxide in the environment. Hallucinogenic drugs appear to cause irreversible rearrangement in neural assemblies. Some of these drugs are highly toxic to neural functions in doses as small as a billionth of a gram per gram of tissue. It is man's central nervous system that differentiates him most markedly from lower forms of life and it is the brain that allows man the quality of intellectual life. The gift of such exquisite nervous systems should be properly appreciated, and they should be studied and protected not only for our own well-being and progress but for the future of mankind as well.

Following large lesions to the brain, such as those which may be produced following the trauma of concussion with bleeding, the circulation in significant regions is impaired. This is followed by increased intracranial pressure due to swelling or inflammation, and later degenerative changes, culminating in necrosis and extensive formation of scar tissue. The scar tissue in some manner appears to act as irritation or "short circuit" for some neuronal nets and may precipitate late epileptic seizures. The degenerative changes are often progressive and they can eventually encompass a much larger volume than the volume initially injured. Thus, the long-term result of trauma or extensive surgery can be the progressive deterioration of brain tissue.

For the future, it appears that surface radiosurgery may be carried out by heavy particles on brain tissue, with such slight scar tissue formation that the radiation lesion may be sealed off and not spread. An example of this is seen in rat cerebellum (Fig. 8b), where a large dose of helium ions caused complete disappearance of several upper layers of nerve tissue. VanDyke has produced focal lesions in primate brains by chemical means. This usually develops into an enlarging necrotic domain. Later, ablating radiolesions were applied with helium ions in the same location. For about 5 years after irradiation, the radiated regions retained their initial size and did not produce later complications such as epilepsy.

Similar methods might be used sometime in the future for removing sections from human spinal cord, as sometimes is indicated when the cord is injured or when tumors occur in it. Further research is needed to ascertain whether two parts of the cord, resected by sharp radiation lesions, can weld together into a functioning cord again. The large-scale involvement of brain tissue in traumatic lesions has made it difficult to study the stepwise nature of the changes that occur during the development of the late lesions.

Laminar lesions [26] made with monoenergetic accelerated particles not only have helped to study the nature of the processes involved, but may also be used in the future as a surgical and therapeutic tool. With appropriate dosage, only a narrow laminar section of the nervous tissue is injured. The impairment of circulation is minimal. Such a lesion is shown in Fig. 8.

The lesion is made by the Bragg ionization peak so that it runs exactly parallel to the top surface of the cerebellum where the monoenergetic particles enter. If the dose is large (several thousand rad), then neurons in the plane of the Bragg ionization peak gradually die. They are eliminated apparently by the microglia. Dendrites and axons may also be injured and can disappear from the plane of the laminar lesion, and axons degenerate if the soma is in or near the plane of the lesion. By the use of special stains, it is possible to trace the path of dead axons into other regions of the nervous system and to anatomically localize some interconnections, as was done by Malis et al. [26]. When lower doses are given, degeneration occurs at a later time. By working with rat nervous systems, it was found that the time for development of late necrosis is related inversely to the dose and to the volume irradiated.

The quantity of importance appears not to be dose (rad) but the total energy transferred to tissue (gram rad). This relationship explains why the low scatter of heavy particles and the ability to deliver a sharply demarcated dose is helpful for the production of local lesions.

It has been believed that nerve tissue in brain cannot regenerate. However, convincing evidence has been obtained by the use of accelerated particles that dendrites and axons do regenerate across a laminar lesion pro-

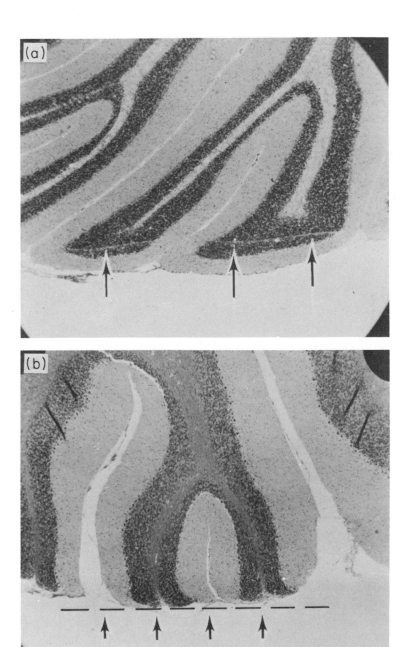

Fig. 8. Exceedingly fine "laminar" lesions produced in rat cerebellum by the use of low-energy monoenergetic helium ions. (a) Lesion in cerebellum about 30 μ wide; the neurons are killed within the lesion but are relatively normal elsewhere. (b) Ablating dose with a monoenergetic particle beam causes gradual liquification of the entire region exposed, while the tissue beyond the reach of the beam remains normal; there is a very small scar, about one cell layer thick. (Courtesy of Dr. W. Haymaker [25a].)

duced by heavy particles, and it is believed that the relative absence of thick scar tissue around the lesion and the presence of good blood capillary circulation across the lesion allows such regeneration [27].

A very puzzling feature in the development of nervous system radiolesions is the manner in which the tissue can recognize that some of its cells have suffered injury. We do not know which particular damage can be recognized and what physical or chemical signals cause glial cells to migrate to the lesion and to attack and phagocytize the injured cells there, particularly since several years can elapse between radiation exposure and development of necrosis. Among the many schemes proposed to explain the process, an interesting one is the autoimmunity theory.

In this theory, the irradiated neurons and other cellular elements become somewhat different from the normal cells of the body and the normal cells respond to this by immune response. If some of the genes of the irradiated cells are altered due to radiation exposure, the cell may start making some proteins different in structure from the proteins normally present. Eventually, perhaps after a long lapse of time, when a sufficient number of these foreign proteins have been made by the irradiated surviving cells, the body recognizes that foreign protein is present and reacts against it by its immunological defense mechanisms, resulting in the process of necrosis. This theory has not been tested in detail, but techniques are becoming available by the use of radioactive labeling and amino acid sequence analysis that may indicate whether such a model may be correct. The model also has been applied to explanation of processes other than local radiolesions. It has been suggested that tissue breakdown due to aging is a process of development of autoimmunity of the body against its own aging cells [28]. In such theories of aging it is also implied that as they age cells develop defects that make them carry out abnormal synthesis.

It is exceedingly important for us to understand the cellular activities of neurons and the manner in which lesions develop in the nervous system and brain. Millions of people would benefit if medical science could stop the development of necrosis or cause regeneration of the injured or diseased nervous system at will.

Lesions in the nervous system have very peculiar properties. It has been shown, for example, by Klatzo et al. [29], that if radiolesions are produced in one side of the cerebral hemisphere, with time a symmetrical lesion develops in a "mirror" location on the other side of the cerebrum. Analogous observations have been made previously with neural lesions of varied origin. It is not understood how such secondary mirror lesions can develop in the brain. It is known that we have two cerebral hemispheres and that their structures are mirror images of each other. The two have symmetrical neural interconnections. It is as though we have two computers available to do the necessary job of controlling computation; one is there as a standby for the other. The interconnections between identical sites of the two sides appear to be such that when lesions cause abnormal electrical functions on one side, the unbalanced electrical action and injury currents somehow cause degenerative injury to the mirror side. It is possible to "disconnect" the two cerebral computers from each other by cutting all the neural interconnections between them. These all go through a very massive, large nerve bundle carrying perhaps a billion fibers called the *corpus callosum*. In animals it is reasonably feasible to electrically isolate one side of the brain from the other by a heavy-ion cutting lesion in the corpus callosum. This is, of course, bloodless. A peculiar aspect of such lesion, shown by Gaffey and Montoya [30], was that it must be at least 2 mm wide; otherwise neural impulses jump across the lesion, even following a large dose of radiation.

When radiolesions interfere with local brain structures, one may expect to observe abnormalities in the function controlled by that part of the brain. The number of studies that need to be undertaken by such methods to fully understand each aspect of brain function on these problems is manyfold. Some beginning has been made already on the "neural circuits" by producing lesions in the "visual" cortex, which is apparently responsible for part of our ability to see and interpret images. Other irradiated regions include the hippocampus, which is believed to relate to the process of learning. When animals are given radiolesions by heavy particles in the hippocampus, their ability to learn is decreased, but the memory of tasks learned before the lesion was made remains intact [31].

We do not understand the basic mechanism of either memory or learning at the cellular or at the tissue level and much more needs to be learned about these processes.

C. Sensation and Motor Action Caused by Radiation Pulses

Some of the sensory nerve endings are quite sensitive to radiation. For example, it has been known for almost 50 years that penetrating radiations are recorded as light flashes in the eye. It was shown in our laboratory that exceedingly small doses of ionizing radiation can produce alterations in the electrical impulse pattern emanating from the frog retina. The doses necessary are so small that a few ionization events can give rise to changes in the electrical pattern [32]. This has also been observed in the dark-adapted human eye, which can perceive a bluish light when ionizing radiation strikes the retina. The small dose necessary for perception can alter the subsequent sensitivity of the retina to light, as evidenced by change in the absolute light sensitivity threshold following a pulse of radiation as small as 0.002 rad [33]. In a few minutes following a small radiation dose light sensitivity apparently returns to normal.

Very recently several astronauts, flying on the Apollo missions to and from the moon, observed an interesting phenomenon of occasional light flashes and streaks; these would appear to the dark-adapted person with eyes open or closed [34]. It was possible to observe somewhat similar flashes of light in the fast neutron beams of the cyclotron. Light sensation at the cyclotron seemed to be due to heavy nuclear recoils from atomic collisions affecting the retina. In space these events appear to be due to fast, heavy cosmic ray particles [35].

It has become interesting to find out whether or not heavy charged particles could be perceived only in visual receptor endings or if radiation can cause active processes elsewhere in the nervous system as well. By using the Berkeley heavy-ion accelerator millisecond pulses of monoenergetic helium ions were directed to various parts of the bodies of small animals. It was observed that such beam pulses can cause reflex action. For example, a single helium ion pulse impinging on the pupil of one eye and penetrating to about 140-μ depth, where the sensory nerve endings are, can cause a "corneal blinking reflex." When the beam penetration is shallower or deeper, the efficiency of inducing such reflexes decreases.

Hug, in Germany, and others attempted to stimulate muscle and nerve action by various types of radiation. According to them [36] the cells of the body are divided into two large groups. Those cells or nerve endings that are sensitive to light can also be easily stimulated by ionizing radiation. Light-sensitive nerves send out electrical signals in response to a relatively small dose of radiation. Light-sensitive muscle, such as the smooth muscle of the heart, contracts. The sensitivity seems to be related to chromophoric substances within the cells that appear to be responsible for the release of chemicals (for example, transmitter substances) which cause membrane changes.

It is essential to have the "threshold" radiation doses delivered in a time smaller than 2 msec. Ordinary radiation is usually not perceived by humans, the intensity is far below the threshold.

Most of the cells of the brain are not sensitive to light and are quite refractory to excitation by penetrating radiations. Nevertheless, it was possible to show that high-intensity pulses of α particles sent to local regions in the motor cortex of rats could elicit movement of certain groups of muscles in the other side of the body of the animals in a manner somewhat similar to what one obtains with electrical stimulation. Such experiments are exceedingly hard to carry out since, in order to find the right location, it seems necessary to anesthetize the animals in advance, and narcotics render the nervous structures quite insensitive.

D. Nerve Cultures

There are many complex and still unsolved problems in connection with an understanding of brain and neural functions that it seems necessary to dissect these problems to their elements. At the core of most of the changes are the response and behavior of individual cells. It seems to be logical to attempt to culture mammalian nerve cells and study the changes induced by radiation and by other agents at the cellular level. Since nerve cells do not divide in the lifetime of animals, such cultures do not proliferate in the same manner as cultures of other body constituents. It is quite possible to keep neurons alive for many weeks away from the body, where the supporting cells, the various types of glia, proliferate and assist the neurons to survive. Several groups have been successful in culturing various parts of the nervous system [37]. Among such efforts are studies of the spinal cord in culture, of cerebellum, and even of the cerebral cortex. We cannot review the many interesting experiments that have been attempted in culture and will deal only with a few of the points. In our laboratory a method was found by which the neurons, following explantation from a newborn mouse or rat, arranged themselves in a monolayer along the surface of a microscope cover glass in such a manner that neurons could be individually studied and their interconnections mapped to some extent. A picture of the cells in such a culture taken by scanning electron microscopy is shown in Fig. 9. We see that each large neuron is surrounded with a number of other cells and that the interconnections (axons and dendrites) are exceedingly complex. Nevertheless, it has been possible either to insert electrodes into neurons or to place them at the membrane of a single neuron (soma) and observe the spontaneous electrical activity of these [37]. It is of interest that even a small piece of neural tissue, when explanted and kept alive, exhibits spontaneous activity which manifests itself in electrical pulses along the axons of neurons and passes with some time delay from one neuron to another via synaptic action. Often several hundred pulses per minute are observed at room temperature. These are not random, but are bunched in a curious, near periodic manner. The pulses are usually not observed in cultures which are not dense with neurons, indicating that the spontaneous activity requires communication between a number of neurons. Two or more cells may be observed simultaneously, and when this is done, it seems quite obvious that interconnections cause coincidences or delayed coincidences between each. The laborious process of mapping such events is in progress. We already know that environmental changes, for example, a change in temperature and in ionic concentration of the cul-

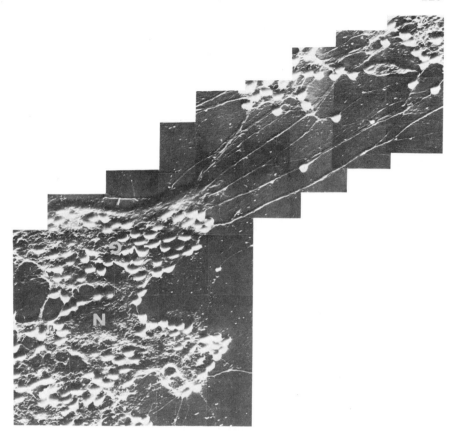

FIG. 9. Neurons from the cerebellum grown in thin layers in the laboratory. This technique allows direct study, at the cellular level, of the electrical and chemical manner in which such cells communicate. The photograph was made with the scanning electron microscope and indicates the cells and their connections. (Courtesy of Dr. Abdel-Megid Mamoon and Dr. Helge Dalen). The large cell marked N is a neuron. The smaller bumps marked G are glial cells and microneurons. The interconnections are neurites (axons and dendrites) as well as glial processes.

ture medium, profoundly alter the spontaneous electrical activity. A search has started for responses of neural nets to temporary environmental changes which could be interpreted as "memory" of previous events. Micromethods are being developed to study the effects of accelerated particles on various parts of the single neurons by the use of microbeams. Such beams can be formed not only with heavy particles but also by the use of laser and ultraviolet light. In particular, we are testing the hypothesis that radiation might be able to affect certain submicroscopic structures in the synaptic region of neurons, thus causing an alteration in their ability for synaptic transfer. There are hopes that one can eventually carry out work with neural cultures not only at the electrical level but also at the biochemical level. It is already possible, by the addition of radioactive tracers, to observe the rate of incorporation of molecules into the genetic apparatus, the formation of nucleic acids, and the synthesis and transport of proteins. Abnormal neural cultures can also be developed. For example, in our laboratory, Mamoon has shown that when the culture from a newborn rat is irradiated [50], it fails to develop myelin sheaths around its axons; thus, we now have normal myelinated and abnormal unmyelinated cultures. The behavior of the unmyelinated cultures resembles the behavior of neural tissue in persons with multiple sclerosis. This is a serious and debilitating disease of unknown cause and origin. There is some hope that more can be learned about it by the cell culture method. Various properties of the brain are being studied in several laboratories and there are even some cultures available from human brain tumors.

By the use of the cultures it should eventually be possible to test quantitatively the various theories for learning and memory. In addition, at least in principle, it would appear to be feasible to grow neural cultures that become very large in volume. As in one of the known classes of developmental abnormalities, "exocephalus," the brain continues to grow and may reach considerable size outside of the skull. Are such large volumes of neural tissue capable of more complex tasks than a normal-size brain? It is a particularly mystifying task for the investigator to work with neuron cultures and for him to know that even though they were once part of the body of an animal whose language he cannot understand, the electrical impulses are clicking on and off day and night, carrying some mysterious message he would very much like to understand.

IX. PARTICLES WITH A HIGH RATE OF ENERGY LOSS

Something should be said now about entirely different properties of the accelerated particles: the "rate of energy transfer." As described at the beginning of this chapter, the transfer of energy from these particles to matter depends very sensitively on the electrical charge of the particles as well as on their velocity. Particles that move relatively slowly have much greater energy transfer than fast particles, and particles that have high charge ionize much more effectively than those with a low charge. In fact, the rate of energy loss varies proportionally to the square of the charge of the particles. This property is of great interest when heavy atomic nuclei are used.

For example, an iron atom has 26 electrons. When the particle is accelerated to high velocities, all these electrons can come off and the iron nucleus is left with 26 charges. The rate of energy loss of such a nucleus is $26^2 = 676$ times higher than the rate of energy loss of a proton of the same velocity. A great deal of study has gone into the effects of such high-energy transfer to tissue. We will not describe this in detail but it has been adequately documented elsewhere. Qualitatively we might describe this effect as high-LET particles being much more effective in killing cells than low-LET radiation, and also much more effective at producing chromosome deletions, translocations, and interchanges. It is believed that most of the effects of low-LET particles are on a single strand of the double-stranded DNA, and the lesions made in the single strands, such as scission, can be repaired by an enzymatic repair mechanism. The high-LET particles can apparently break both strands of DNA and thus interrupt the continuity of the genetic material. Such damage is much more serious than single-strand damage, which is often repaired by the cell. It is not yet known whether a double-strand repair mechanism exists.

X. HEAVY PARTICLES IN CANCER THERAPY

Radiation therapy of cancer is an art in a state of very detailed technical development. Even in its present advanced stage it has great limitations and much improvement is desired. One must realize that many classes of cancer cause metastases and disseminate in the body so rapidly that local therapy is not useful, and one has to resort to other methods, such as the pituitary irradiation described earlier in this chapter. In other situations, tumors are well localized, but it is a difficult problem to ascertain where they are, since tumor tissue is soft and often will not produce very definite shadows in conventional x-ray roentgenography.

The methods of nuclear physics also have been applied to this problem; for example, in the development of the "γ-ray camera," which in some instances can detect the presence and extent of tumors by the uptake of radioisotopes [38]. Radiation therapy with heavy particles is desirable under certain special conditions [39, 40]. Initial applications of direct proton radiotherapy were made in Uppsala on metastatic tumors of the uterus [21, 41] and in our laboratory with helium ions on selected cases of brain and mammary tumors. There are three distinct classes of tumors where the use of heavy particle beams may be beneficial:

1. The tumor is localizable with diagnostic methods but is in a site which is surgically inaccessible. Some brain tumors fall into this class.

2. The cells comprising the tumor are radioresistant to conventional radiations but sensitive to high-LET radiation.

3. A tumor produces necrotic centers. In such centers oxygen supply becomes deficient and anoxic tumor cells can easily outpace the normal cells in growth since the latter require oxygen. The property of x rays, γ rays, and other low-LET radiations is that they become quite ineffective in anaerobic environment, whereas high-energy radiations are nearly equally effective whether the environment is oxygenated or anaerobic. By use of the Bragg ionization curve and heavy accelerated ions, suitable depth–dose distributions can be achieved, together with an appropriately high-LET distribution in the tumor. Figure 10 gives an example of depth–dose distribution for heavy particles in a stimulated therapeutic exposure.

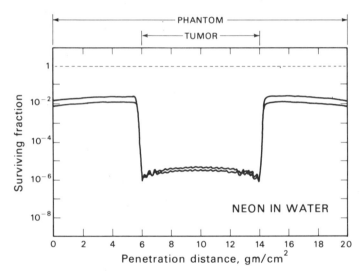

FIG. 10. Survival ratios calculated for a hypothetical therapeutic situation. The lower curve is in the presence of oxygen, the upper one in anaerobic condition. Conventional radiations would lead to thousandfold greater survival of anoxic tumor cells than shown here. (For more details see Tobias *et al.* [39].)

XI. SPECIALIZED TECHNIQUES

Accelerated particles can be used for so many interesting studies that in the space available here one cannot do equal justice to each of them. Nevertheless, we wish to mention two techniques which are in the infancy of their development at present. Both of these could turn out to be important tools for biological and medical science.

A. Heavy Ion Microprobe and Microscope

It has been known for many years that microscopy with heavy particles has theoretically a much higher resolving power than microscopy with electrons. Yet electron microscopy has developed a very high degree of perfection. Heavy-ion microscopy has not yet been generally utilized. The explanation for this appears to be that the techniques for building heavy-ion microscopes are similar to techniques used for heavy-ion accelerators, and it is only recently that these have been mastered sufficiently well to seriously think of building heavy ion microscopes with a reasonably high resolution. It turns out that recent advances in the focusing of heavy particles, particularly the quadrupole–octupole focusing methods, will be very useful since these allow correction of the spherical and chromatic aberrations of ordinary electrostatic or magnetic lenses. The resolution of a particle depends on its wavelength, and the wavelength is inversely related to the mass of the particle. Since the proton mass is 1846 times greater than the electron mass, the theoretical resolution attainable by protons is better by the same factor. We are very far from being able to use such high resolution, however. It is not generally known that the French investigator Crouchet [42] built a preliminary form of a proton microscope almost 20 years ago, and that this effort had to be terminated when the investigator developed a chronic disease. An ion microprobe was built [43] recently in England for surface elemental analysis. At Berkeley work has been started to develop microbeams of accelerated particles that probe the surface of the specimen and analyze certain secondary processes. Among the secondaries are characteristic x rays of the elements from the surface, recoil nuclei from the surface, and finally, the scattered primary particles. It has been shown that one can expect a cleaner characteristic x-ray spectrum from surface irradiation than with electrons. It has also been shown that when the particle beams are used for scanning, a smaller beam probe size can be expected than that with electrons since the electrostatic repulsion of heavy particles is less than that of electrons of comparable velocity and the heavy particles scatter less in the target material. In addition to scanning microscopy, heavy particles can be used for surface etching when the scanning beam is made sufficiently intense to cause vaporization of some of the atoms or molecules from the surface. A striking picture of a red blood corpuscle ablated by heavy-ion scanning is reproduced in Fig. 11. It reveals structure in the

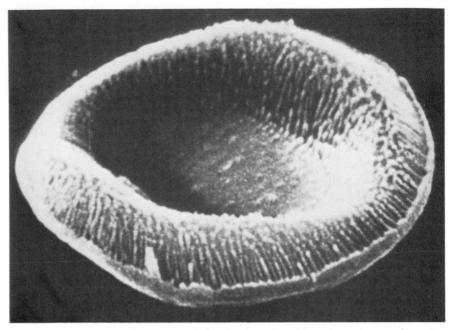

FIG. 11. The appearance of a red blood corpuscle without its outer membrane as photographed with the scanning electron microscope, following surface etching by a helium ion beam. (Courtesy of S. M. Lewis [43].)

frozen red blood corpuscle that has not been previously seen. By using different monoenergetic heavy ions with different penetration and intensity it should be possible to control the amount of heat released locally in the specimen and cause surface etching with considerable control. Such a technique may be important in unraveling the biological structures underneath the surface. Heavy accelerated particles can be used to implant the accelerated atoms to some layer below the surface. This technique has not yet been utilized in biology, but it has given some interesting results in the study of metals [44]. It is believed that the intricate structure of biological membranes and of artificial membranes built to mimic the properties of biological membranes might be influenced by implanted atoms due to the finite range of such particles; the implantation can occur at a definite microscopic depth.

A microprobe is being constructed in our laboratory. By this instrument we hope to obtain a beam under 100 Å in diameter which can be used for probing and scanning of biological surfaces.

B. Laminar Heavy-Ion Roentgenography

It has been known for some time that heavy-particle beams are suitable to detect small changes in density or small flaws in the matter they cross. This is due to the fact that the range of the monoenergetic particles is a very sensitive function of the thickness and composition of the matter they cross. By measuring deviations from the expected range penetration, flaws in material can be detected [45]. If a particle crosses an object and emerges on the other side, the residual range (range of the particle left after crossing the object) will vary precisely according to the thickness of the body itself. We speak here not of thickness in terms of inches, but rather in terms of the "stopping power," which depends on the density of the atoms and their atomic number. Andersson of the University of Chicago suggested some time ago that the resolution attainable is much higher than that with x-ray roentgenography and Koehler at Harvard University recently has actually demonstrated that proton roentgenography is feasible [46]. It was realized recently that measurement of the residual range of particles after they cross an object can be used to calculate the density distribution inside of the object, which is something that one usually cannot obtain in x-ray roentgenography. In conventional roentgenography we usually obtain a shadowgram of the object, whereas ion laminography can produce a density distribution in the plane of the beam. In order to obtain this distribution one measures the stopping power of the body in many places and in many directions [47].

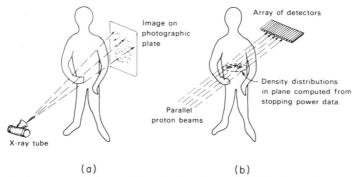

(a)   (b)

FIG. 12. A proposed diagnostic technique, heavy-ion laminography. In contrast to X-ray diagnostic pictures (a), which give a projected image, the laminography technique (b) will produce an image of the density distribution of tissue in the plane in which the heavy ions intersect the body.

Experiments have been started at Berkeley to prove the feasibility of this method. It appears that it will be possible within a period of about 2 sec to obtain stopping power measurements in about 40 places across the human body and in 40 different directions, giving 1600 pieces of data. Using Fourier transforms and computer handling, in theory, it is possible then to plot the density distribution within the body (see Fig. 12). Although this proposed technique appears to be theoretically sound, it is anticipated that the experimental and computational complications are such that it may take several years to develop it to a high degree of perfection.

XII. THE FUTURE

We tried to give some idea of the many interesting and important biological and medical problems that remain unsolved, particularly in connection with the brain, and of some heavy-particle techniques. The wider, therapeutic use of these techniques would surely have beneficial effects. Most of the ideas were developed over the past 20 years in relatively modest programs using accelerators available to physicists at various laboratories. The list of these laboratories is continuously increasing. We mention the Uppsala and Harvard cyclotrons. The 300–600 MeV proton cyclotron of the National Aeronautics and Space Administration is currently being prepared for biomedical investigations. The cyclotrons at Harwell, England, and Orsay, France, have been used in specialized investigations and the CERN synchrocyclotron in Geneva is currently the site of pretherapeutic studies with negative π mesons. In the Soviet Union considerable recent interest has developed, chiefly due to the work of Dzhelepov and Gol'din [48], and a number of patients with cancer have already been treated.

The work has progressed to such a complexity that it seems worthwhile to think of the parameters needed for biology and medicine and of ways in which machines could be made available for them. The main criteria in developing design parameters were to have beams available with sufficient penetration to be able to reach most parts of the human body from the outside, to have beams available with a variety of atomic numbers so that high-LET effects could be studied, to make the energy resolution of the beams precise, and to have variable energy so that laminar lesions can be produced at any desired depth. It turns out that no existing accelerators fulfill these criteria.

The acceleration of ions much heavier than protons or α particles requires special consideration. When atoms are ionized in collisions, the state of

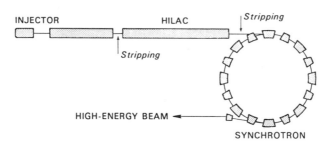

INJECTOR   HILAC   Stripping

Stripping

HIGH-ENERGY BEAM

SYNCHROTRON

FIG. 13. Schematic diagram of a proposed heavy-ion synchrotron. The ionized particles are first accelerated in a linear accelerator to travel at speeds of about the same magnitude as the velocity of their remaining electrons. By passing through a thin layer of gas or metal, their remaining electrons are stripped off. The stripped particles then are further accelerated in a synchrotron to the desired higher energies.

ionization (that is, the number of electrons they can lose) depends on the velocity of the atoms. In the ion plasma of a conventional ion source the thermal velocity is relatively low and one can expect atoms that have lost not more than one, two, or three electrons. Yet for the efficient acceleration of heavy nuclei, it is desirable to have these nuclei stripped of all of their electrons. It turns out that there are two generally feasible methods for this.

The first, that of gradual acceleration of the ions followed by electron stripping, has been tested in the laboratory sufficiently so that we are certain an accelerator could be built that would deliver the types of beams wanted. A proposal has been submitted to the Atomic Energy Commission for such a machine. The beams for this machine start with the Berkeley heavy-ion linear accelerator (HILAC), which is being modified at the time of this writing. The particles are injected with a few electric charges, accelerated, then stripped in two places in the linear accelerator tank until they emerge with an energy of about 8.5 MeV/nucleon. It was shown that this beam could be stripped again and injected into a heavy-ion synchrotron, which would allow variable-energy beams up to 500 MeV/nucleon—the energy desirable to have for medical application.* The general layout of such a machine is shown in Fig. 13. This machine could have fast accelerated particles of all the elements of the periodic table, except that the very heavy elements (those above calcium) would have somewhat less kinetic energy than the light ones since they will not be completely stripped before entering the synchrotron. The low-energy heavy-ion beams and the high-

_____

* Note added in proof: The existing Berkeley Bevatron is a suitable machine to accelerate ions received from the Hilac to appropriate energies.

energy synchrotron beams can be obtained and used simultaneously: low-energy nuclear chemistry will use 99.9% of the available beam. The remaining 0.1% will be sufficient for biomedical investigations at high energy.

The second method is based on a new principle of acceleration. This is the "electron ring accelerator" (ERA). Some aspects of this originally were conceived by the Russia scientist Vechsler about ten years ago. Experimentally, the electron ring accelerator principle is currently being tested in the Soviet Union and in the United States [49]. The principle of this machine involves using fast electrons for the acceleration of heavy positive particles by pulling the ions along in an electron plasma. Accelerator techniques are well known which can produce electron beams of about 20 MeV relatively easily. Strong magnetic fields in a "magnetic bottle" are used to hold and shape the plasma of electrons while neutral atoms are injected into the plasma. The atoms soon become ionized by collisions with electrons. They then are surrounded by many electrons (perhaps 1000 electrons around each heavy ion), after which the electron plasma is extruded from the magnetic bottle, pulling the heavy ions along.

Since in the ERA the acceleration is magnetic, the greater efficiency of this accelerator (if it can be shown to work reliably) makes it preferable to conventional accelerators. Russian scientists have claimed to have already obtained nitrogen ions of about 4 MeV/nucleon in just a few-centimeter path of acceleration. Beams of 500 MeV/nucleon might be possible with a linear acceleration distance of under 50 ft. This method is also suitable for the acceleration of protons to multibillion-volt kinetic energies.

Accelerated particles have opened up new avenues in research in biology and medicine, yielded new therapeutic methods for cancer and other diseases, and could make exceedingly important contributions in the future. In order to progress further, it seems essential to use accelerators that have the precision properties which are shown to be needed for this endeavor. Perhaps such accelerators always appear expensive when they are first built. It has often happened that the uses derived from the results of research have resulted in benefits that repaid the initial cost many times over. If the heavy-ion accelerator were to be used extensively in cancer therapy, it can be shown that the cost per patient might be less than the cost of conventional methods. This is due to the fact that a single accelerator can provide facilities for the treatment of a number of patients in different rooms at the same time, perhaps as many as 10 or 20 individuals. The treatment period for each can be kept exceedingly short, on the order of a few minutes for each patient, due to the high beam intensity. Currently, in large hospitals radiation therapy is performed by providing multiple numbers of accelerations, cobalt sources, and x-ray machines in a number of separate installations. Thus, the large beam capacity of heavy-ion accelerators, when used with an appropriate number of exposure rooms, might make these machines economically as feasible as the use of many smaller units.

GLOSSARY *

Acromegaly—A serious and debilitating disease where, due to a tumor in the pituitary gland, excessive growth hormone is secreted. Visible symptoms are progressive enlargement of hands, face, feet, and thorax.

Afferent nerve fibers—Nerves carrying electrical messages from peripheral and sense organs to the central nervous system. These messages can be somatic, visceral, or proprioceptive.

Autoimmunity—When the body develops immunity against some of its own cells and attacks them as though they contained foreign substances. It may be the natural consequence of disease or of deleterious changes caused by external agents.

Axon (of a neuron)—A nerve trunk starting from the soma and, usually after branching, ending in synapses. It conducts action potentials.

Blood–brain barrier—Membranous structure separating neural tissues from the circulation and therefore the rest of the body. It has selective barrier action for many ions and molecules. The rate of entry of potassium ions from circulating blood to spaces containing nerve cells is limited by the blood–brain barrier.

Carcinogenesis—The mode of initiating, causing, or originating cancer.

Congenital malformations—Deformities and diseases that exist in the newborn. Often these are the result of deleterious effects of external agents during gestation.

Corpus callosum—Large bundle of nerve fibers connecting the two cerebral hemispheres together.

Cytomembranes—Laminar and mebranous structures seen in the cytoplasm of cells, particularly of secretory cells, with the aid of the electron microscope.

Dendrites—Neural processes.

Efferent nerve fibers—Nerves carrying electrical messages from the brain and spinal cord to peripheral parts of the body.

Endocrine system—System of hormone secreting glands.

Estrogenic hormone—Generic term for substances which produce estrus and development of secondary female sexual characteristics.

* For more precise definition of words also consult a standard medical dictionary.

Glia (neuroglia)—Cells and fibrous substance surrounding and supporting neurons in the central nervous system. Morphologically, several such cell types are known —e.g., astroglia, oligodendroglia, and microglia.

Gonadotropins—Pituitary hormones regulating the reproductive organs.

Hippocampus—A portion of the brain forming part of the rhinencephalon near the floor of the lateral ventricle.

Homeostasis—Control action centering in the hypothalamus, causing normal body organs to perform within normal limits.

Hypophysectomy—Surgical removal of the hyphysis. *Radiation hypophysectomy* refers to arresting or modifying the secretions of the hypophysis by radiation.

Hypophysis, or pituitary gland—A small, two-lobed gland at the base of the brain, lying in the sella turcica, a bony cavity near the geometrical center of the skull. Under the neural and humoral influence of the hypothalamus, the hypophysis secretes hormones that regulate many of the physiological functions of mammals. Growth (somatotropic hormone), development of sex organs (follicle stimulating hormone), the adrenal glands (ACTH), the thyroid (TSH), milk production (LH), water balance (antidiuretic hormone), blood pressure (vasopressin), and other body functions are under the influence of the pituitary gland.

Hypothalamus—Located below the thalamus and above the pituitary, the hypothalamus contains important nerve centers ("nuclei") for autonomic control of many lower functions of the body. Its control functions are exercised either via nerve impulses or by neurosecretion of humoral substances. The secretions of the anterior and posterior pituitary gland are under direct control of the hypothalamus.

Intracranial pressure—Hydrostatic pressure developed by tissues of the brain, which are enclosed in the rigid cranium. This pressure is the result of metabolic and osmotic action. When abnormally high, it can cause headaches and even loss of consciousness.

Malignant tumor—Proliferating and disseminating neoplasm capable of causing fatality either due to its toxicity or its invasiveness.

Metastasis—The appearance of groups of cancer cells in parts of the body remote from the seat of the primary tumor.

Myelin—multilayered membranous structure sururounding many axons. Its layered structure can be seen in the electron microscope. The membranes are formed from lipid and protein; they alter the electrical conduction properties of nerves quite markedly. In normal brain, neurodendroglia produce myelinization; these cells literally wrap themselves around neurons.

Nanogram—10^{-9} gm; one billionth of a gram.

Neoplasm—Tumor with proliferating cells.

Phagocytosis—A complex process whereby foreign particles, microbes, or sick cells are removed from the tissue by their ingestion into specialized cells—in the case of brain tissue, probably microglia.

Pituitary—See *Hypophysis.*

Rad—Unit of radiation dose. It corresponds to 100 ergs of energy, in the form of ionization and excitation transferred to 1 gm of tissue.

Sella turcica—A saddle-like prominence on the upper surface of the sphenoid bone; it surrounds the pituitary gland.

Soma—The main cell body of a neuron. Axons and dendrites project from it.

Stem cells—Cells in various specialized organs that originate the development of differentiated cells with specific biochemical functions. The division and differentiation of stem cells is believed to be under external (e.g., hormonal and neural) control.

Synapse—A microscopic contact region where a nerve impulse is transmitted from one neuron to another.

Tumor—Abnormal growth.

Visual cortex—The part of the cerebral cortex most responsible for seeing and interpreting images, it is located in the occipital lobes.

ACKNOWLEDGMENT

Some of the work described in this chapter is the result of twenty-five years of collaboration between physical, biological, and medical groups at Donner Laboraotry and Lawrence Radiation Laboratory. The authors are indebted to Drs. John Lawrence and James Born for their cooperation and for fruitful discussions in the course of the preparation of this manuscript.

REFERENCES

1. C. A. Tobias, J. H. Lawrence, J. L. Born, R. K. McCombs, J. E. Roberts, H. O. Anger, B. V. A. Low-Beer, and C. B. Huggins, Pituitary irradiation with high-energy proton beams: a preliminary report, *Cancer Res.* **18,** 121–134 (1958).
2. J. H. Lawrence and C. A. Tobias, Heavy particles in therapy, *In* "Modern Trends in Radiotherapy" (Thomas J. Deeley and Constance Wood, eds.), pp. 260–276. Butterworth, London, 1967.
3. P. Steward, Stopping power and range for any nucleus in the specific energy

interval 0.01–500-MeV/amu in any nongaseous material, Ph.D. thesis, Univ. of California, Lawrence Radiation Lab. Rep. 18127, Univ. of California (1968).

4. G. Litton, Penetration of high-energy heavy ions, with the inclusion of coulomb, nuclear, and other stochastic processes. Ph.D. thesis, Univ. of California, Lawrence Radiation Lab. Rep. 17392 (August 1967).

5. H. D. Maccabee, M. R. Raju, and C. A. Tobias, Fluctuations of energy loss by heavy-charged particles in thin absorbers, *Phys. Rev.* **165,** 469–474 (1968).

6. M. Raju, J. Lyman, T. Brustad, and C. A. Tobias, Heavy charged particle beams, *In* "Radiation Dosimetry" (F. H. Attix and E. Tochilin, eds.), Volume 3, 2nd ed., Chapter 20, pp. 151–193. Academic Press, New York, 1969.

7. M. Singer, "The Human Brain in Sagittal Section," p. 81. Thomas, Springfield, Illinois, 1954.

8. F. H. Netter, The CIBA collection of medical illustrations, *CIBA Pharm. Prod.* **1** and Suppl. (1953).

9. C. A. Tobias, D. C. VanDyke, M. E. Simpson, H. O. Anger, R. L. Huff, and A. A. Koneff, Irradiation of the pituitary of the rat with high energy deuterons, *Am. J. Roentgenol. Rad Ther. Nucl. Med.* **72,** 1–21 (1954).

10. J. H. Lawrence, C. A. Tobias, J. A. Linfoot, J. L. Born, J. T. Lyman, C. Y. Chong, E. Manougian, and W. C. Wei, Successful treatment of acromegaly: metabolic and clinical studies in 145 patients, *J. Clin. Endocrinol. Metabolism* **31,** 180–198 (1970).

11. R. N. Kjellberg, A. Shintani, A. G. Frantz, and B. Kliman, Proton-beam therapy in acromegaly, *New England J. Med.* **278,** No. 13, pp. 689–695 (March 1968).

12. W. Haymaker, E. Anderson, and W. Nauta, eds., "The Hypothalamus." Thomas, Springfield, Illinois, 1969.

13. A. Anderson, J. Garcia, J. Henry, C. Riggs, J. E. Roberts, B. Thorell, and C. A. Tobias, Pituitary and hypothalamic lesions produced by high energy deuterons and protons, Lawrence Radiation Lab. Rep. 3737, Univ. of California (1957); *Rad. Res.* **7,** 299 (1957) abstr.

14. C. J. Shellabarger, G. E. Aponte, E. P. Cronkite, and V. P. Bond, Studies on radiation induced mammary gland neoplasia in the rat, VI. Effect of changes in thyroid function, ovarian function and pregnancy, *Rad. Res.* **17,** 492–507 (1962).

15. R. Luft and H. Olivecrona, Experiences with hypophysectomy in man, *J. Neurosurg.* **10,** 301–316 (1953).

16. C. Huggins, Endocrine induced regression of cancers, *Science* **156,** 1050 (1967).

17. D. C. VanDyke, M. E. Simpson, A. A. Koneff, and C. A. Tobias, Long-term effects of deuteron irradiation of the rat pituitary, *Endocrinology* **64,** 240–257 (1959).

18. J. Furth, N. Haran-Chera, H. J. Curtis, and R. F. Buffet, Studies of the pathogenesis of neoplasms by radiation, I. Pituitary tumors, *Cancer Res.* **19,** 550 (1959).

19. K. H. Clifton, E. L. Gadsen, and R. F. Buffet, Dependent and autonomous mammotropic pituitary tumors in rats, *Cancer Res.* **16,** 608–616 (1956).

20. A. Anderson, J. E. Roberts, B. Thorell, and C. A. Tobias, Arrest and growth in young rats following hypothalamic deuteron irradiation, Lawrence Radiation Lab. Rep. 3738, Univ. of California (1957); *Rad. Res.* **7,** 300 (1967) abstr.

21. S. Graffman and B. Jung, Clinical trials in radiotherapy and the merits of high energy protons, *Acta Radiol.* **9,** 1–23 (1970).

22. B. Andersson, B. Larsson, L. Leksell, W. Mair, B. Rexed, and P. Sourander,

Effect of local radiation of the central nervous system with high energy protons, *Intern. Symp. Responses Nervous System to Ionizing Radiation* (1962).

23. P. Blanquet, Hypothalamus and thyroid, *Advan. Biol. Med. Phys.*, **8**, 225–306 (1962).

24. I. Klatzo, J. Miquel, C. Tobias, and W. Haymaker, Effects of alpha particle radiation on the rat brain including vascular permeability and glycogen studies, *J. Neuropathol. Exp. Neurol.* **20**, No. 4, 459–483 (October 1961).

25. F. D. Walker and W. J. Hild, Neuroglia electrically coupled to neurons, *Science* **165**, 602 (1969).

25a. W. Haymaker, *In* "Effects of Ionizing Radiation on the Nervous System," p. 309, International Atomic Energy Agency, Vienna ,1962.

26. L. I. Malis, J. E. Rose, L. Kruger, and C. Baker, Production of laminar lesions in the cerebral cortex by deuteron irradiation, *In* "Response of the Nervous System to Ionizing Radiation" (T. J. Haley and R. S. Snider, eds.), pp. 359–368. Academic Press, New York, 1962.

27. L. Kruger and C. D. Clemente, Anatomical and functional studies of the cerebral cortex by means of laminar destruction with ionizing radiation, *In* "Response of the Nervous System to Ionizing Radiation, II" (T. J. Haley and R. S. Snider, eds.), pp. 84–104. Little, Brown, Boston, Massachusetts, 1964.

28. H. J. Curtis, Radiation and aging, *Symp. Soc. Exp. Biol.* **21**, 51–64 (1967).

29. I. Klatzo, J. Miquel, C. A. Tobias, and W. Haymaker, Effects of alpha particles on rat brain including vascular permeability and glycogen studies, *J. Neuropathol.* **20**, 459–483 (1961).

30. C. T. Gaffey and V. J. Montoya, Bioelectric sensitivity of *Corpus callosum* to cyclotron-accelerated alpha particles. Lawrence Radiation Lab. Rep., Univ. of California (in preparation); *Rad. Res.* **31**, 560 (1967) abstr.

31. R. L. Schoenbrun and W. R. Adey, Space flight related stresses on the central nervous system, *Rad. Res.* **7**, 423–438 (1967).

32. L. E. Lipetz, An electrophysiological study of some properties of the vertebrate retina, Ph.D. thesis, Univ. of California, Lawrence Radiation Lab. Rep. 2056 (January 1953).

33. K. Motokawa, T. Kohata, M. Komatso, S. Chichiba, Y. Koga, and T. Kasai, A sensitive method for detecting the effect of radiation upon the human body, *Tohoku J. Exp. Med.* **66**, 389–404 (1957).

34. Manned Spaceflight Center, Medical Division, NASA, Houston, private communications 1970.

35. C. A. Tobias, T. F. Budinger, and J. T. Lyman, Radiation induced phosphenes observed in fast neutron and X-ray beams by human subjects, Lawrence Radiation Lab. Rep. 19686, Univ. of California (August 1970).

36. H. Gangloff and O. Hug, The effects of ionizing radiation on the nervous system, *Advan. Biol. Med. Phys.* **10**, 1–73 (1965).

37. W. Schlapfer, A. Mamoon, and C. Tobias, Spontaneous Bioelectric Activity of Cultured Neurons, presented at Third Intern. Biophys. Congr., Aug. 29–Sept. 3, Cambridge, Massachusetts (1969).

38. H. O. Anger, Radioisotope cameras, *In* "Instrumentation in Nuclear Medicine" (Gerald Hine, ed.). Academic Press, New York, 1967.

39. C. A. Tobias, J. Lyman, and J. H. Lawrence, Some considerations of physical and biological factors in radiotherapy with high LET radiations, *Progr. Atom. Med.*, **3** (in press).

40. C. A. Tobias and P. W. Todd, Heavy charged particles in cancer therapy, *U.S. Natl. Cancer Monogr.* No. 24, Radiobiol., Radiotherapy, Natl. Cancer Inst., Bethesda, Maryland, 1–21 (1967).

41. J. Naeslund, S. Stenson, and B. Larsson, Result of proton irradiation on V × 2 carcinoma in the rabbit, *Acta Obst. Gynecol. Scand.* **38**, 1–23 (1959).

42. G. Crouchet, Contribution à L'étude des sources solids ioniques—Application à l'emission electronique secondaire et à la microscopie ionique par emission, Ph.D. thesis, Faculty of Sciences, Paris (1953).

43. S. M. Lewis, J. S. Osborn, and P. R. Stuart, Demonstration of an internal structure within the red blood cell by ion etching and scanning electron microscopy, *Nature* **220**, 614–616 (1968).

44. S. Matsui, H. Kawakatsu, H. Yamasaki, and K. Kanaya, Observation of colour images recorded by ion beam bombardment, *J. Electron Microsc.* **14**, 290–296 (1965).

45. C. A. Tobias, T. L. Hayes, H. D. Maccabee, and R. M. Glaeser, Heavy ion radiography and microscopy, Biomedical Studies with Heavy-Ion Beams, Lawrence Radiation Lab. Rep. 17357, Univ. of California (March 1967).

46. A. M. Koehler, Proton radiography, *Science* **160**, 303 (1968).

47. C. A. Tobias, R. M. Glaeser, J. T. Lyman, T. F. Budinger, Laminar proton roentgenography, Lawrence Radiation Lab. Rep. 20637, Univ. of California (in preparation) 1970.

48. V. P. Dzhelepov and L. L. Gol'din, The use of the existing charged heavy-particle accelerators and the possibilities of creating new domestic ones for radiation therapy. Presented at Symposium on Problems in the Development of Radiation Therapy Techniques in Oncology, Moscow, Institute of Experimental and Clinical Oncology, April 1969. UCRL translation #1422, September 1970.

49. E. A. McMillan, Symposium on electron ring accelerators, Lawrence Radiation Lab. Rep. 18103, University of California (1968).

50. A. Mamoon, Effects of ionizing radiations on myelin formation in rat brain cultures, Ph.D. thesis, Lawrence Radiation Lab. Rep. 19481, University of California, December 31, 1969.

51. C. W. Welsch, H. Nagasawa, and J. Meites, Increased incidence of spontaneous mammary tumors in female rats with induced hypothalamic lesions. *Cancer Res.* **30**, 2310–2313. September 1970.

6

Possible Uses of Densely Ionizing Radiations in Radiotherapy

V. P. Bond   *Medical Department*
G. M. Tisljar-Lentulis*   *Brookhaven National Laboratory*
                        Upton, New York

* Present address: Strahlenbiologisches Institut der Universität München, Munich, West Germany.

I. INTRODUCTION

The application of ionizing radiation for therapeutic purposes has been well established since x rays were first used about 70 years ago. Considerable improvement has been achieved over the years with respect to type of radiation, energy, intensity, and irradiation technique. Various neoplastic diseases have been attacked successfully by radiation therapy used independently or in connection with conventional surgery or chemotherapy. It is also possible to influence favorably certain malfunctions in a patient by irradiation of organs, the functions of which have a strong impact on the physiological system. Thus, irradiations of the pituitary gland have proved to be beneficial in certain cases of metastatic carcinomas, diabetes, and acromegaly.

Although conventional radiotherapy has been used with remarkable success in the control of neoplastic disease, a number of patients do not respond in a satisfactory manner. There are several reasons for this less than satisfactory response. Some tumors have metastasized when first seen by the radiotherapist, and it frequently is impossible to treat all of the resulting lesions. Also, the location of many tumors is such that a radiation dose sufficient to control the tumor cannot be delivered without at the same time delivering doses to normal tissues that will cause unacceptable damage.

The relatively poor blood supply to portions of tumors, resulting in regions containing hypoxic cells, is thought to impose a severe limitation on the control of tumors with conventional x or γ radiation. The reason stems from a basic fact of radiobiology, namely, that cells are relatively resistant to x or γ radiation when in the hypoxic state. Cells exposed to x or γ radiation in a nitrogen atmosphere may be as much as three times more resistant to radiation than the same cells exposed in air or in a pure oxygen atmosphere. Rapid regrowth of the tumor may occur from these protected hypoxic cells, even though large numbers of aerated tumor cells may have been eradicated.

A number of approaches have been suggested and tried to correct the difficulties arising from the presence of hypoxic cells. Patients have been exposed to oxygen under hyperbaric conditions, and pure oxygen at 3 atm pressure frequently has been employed. Results to date are equivocal, and certainly no dramatic increase in cure rate has been demonstrated. The explanation may lie in a failure of oxygen transport, even under hyperbaric conditions, to the severely hypoxic regions. Intravascular hydrogen peroxide has been tried, with equivocal results to date. The approach of rendering anoxic all tissues in the region of the tumor to be treated, and thus making all normal and all tumor cells equally hypoxic and equally resistant to radiation, is attractive. The anatomical locations where this approach is feasible at present are limited, and thus the approach has not been widely exploited.

An attractive approach with some promise of ameliorating the problems resulting from hypoxic cells in tumors lies in the use of densely ionizing or high LET radiations (see definition below). As opposed to x or γ radiations (low LET), densely ionizing radiations show less dependence, in their cellular effects, on the degree of oxygenation of the cells. With very high LET radiations, the degree of effect in a given cell population is indistinguishable, whether the cells are highly oxygenated or severely hypoxic. This and other effects of densely ionizing radiation, detailed below, render them attractive as possible radiotherapeutic tools and has spurred a great deal of investigation into the effects of these radiations and how they might be utilized in the treatment of neoplastic disease.

The quality of a given radiation is usually characterized by its linear energy transfer (LET), or its LET spectrum. The concept of the linear energy transfer is closely related to, but not identical with, the rate of energy loss dE/dx. The term LET refers to the energy absorbed in the immediate neighborhood of the track of the ionizing particle. Principally for this reason, LET values in general are smaller than the rate of energy loss of the particle. For biological effects it is important to consider the ionization not only in the particle track core but in the secondarily produced δ rays generally associated with the ionization track as well. If there is no appreciable δ ray production, or if the biological volume under consideration is large enough to guarantee total δ ray absorption in that volume, LET becomes LET_∞ and is identical with dE/dx. Thus, LET_∞ corresponds to an unlimited inclusion of δ ray energy. For radiobiological purposes, however, it is more appropriate to consider a cylinder around the particle track which includes only δ rays of energies up to 100 eV. If δ rays of more than 100-eV energy* are produced, they are dealt with as if they were independent particles with their own independent LET. The energy of these "independent" particles represents the difference between dE/dx and LET of the primary particle. The greater the energy of the secondaries, the larger will be the percentage of the energy of the primary particle that will be lost with respect to local absorption, and the greater will be the discrepancy between dE/dx and LET. The degree of the discrepancy varies with type and energy of the primary particle as a result of differences in the nature of δ ray production. With decreasing energy of the primary particles, a rise in LET similar to that seen for dE/dx is observed (Fig. 1).

* Sometimes a "cutoff" energy of 200 eV is arbitrarily chosen.

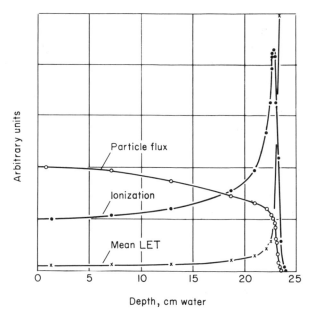

FIG. 1. Integral range curve, ionization and mean LET of 187-MeV protons in water. Diameter of the beam aperture: 2 cm [from B. Larsson, *Brit. J. Radiol.* **34**, 143 (1961)].

The energy deposited per unit length, LET, is usually expressed in terms of kilo electron volts per micron (keV/μ). Some authors prefer million electron volts square centimeter per gram (MeV·cm²/g).

The LET of conventional x rays (low LET radiation) ranges in the order of 1–3 keV per micron of water. With densely ionizing radiations the LET may be several hundred kilo electron volts per micron. The borderline between low and high LET radiation is difficult to draw and has not been uniformly defined. Acceptable values are those suggested by Fowler [1], who refers to values up to about 5 keV/μ as being "low LET," and "high LET" above approximately 60 keV/μ.

All radiations are delivered in a spectrum of LET's, and the variety of spectra encountered in radiations commonly used is quite extensive. It is common to refer to the mean LET of a radiation, and this represents at present the best single parameter in terms of which to characterize the quality of a given radiation. However, it is recognized that the mean LET, and perhaps even the concept of LET, may not be adequate to allow quantitative predictions of the degree of biological effects. Newer approaches to characterizing the microdeposition of radiation energy are being developed [2]; however, these as yet cannot be generally applied. Thus, the concept of LET, and the mean LET, must be regarded as a useful, but an only approximate approach to characterizing quantitatively the micro-deposition of energy from ionizing radiations.

The effects of high LET radiations, in addition to the relative independence of oxygen tension, differ in other important respects from those of low LET radiations. The relative biological effectiveness (RBE)* of high LET radiations is greater than that of x or γ radiation, that is, with the same dose (same number of ergs per gram) to a given tissue, the degree of effect will be greater for high LET than for low LET radiations. Another striking difference between the effects of high and low LET radiation is seen in the dose–rate dependence of their effects. The effects of low LET radiation are markedly dose–rate (or dose–fractionation) dependent, with a higher dose required to yield the same degree of effect as the dose rate is reduced. The phenomenon is due to recovery† of cells from sublethal damage. No such dose–rate dependence is seen with very high LET radiation, that is, no cell recovery is possible. The distributions of dose in depth in tissue that can be achieved with high LET radiations are, in general, different from the distributions obtainable with low LET radiations, and with some high LET radiations it is possible to deposit energy in highly discrete anatomical locations. These differences in effect between high and low LET radiations, and their possible advantages and disadvantages with different applications in radiotherapy, are discussed below.

II. CELLULAR RESPONSES TO EXPOSURE TO HIGH LET RADIATIONS

A. Relative Biological Effectiveness (RBE)

The degree of biological radiation damage is not only a function of the absorbed dose but depends strongly on the way energy is transferred, that is, on LET. It can be appreciated intuitively that the thin line of sparsely located ions produced by β particles should affect tissue in a different way than the bulky clusters caused by α particles of heavier ions. Several "hits" from low LET particles might be necessary to inactivate a cell beyond

---

* The ratio of the dose of low LET (standard x or γ radiation) required to yield a given degree of a biological effect, to the dose of high LET radiation to yield the same degree of the same biological effect, is known as the RBE of the high LET radiation.

† "Recovery" is frequently employed to indicate the ultimate survival of a single cell after it has sustained sublethal damage from low LET radiation. "Tissue repair" usually refers to restitution of a cell population or of a tissue. It may involve cell recovery, but involves, in addition, the division of surviving cells.

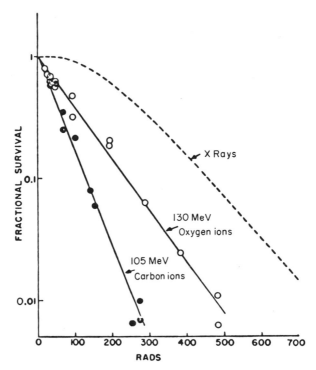

FIG. 2. Effects of heavy ions on human cancer cells. Sublethal damage and recovery indicated by the shoulder of the x-ray curve does not take place in case of carbon and oxygen ions (single-hit exponential response curves). The biological effectiveness of the ions in comparison to x rays (RBE) is a function of dose. At the 200-rad level, carbon ions are about three times as effective as x rays. Oxygen ions of 130-MeV energy are less effective than 105-MeV carbon ions because part of the energy they lose is wasted [from R. A. Deering and R. Rice, *Rad. Res.* **17**, 774–786 (1962)].

repair, but only one hit of a high LET particle might suffice to cause the same effect. Thus, in general the biological effectiveness increases with increasing LET. X rays of 250 kVp or ⁶⁰Co γ rays are usually chosen as a (low LET) reference radiation.

The differences in effect between high and low LET radiation can perhaps be shown most easily by comparing the "survival curves" resulting from the irradiation of cell populations in tissue culture. In Fig. 2 the types of curve are shown where fractional survival is plotted as a function of radiation dose in rads. The curve obtained with x rays can be divided conveniently into two regions, the so-called "shoulder" portion, and the portion that appears to be exponential. The shoulder region implies sublethal cell damage and recovery which will be discussed below. The apparently exponential portion implies that the region of $(n-1)$ "hits" has been reached, and that only a single additional "hit" is required for lethality. Extrapolation of the exponential portion of the curve to the ordinate yields the "extrapolation number." No shoulder region is seen with high LET radiation; the curves are exponential within limits of error (Fig. 2, curves for carbon and oxygen ions). The extrapolation number is one.

The concept of RBE can be seen easily from Fig. 2. If one considers the level of 50% survival with x rays and carbon ions, then the ratio of the dose of x radiation required for this effect (250 rads), to that of high LET carbon ion irradiation to produce the same effect (50 rads) is the RBE, and is equal to 5. It will be apparent, however, that the RBE is a function of dose, or of level of degree of biological effect. At high doses (high degree of effect), the RBE is relatively low and in general does not vary appreciably with dose. At lower doses, however, x rays become

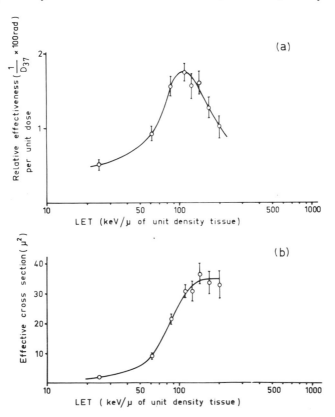

FIG. 3. Relative effectiveness per unit dose (a) and per particle (b) of α particles for inhibition of clone formation of human kidney cells in culture.

markedly inefficient because of sublethal cell damage and subsequent recovery, and thus the RBE increases with decreasing dose or level of effect.

With radiations of moderately high LET, the curve is exponential (Fig. 2, curve for carbon ions). With further increase in LET, the slope may decrease (curve for oxygen ions). This is presumably because a region of "overkill" has been entered, and part of the energy deposited will be wasted. Thus, the LET–RBE relationship curves pass through a maximum (Fig. 3). The RBE maxima obtained with different biological systems do not necessarily coincide, although most maxima fall between 50 and 250 keV/μ. It has been observed that the maxima are located, at least in some systems, at about the LET value where the dose effect curve becomes a single exponential [3].

B. Cell Recovery, Dose Rate Effects

If the total time for delivery of a given dose of low LET radiation is increased, either by decreasing the dose rate or by delivering the radiation at the same dose rate but in fractions separated in time, the degree of the biological effect is reduced (the dose required to cause the same degree of effect is increased). This is due to sublethal damage from which recovery is possible, which was previously referred to in Section II.

The phenomenon of cell recovery has been studied extensively, and the quantitative relationships between the time intervals separating dose fractions and the degree of recovery have been worked out. If the surviving cells from a dose of radiation on the exponential portions of the curve are allowed to "recover" for several hours and are then subjected to additional radiation, a shoulder on the curve indistinguishable from that seen in the irradiated pristine population will reappear. Cell recovery undoubtedly plays an important role in radiotherapy with low LET radiations, although the exact contribution is not clear. The fact that protracted or fractionated exposures appear to be more suitable in radiotherapy than are single exposures may involve the cell recovery phenomenon, although it may well involve cell population repair as well. There is evidence to suggest that feedback mechanisms allow normal tissue to repair more rapidly than would tumor tissue which presumably lacks feedback mechanisms. Thus, with fractionated exposure the tumor tissue would be more damaged than normal tissue because of the intervening repair of normal tissue.

As a result of the characteristic differences in the cell survival curves

(Fig. 2), effects of fractionation or dose rate would be expected to be considerably less with high LET radiations than with x or γ radiation. The degree to which this might be expected to influence the efficacy of radiotherapy is known to depend on a number of factors [4, 5]. Precise data, not now available, on the relative response of normal and tumor populations in question must be known before accurate evaluation of the overall effect would be possible.

C. The Oxygen Effect

It was observed by Gray [6] that cells exposed in the relative absence of oxygen are more resistant to radiation than are the same cells exposed in oxygen. The ratio of doses to reduce a cell population to a given level (usually to 50%, or to $1/e$), when exposed in the absence of oxygen as opposed to exposure in an oxygen atmosphere, is referred to as the oxygen enhancement ratio, or the OER. The maximum OER obtainable is approximately 3.0. Oxygen enhancement ratios versus LET are shown in Fig. 4. With low LET radiation (x rays) the OER is of the order of 2.7 or 2.8. At 5.6 keV/μ the ratio is still about 2.6, and about 2.0 at 60 keV/μ. A ratio of one is not approached until LET values as high as 165 keV/μ are achieved. Thus, relatively high LET is indeed required if the "oxygen effect" is to be eliminated entirely.

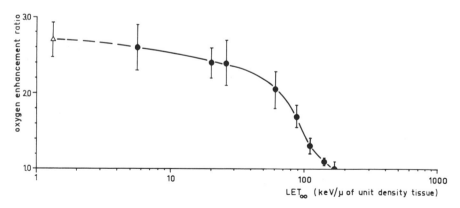

FIG. 4. Effect of oxygen on the effectiveness of ionizing radiations for proliferative capacity of human kidney cells. When the energy loss of the radiation (LET$_\infty$) is larger than about 100 keV/μ of wet tissue, then anoxic cells are equally affected as oxygenated cells (the oxygen enhancement ratio is reduced to unity).

III. HIGH LET RADIATIONS IN RADIOTHERAPY

A. Fast Neutrons

The first densely ionizing radiations used in radiotherapy were fast neutrons, and patient trials were carried out in the late 1930's before the oxygen effect and the lack of cellular recovery following exposure to high LET radiations were appreciated [7]. After a careful series of radiobiological experiments in which the RBE of the neutrons used (maximum energy of about 20 MeV; mean energy of approximately 7–8 MeV) was determined both in animals and on the skin of the human being. The doses to be used in patient treatment were calculated on the basis of the RBE values so determined, and sizable series of patients were treated. This approach to radiotherapy was discontinued because very severe skin reactions resulted.

The reasons for these early failures are now appreciated, and the poor results in terms of severe damage to the skin were undoubtedly due to the differences in recovery (and perhaps repair) rates of tissues following exposure to high versus low LET radiations. The treatments were carried out, as was and is customary with low LET radiations, in fractions separated by days. The doses of x and γ radiation delivered in a fractionated pattern were scaled up over what might be given in a single exposure, on the basis of the well-documented recovery rate (dose rate of fractionation dependence) of human tissues, particularly the skin. The RBE values determined for the neutrons to be used were obtained with single exposures to x rays and neutrons, and the dose of neutrons to be given in a fractionated pattern was scaled up on the basis of dose rate or fractionation factors determined from studies with x ray γ radiation. Because of the lack of recovery or repair from high LET radiations, the RBE values used in calculating the neutron dose should have been higher than those determined with single exposures. This resulted effectively in overdosing the patients, with the resultant serious skin damage observed.

Interest in the possible use of fast neutrons was rekindled with the appreciation of the oxygen effect and its possible bearing on the efficacy of radiation therapy. As a result, a large amount of radiobiological experimentation has been done, and actual patient trials are being started in England.

Fast neutrons are attractive for radiotherapy because they will deliver relatively high LET radiation at appreciable depth in tissue. Fission spectrum neutrons yield central axis depth–dose curves roughly similar to those obtained with 250 kVp moderately filtered x rays. The degree of penetration improves, of course, with increasing energy. The dose to tissue from neutrons in the few MeV energy range is delivered almost entirely by recoil protons from knock-on collisions with hydrogen, with a minimal contribution from γ radiation. As the kilo electron volt range is approached, appreciable thermalization may occur in tissues with a resulting marked increase in the gamma contribution to dose from capture by hydrogen. At approximately 15 MeV and higher, the contribution from heavy nuclei other than hydrogen may be appreciable. The RBE varies with neutron energy, and this relationship for one system of mammalian cells is shown in Fig. 5.

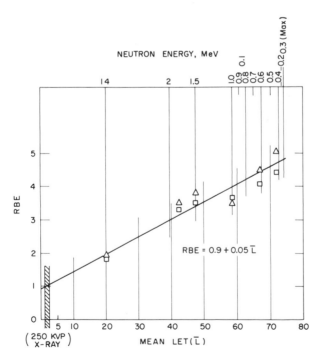

FIG. 5. RBE of monoenergetic neutrons for spleen and thymus weight loss in the mouse. The values of mean LET are calculated.

It was thought initially that the OER, like the RBE, should be a fairly strong function of neutron energy, and there is little doubt that this could be shown to be the case if one dealt with relatively narrow bands of neutron energy. Practically, however, one deals almost entirely with broad spectra of neutrons, and recent work has shown that the OER remains remarkedly constant with different incident neutron spectra, with average energies ranging from the keV regions up to 14.7 MeV, and perhaps even higher. It was thought initially that one would have to compromise between the most desirable OER (obtainable with low-energy neutrons) and optimal penetration (obtainable with high-energy neutrons), and it was thought that spectra with a mean energy of around 6 MeV might represent a suitable compromise. The demonstration of the relative invariance of OER with incident neutron spectrum has now shifted emphasis to 14.7 MeV or higher energies which yield the most desirable depth–dose pattern and the highest tumor-to-normal tissue dose ratio.

Studies on patients are not as yet sufficiently advanced to allow a statement with respect to how successful the treatments may be. The oxygen-enhancement ratio obtainable with fast neutrons, 1.6, is of course more favorable than the 2.5–3.0 obtainable with x or γ radiation, but is not ideal. In addition, while fast neutrons are indeed penetrating, the collimation possible and the resulting depth–dose patterns are less suitable than those obtainable with x or δ radiations. In general, for a given dose to the tumor, more normal tissue must receive a higher dose than is necessary with x or γ radiation. The degree to which this may balance the gain from the more favorable OER remains to be seen. At any rate, the approach definitely deserves a clinical trial. If clear-cut improvement is seen, it will serve to stimulate the use of other approaches to selectively irradiating tumors with very high LET radiations.

In the discussions above, the use of fast neutrons from sources outside of the body has been discussed. With the advent of the availability of transuranic elements, an avenue is open for the possible use of fast neutrons for interstitial therapy similar to that accomplished with radium needle implantation. A number of transuranic elements fission spontaneously, and can be made into needles similar to radium needles. A suitable element in terms of half-life is californium (^{252}Cf). This approach might be quite useful with certain types of tumors, such as neoplasia of the cervix or of the head and neck region. In principle, with implantation therapy, one would avoid damage to intervening tissues that must be irradiated with the use of external beams. The fast neutrons resulting from the spontaneous fission constitute high LET radiation, and therefore should aid to a degree problems resulting from the presence of hypoxic neoplastic cells in tumors.

B. Thermal Neutrons (Neutron Capture Therapy)

Thermal or epithermal neutrons in themselves are not at all attractive for purposes of radiotherapy. The thermal neutron flux from a beam of thermal neutrons impinging on tissue falls off quite rapidly, with a half-value of 2 cm or less. The dose to tissues is produced almost exclusively by recoil 0.6-MeV protons from the $^{14}N(n, p)^{14}C$ reaction (high LET radiation), and from hydrogen capture yielding a 2.2-MeV γ (low LET radiation). Thus, at depths of even a centimeter, the bulk of the dose is delivered by low LET radiation, with a resulting high OER. Irradiation of a large tissue mass with thermal neutrons may be thought of as a rather exotic method of exposing the tissues to high-energy γ radiation.

Thermal neutrons may be of use theoretically, however, in the approach known as neutron capture therapy, involving a "capture element" with a high cross section for thermal neutrons. The capture elements most commonly employed have been ^{10}B, although ^{6}Li and higher Z fissionable nuclei represent theoretical possibilities. If such a capture element could be incorporated into a compound that could in turn be incorporated into a tumor in amounts far exceeding that in adjacent normal tissues, and if both normal and tumor tissues could then be exposed to thermal neutrons, the tumor would be selectively irradiated. With ^{10}B, most of the energy is liberated in the form of an α particle and a ^{7}Li nucleus. The range of these very high LET particles is of the order of microns in tissue, and thus in principle one might attain highly selective exposure of the tumor with sparing of the normal tissues.

Therapeutic trials of this new clinical approach were initiated at Brookhaven and at the Massachusetts General Hospital in Boston in the late 1950's and early 1960's. Inorganic ^{10}B compounds were used, and the principal tumor employed was glioblastoma multiforme of the brain. Selective deposition in the tumor area was expected because the so-called blood–brain barrier limits severely the amount of this foreign substance that can gain access to the normal brain. The blood–brain barrier apparently is impaired in tumor tissue, and the material accordingly can gain access to neoplastic tissue. The medical reactor at Brookhaven was designed specifically for this type of therapy, and patients were treated both in the large graphite reactor and in the medical research reactor at Brookhaven. The results were disappointing in that survival time in the treated patients was minimally prolonged, if at all. Serious complications in the form of severe skin damage and brain edema were encountered. Therapeutic trials were discontinued in 1962, and a review of the principal reasons for the complications was instituted to explore if improvements could be effected.

The principal reason for the complications probably was ascribable to the poor penetration of a thermal neutron beam in tissue. Thus, tumor tissue located at 8 cm in depth receives a fluence of thermal neutrons from 20 to 60 times less than that of the intervening skin, depending on the precise geometry. A very large ratio of boron in tumor to that in skin would be required to overcome this severe disadvantage. Another factor relates to the amount of boron remaining in the blood stream at the time the gross ratio of boron in the tumor to that in normal brain appears to be favorable. The ratio varies in time after intravenous injection of the boron compound, and appears to be most suitable at approximately $\frac{1}{2}$ hr after injection. At all times when this favorable ratio exists, however, the absolute concentration of boron in the blood is higher than that of either the tumor or normal tissue. Thus, if exposure to thermal neutrons is effected at the time of the most suitable ratio, the walls of the blood vessels in both normal and tumor tissue will receive a dose of radiation comparable to or greater than that of the tumor. This may well account for the hemorrhages that were seen in the brain of patients so exposed.

A number of attempts have been made to correct the difficulties encountered, and attempts to improve both the physical or dosimetric aspects of the procedure and the degree of localization of capture elements in tumor tissues have been made. In order to overcome the high degree of attenuation of thermal neutrons in tissues, "epithermal" or higher energy neutrons have been investigated. So-called epithermal neutrons may be obtained from the thermal port of a reactor by placing a filter of ^6Li or cadmium in the thermal neutron beam. With the effective removal of thermal neutrons, the remaining beam is composed of neutrons of higher energy that can be allowed to thermalize in tissue. The depth–flux pattern obtainable with such neutrons is much more satisfactory than that obtained with thermal neutrons alone, and the thermal neutron flux at several centimeters in depth may even be appreciably higher than at the skin surface. With the epithermal beam from the reactor, the flux at 3 cm depth is approximately 1.5 times that at the surface, and the flux at 6 cm is equal to that at the surface.

In order to obtain better relative concentrations in neutron capturing substances in neoplastic tissue, a large number of investigations have been initiated in recent years [8, 9]. It has been noted that serum protein has a tendency to localize in tumors. Boron can be incorporated into proteins in the form of arseno–azo proteins, and boron concentrations in the order of 40 $\mu g/g$, comparable to those obtained within organic boron compounds, appear to be possible. These studies have been done in animals; it is not yet clear the degree to which such an approach may be useful in the human being. Also it is not clear exactly where the foreign protein is deposited. Inorganic boron compounds gain entrance into the cell; proteins containing boron may be in the cell, or may be localized in the supporting structures between cells. Attempts to incorporate boron into other compounds such as uracil have been made with no success reported to date.

Further remarks are in order with respect to neutron capture therapy and possible advantages in radiotherapy from the use of densely ionizing radiation. The recoil α and ^7Li nuclei from neutron capture of boron constitute very high LET radiation. Thus, it might be thought to be advantageous with respect to the problem of hypoxic cells in tumors. A question remains, however, as to whether or not the boron, and therefore the densely ionizing recoil particles from thermal capture reach the hypoxic cells. The blood supply to these areas is quite poor, and there is a real question as to whether even oxygen, under hyperbaric conditions, finds its way to these areas in any appreciable concentration. Thus, it is doubtful that the procedure, at least with a single exposure, will contribute to the solution of the oxygen problem. There is some evidence that, with repeated exposure, the blood supply may improve as a result of the net decrease in the mass of tumor cells. Thus, with repeated exposures, boron distribution might be expected to be more uniform. On the other hand, the distribution of oxygen could be expected to be more uniform as well. Procedures have recently been developed to determine the microdistribution of boron in tissue by observation autoradiographically of the recoil heavy particles from capture. By this technique it should be possible to evaluate whether or not the high LET radiations from neutron capture therapy will be of use with respect to the oxygen problem.

C. High-Energy Proton and α Particles

Proton and α particle beams are treated separately from beams of higher Z nuclei for two principal reasons. To date the only charged particle beams that have been available with energies sufficiently high to achieve the several centimeters penetration in tissue required for depth radiation therapy have been of protons, deuterons, and alpha particles. In addition, these particles when accelerated to energies necessary for adequate penetration represent low LET radiation through most of their path in tissue, and thus they are minimally attractive with respect to the LET characteristics.

The potential advantages of proton and α particle beams can be summarized as follows: they represent a high-precision instrument allowing the exact localization of energy deposition in tissue. Because there is minimum lateral scatter, the separation between irradiated and nonirradiated tissues can be quite distinct. Beams of almost any diameter can be obtained. At the end of the depth–dose curve in tissue from such beams (the "Bragg curve") there is increased energy deposition (Fig. 1), and thus the dose at the end of the range can be appreciably higher than at the entrance or skin surface. The width of the curve at half-height is only of the order of 1 or 2 cm, however, and thus advantage of this increase in dose at the end of the range can be utilized only if the target biological structure is of similar dimensions. Most tumors that present radiotherapeutic problems are much larger in size. Broadening of the "peak" by differential filtration or by back-and-forth scanning of the beam can be done only at the expense of lowering the differential dose advantage.

The LET at the end of the range of proton or α particles increases and accordingly the RBE goes up. Thus, the "effective dose" (dose $\times$ RBE) advantage in the Bragg peak is greater than the physical, or absorbed dose advantage. High LET's are achieved only at the very end of the range, however, and thus the differential in terms of "effective dose" (absorbed dose times RBE) is still inadequate to allow utilization of the dose differential for therapy of sizable tumors.

With increasing LET at the end of the range, the oxygen enhancement ratio would be expected to decrease. However, the OER, as with the RBE, changes appreciably only at the extreme end of the range, and therefore these beams are of no practical value with respect to overcoming the oxygen problem.

Charged particle beams in radiotherapy were first utilized at the University of California [10, 11] and protons, deuterons, and α particles have been employed for this purpose. Additional studies have been done using the cyclotrons at Harvard and of the Gustav Werner Institute at Uppsala in Sweden.

Initial therapeutic trials involved exposure of the pituitary as a form of palliative therapy in patients with cancer of the breast. It was known that ment in an appreciable number of such patients whose disease was not obliteration of the pituitary or its removal resulted in significant improve-amenable to other types of treatment. The beam was ideally suited for this purpose, since the pituitary is a small gland of the order of 1 cm in diameter, located in the central region of the cranium. Its precise location is easily determinable in the intact cranium, by reference to adjacent bony landmarks that are readily observable radiographically. Thus, by stereotactic procedures the organ could be precisely localized, and the beam could be aimed precisely at the target area. Maximum dose to the organ with minimal dose to adjacent normal structures and the skin was obtained by rotational techniques, such that the pituitary was at all times in the beam while a given skin area was in the beam only during the brief time the scan traversed that position.

Results were comparable to those obtained either by surgical extirpation of the gland or by destruction of the gland by direct injection of radioactive isotopes into the region of the gland. The procedure, of course, carried no surgical risk, and it can be used on patients in whom any procedure involving a surgical intervention would be precluded. Initial difficulty was encountered in the form of injury to cranial nerves whose paths lay closely adjacent to the pituitary. These difficulties appeared to have been largely overcome by improvements in methods of beam localization, and by reduction of the total dose administered to the gland.

Irradiation of the pituitary gland is also beneficial in other diseases, including diabetic retinopathy and acromegaly. Definite benefit in patients so treated has been achieved at Berkeley.

In essentially all of the above-quoted studies, the beam has been allowed to traverse the entire cranium, and little effort has been made to take advantage of the "peak end" of the Bragg curve. In principle, a larger relative dose could be delivered to the pituitary if the beam were directed such that the Bragg peak would coincide with the location of the gland. An even greater differential could be obtained by additional scanning and cross-fire techniques.

The possible use of the Bragg peak has been suggested for other diseases in which it is desirable to damage only a very small volume of the brain. Such diseases include Parkinsonism. At Berkeley two patients have been irradiated to date for this disorder with apparently appreciable improvements in one.

The beam has been used also for the treatment of intracranial and other tumors. The beam has the advantage over surgery with respect to intra-cranial neoplasms, in that apparently healing from radiation damage occurs without scar formation. Use of these beams for therapy of neoplasia has not been extensive, and it remains to be seen what place their use will occupy in the overall armamentarium against neoplasia.

With the use of the beams, precise localization of the lesion and of the beam as it passes through tissue is mandatory. This requirement is less stringent with x and γ radiation, since deposition of energy from these beams is relatively diffuse. Cranial lesions lend themselves to this type of approach, since regions of the brain and tumors frequently can be localized quite precisely relative to bony landmarks. With tumors in soft tissue, the problem is compounded appreciably.

D. Beams of High-Energy Heavy Ions

If heavy ions could be accelerated to energies permitting penetration of many centimeters in tissue, a number of advantages could accrue in principle. Because of the possibility of precise focusing, and because of minimal lateral scattering, energy could be deposited in precisely defined tissue volumes, as is possible with protons or α particles. The principal advantage with respect to radiotherapy would involve the capability of irradiating tumors in depth with very high LET radiations, perhaps allowing one to deal effectively with the problem of hypoxic neoplastic cells within tumors.

Radiobiological experiments have been done with beams of heavy nuclei accelerated to only a few million electron volts per nucleon, and allowing studies only with cells in culture. Such studies indicate that the oxygen effect is markedly reduced with accelerated boron and carbon nuclei, and that it is essentially eliminated with accelerated nitrogen, oxygen, neon, and argon nuclei. These beams thus may represent a very attractive approach to overcoming the problem of hypoxic tumor cells. However, beams with sufficient energy to allow useful penetration are not now available, but are expected to be available in the future. A large synchrotron, the Omnitron, has been planned in Berkeley, designed to accelerate heavy nuclei to energies up to 500 MeV per nucleon.

The degree to which one might expect a Bragg peak at the end of the range of very fast high Z nuclei is not clear. Heavy ions with a range of the order of 10 cm of tissue have a high probability of causing nuclear interactions, and contributions of secondary particles originating from such events have to be considered in the total ionization curves of heavy ion beams. An appreciable number of the secondary particles are produced in stripping reactions, which have essentially the same velocity as the primary particles, and penetrate deeper because of their smaller charges. For primary particles of charge 20 and a range of 10 cm in tissue a broadening of the ionization "peak" to some 15 cm or more has been estimated. Another effect which must be expected to have a strong impact on the ionization curves of heavy ions is the increasing rate of electron pickup when the velocity of the ions approaches low values [12, 13]. In Fig. 6a is shown how the range extension, which results from the increasing neutralization of the charge, increases with the mass of the ions. The deviations of the effective charge from the full charge of the particles, as a function of the reduced velocity, are shown in Fig. 6b. The rate of energy loss—after having increased with decreasing particle energy—decreases again with further decrease of energy because of the electron pickup. This reduction in energy loss results

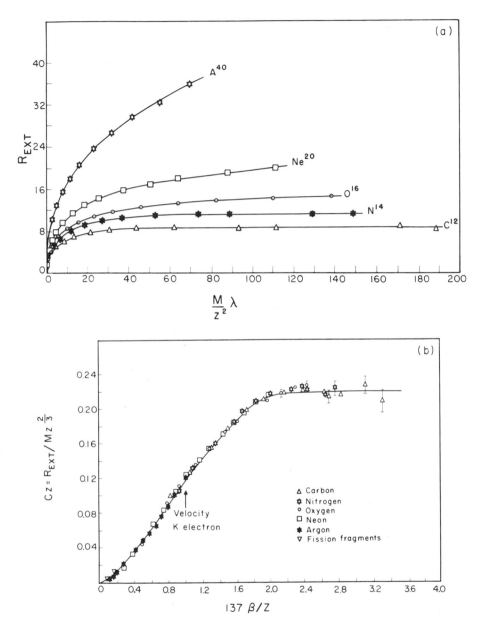

FIG. 6. (a) Range extension of heavy ions caused by partial neutralization of the ion charge; plotted on the abscissa is the depth the ions would reach if no neutralization took place. (b) Effective charges Z^*/Z which correspond to observed values of dE/dx, plotted vs. the reduced velocity of the ions $\beta/Z^{2/3}$.

in elevated values of residual range, and thus a smear of ending tracks is to be expected rather than a well-defined range. Also, it is not clear to what degree LET may cease to track dE/dx as heavy ions are accelerated to very high energies. Higher energies of the primaries produce more energetic δ rays with a net decrease in LET. It is probable, however, that Z and energy can be chosen appropriately, such that the LET in the region of a tumor in depth would be high indeed, and the OER considerably lower than the 1.6 achievable with fast neutrons.

E. Negative π Mesons

Several years ago Fowler and Perkins [14] suggested the radiological application of negative pions. This idea is based on the observation of stars (clusters of tracks looking like stars as a result of nuclear disintegration) which are produced when a π meson is captured by one of the nuclei of the absorber material (Fig. 7). An energy of about 40 MeV is necessary to break up the disintegrating nucleus. The remainder of the pion rest mass energy of 140 MeV is released in the production of several charged particles and 2–3 neutrons per capture process. In an "average" capture process, more than 50 MeV are carried away by fast neutrons, and a total energy of 20–30 MeV is made available locally in the form of protons, α particles, and heavier ions such as Li, B, and Be nuclei. This local energy release would enhance the "Bragg peak" * effect, since the probability of a capture taking place is appreciably higher for slowed-down pions in the peak area than for fast pions in the surface area of the absorbing body. The heavy ions and low-energy α particles produced represent high LET particles with the biological advantages mentioned above.

Several machines in the United States are capable of producing π^- mesons; however, the intensities are low. Dosimetric studies can be done; however, only limited radiobiological studies are possible. One machine in Zurich [15] may be used to produce high-intensity beams; a "meson factory" planned for the Los Alamos Scientific Laboratory would provide a high-output source. To date dose rates of order of 0.1 rads/min or less have been available [16].

It is difficult to produce beams without considerable contamination with μ mesons and electrons. The electron contamination can be reduced by using proton energies of the order of 500–600 MeV, which produce charged mesons energetic enough for biomedical work but avoid the large production

* The total ionization–depth curve obtained with π^- mesons is not strictly a dE/dx vs. x, or Bragg curve because it involves nuclear processes. However, for convenience it is referred to loosely as a Bragg curve.

of π^0 mesons with their fast decay and γ-electron cascades. Problems of beam intensity and contamination arise, furthermore, from the short mean survival time of charged pions, of about 10^{-8} sec. A flux of 70-MeV particles, with a range of 15.6 cm in water, is reduced by a factor e if the flight path from the π-producing target to the first counter or biological sample is

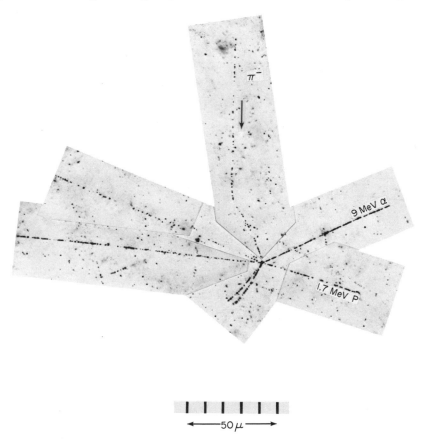

FIG. 7. Star produced by a negative π meson. The emulsion had been soaked in water prior to exposure in order to measure pion capture by oxygen nuclei [photograph courtesy of P. H. Fowler, *Proc. Phys. Soc.* **85**, 1051 (1965)].

8 m long, and an appreciable number of muons and electrons from pion decay inevitably populate the beam. The main difficulty, however, results from the interaction of the beam with the biological absorber itself. Additional decay, multiple scattering, and early capture cause further pion losses and reduce the number of pions reaching the Bragg peak area to an estimated 40% of the flux at the absorber surface.

Calculations of the ionization which can be expected from the energy losses of such a beam in the target area [17] result in relative Bragg peak heights of 2–3, which is in good agreement with measured values (Fig. 8).

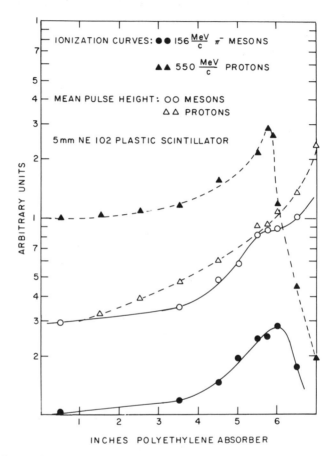

FIG. 8. Comparison of 70-MeV negative π mesons with protons of the same range in water (150-MeV energy). The relative height of the ionization peak is practically the same for both particles. However, the FWHM (full width at half maximum) width of the pion peak is considerably larger. Plotting of the mean pulse height versus depth shows the effects of pion captures in the region of the ionization peak.

The α particles emitted after pion capture at the end of the π^- range have an initial mean energy of 5–10 MeV, corresponding to about 50–100-keV/μ local energy deposition at the beginning of their tracks. The highest value of their LET$_\infty$ when slowing down amounts to about 300 keV/μ. The respective values for 5–10-MeV heavy ions mentioned before are appreciably higher, with peak values of more than 1000 keV/μ. These high LET components are largely responsible for the fact that, for certain biological indicators, appreciably greater effects have been demonstrated in the area of stopping pions than in the neck of the ionization curve [18].

In comparing the possible use of π mesons versus accelerated high Z particles, the two would appear to be complimentary. The π mesons may have an advantage over high Z beams with respect to geometry in that the mesons may provide a sizable region near the end of the range in which the absorbed dose and "effective dose" are appreciably higher than in the plateau portion of the absorption curve. The OER determined in the end of the range portion of the pion absorption curve will be lower than for x or γ radiation, but definitely will not approach 1.0. With high-energy beams of high Z particles, it is doubtful that a large advantage will accrue from the "Bragg curve" with respect to differential dose deposition in the peak region relative to the plateau portion of the absorption curve. On the other hand, it is quite probable that energy can be deposited in sizable tumors with radiations of sufficiently high LET such that the OER will be lower. Thus, these two approaches to radiotherapy have different attributes and possibilities and both deserve further exploration.

In summary of the attractive features of a π^- beam for use in radiotherapy, the beam appears to have advantages over currently employed low LET beams in terms of combining a relatively (though not optimally) low OER limited to the peak portion of the depth–dose curve, and a depth–dose pattern more favorable than those now available. The favorable OER in itself, however, does not constitute strong grounds for making available a high-intensity source of π mesons. The reasons for this are as follows. Other currently available sources of high LET radiation, namely, fast neutrons, should be adequate to evaluate the importance of the problem of hypoxic tumor cells in radiotherapy and the degree to which high LET radiations may help to overcome the problem. Also, fast neutron sources can be made available which compare favorably in terms of central axis depth–dose curve, isodose contour patterns, and the dose rates with those of currently used ^{60}Co therapy machines. Thus, should it be assumed or proved that hypoxic tumor cells do indeed constitute a severe limitation in radiotherapy that can be ameliorated by use of high LET radiations, fast neutron beams can be used.

The degree to which the depth dose pattern obtainable with a π^- beam would constitute an advantage over currently available sources has not yet been adequately evaluated. The advantage depends on the location of the limiting normal tissue in radiotherapy that prevents the delivery of an adequate dose to the tumor. If this limiting normal tissue is that immediately adjacent to the tumor and unavoidably exposed in any beam, then negative

π mesons would provide little if any advantage over fast neutrons. If, however, the limiting normal tissue is frequently that at the entrance (skin), or tissues intermediate between the entrance and those immediately adjacent to the tumor, then the π meson beam might provide for a distinct advantage in radiotherapy. Also, should the minimal exit dose obtainable with the π^- beam permit the delivery of significantly increased dose to a number of tumors, with sparing of otherwise unavoidable damage to vital structures such as the eye, bowel, or spinal cord located "behind" the lesion, then the new modality would be attractive.

Should damage to entrance, intermediate, or exit tissues be a severely limiting factor in a large number of cases, then it would be of considerable importance to develop a high-intensity π^- beam. The relatively low OER should be taken into account additionally. The low OER in the peak region of the π meson depth–dose curve, while not decisive in itself in terms of justifying a high-intensity beam, would tend to favor its development if the assumption is made that hypoxic tumor cells constitute a serious problem in radiotherapy.

With respect to a beam of negative π mesons for radiobiological studies, more extensive and definitive investigations are needed if such a beam is to be considered seriously for radiotherapy. With respect to radiobiological studies not directed to an evaluation of the properties of the beam in connection with radiotherapy, probably only very limited use of the beam would be made. It might have some use in mammalian radiobiology in providing the capability of depositing high LET radiation somewhat selectively in well circumscribed tissue volumes and depth. However, the beam represents a mixture of radiations of widely varying energy and LET. It would thus seem to be unsuitable for most studies in radiobiology, in which radiations having discrete energy and LET are desired.

IV. ILLUSTRATIVE RADIOBIOLOGICAL EXPERIMENTS

A number of factors must be considered in the evaluation of whether or not a particular type of radiation will be advantageous in radiotherapy, compared to the more conventional x rays, electrons, or γ rays such as those from ^{60}Co. Some physical factors include the spectrum of energy and penetration characteristics, which allow calculation or measurement of the "advantage factor." * An evaluation of the quality of the radiation, usually expressed in terms of the spectrum of stopping powers or of LET's

* The ratio of the dose received by the target tumor tissue, to the maximum dose unavoidably delivered to normal tissues in the process of the irradiation.

allows estimate of values for the relative biological effectiveness (RBE) of the radiation, and of the oxygen enhancement ratio (OER). Definitive determination of those values depends on radiobiological experiments.

Radiobiological experiments to yield so-called "dose–effect" curves are necessary to provide pertinent information on cell survival, not only in the tumor but in normal tissues that may be affected as well. Knowledge of the effects of radiations on tissues, as determined by the techniques of pathology and histology is necessary. The RBE of the radiation proposed for use in radiotherapy must be determined in a number of biological systems as well as the OER. The RBE and OER must be determined as a function of dose rate, since the dose rate dependence differs as a function of radiation quality.

In the following paragraphs illustrative examples of radiobiological experiments that shed some light on the above factors that enter into the determination of whether or not a given radiation may be advantageous in radiotherapy are outlined briefly. It is rarely possible to evaluate in the system of most interest, that is, human normal and neoplastic tissues, and it is therefore necessary to work with a number of other systems. If results from a number of systems are generally similar, added confidence is provided in applying the numbers obtained in radiotherapy. The radiobiological experiments listed are by no means inclusive, nor are they necessarily the ones most pertinent in deciding whether or not a given beam would be suitable for radiotherapy. They are intended as illustrative only.

A. Fast Neutrons

RBE values for fast neutrons of different energy spectra have been determined, using chromosomal aberrations in human leukocytes as the criterion of effect [19] for neutrons of mean energy of approximately 14, 2.5, and 1. RBE values of 2.6, 3.1, and 5.6, respectively were obtained. In separate experiments using "fission neutrons," an RBE of about 5 was found for spermatogonial and oocyte killing in the mouse. Monoenergetic neutrons of relatively high LET, obtained with the Van de Graaff generator in which protons were used to bombard a water cooled tritium target were used to determine the RBE of neutrons as a function of energy, and of dose rate [20–23]. For thymus weight reduction, RBE's ranging from 4.3 to 3.0 were obtained for neutrons ranging from energies of 0.4 to 1.8 MeV, respectively. Corresponding RBE values for spleen weight reduction were 4.6 and 3.2, respectively. Spermatogonia depletion gave RBE values of 5.5–3.4.

A series of experiments was done at the Hammersmith Hospital in London, making use of the neutron spectrum produced by deuterons striking a beryllium target in the Medical Research Council Cyclotron. The neutron spectrum extends in energy from 0 to approximately 17 MeV, with a mean of approximately 6 MeV (mean LET about 20 keV/μ in tissue). For erythema of pig skin, an RBE of 2.5 (comparison radiation, 8-MeV x rays) was found for single doses, and a value of 3.3 for doses delivered in six fractions over 17 days [3]. Of great importance, and probably responsible for earlier failures in attempts to use fast neutrons in radiotherapy, is the fact that a dose of fast neutrons given in two fractions yields essentially the same degree of effect as does the same dose delivered in a single exposure. With fractionated exposure to x rays, an appreciable increase in the total dose necessary to yield the same degree of biological effect is required. Experimental evidence of these relationships in the mouse are shown in Fig. 9 [24]. The reduced dependence on dose fractionation of the effects caused by fast neutrons as opposed to x rays results in an increase in the RBE with decreasing dose rate, a pertinent consideration in the clinical application of such particles.

A further example of a difference in degree of effect with respect to neutrons and x rays, and as a function of dose rate is shown in Fig. 10

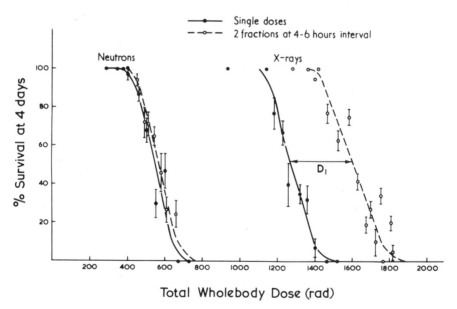

FIG. 9. Effect of dose fractionations on irradiations with low LET (x rays) and high LET radiation (fast neutrons). Two doses separated by a few hours hardly affected the number of mice surviving four days after total-body neutron irradiation. In the case of x rays, the summed fractionated dose had to be about 30% higher in order to cause the same effect as a single dose.

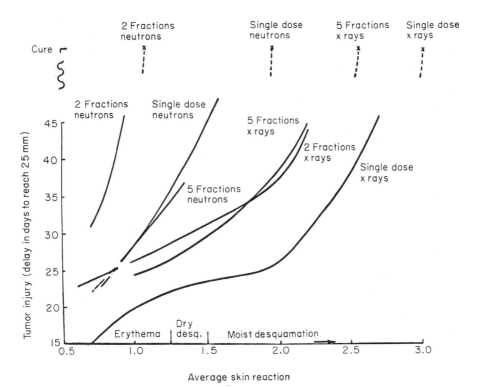

FIG. 10. Optimizing radiation effects on neoplasms: two fractions of fast neutrons appeared to be most effective in causing a delay in the growth of rat tumors, at lowest skin reaction. In this case complete regression of the tumors appears possible with not more than erythema effects [from J. F. Fowler, in "Biological Interpretation of Dese from Accelerator produced Radiation." Conf. 670305, 128 (1967)].

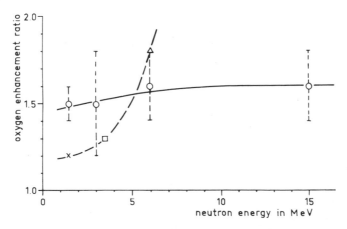

FIG. 11. Oxygen effect as a function of fast neutron energy; for survival of human kidney cells in culture, only a very small increase of the OER has been observed for energies from 1.5 to 15 MeV.

[25, 26]. The response of a fibrosarcoma grown in rats, in terms of the time required for the tumor to reach 25 mm in diameter, is plotted as a function of skin reaction. x Rays and fast neutrons were administered as a single dose, or in doses divided in 2 and 5 fractions. The plot shows the advantage obtainable from neutrons, particularly with dose fractionation. For x rays the most favorable ratio of tumor injury to skin effect was obtained with 2 or 5 fractions. With neutrons a better differential effect as compared with x rays was seen with all degrees of fractionation; however, the most favorable results were obtained with two fractions.

Because the OER depends on LET, with values approaching unity when the LET exceeds about 100 keV/μ (Fig. 4), an OER dependence on the energy of fast neutrons would be expected. Experiments with human kidney cells in culture, however, have provided strong evidence that there is very little dependency of the OER on neutron energy. In Fig. 11 are shown values between 1.5 and 1.6 for monoenergetic neutrons of 3 and 15 MeV, respectively, and for neutron spectra of maximal intensities at 1.5 and 6 MeV. Similar contributions to the dose by the high LET component of the four different LET spectra may be the cause of this effect. The lack of appreciable variation of the OER with neutron spectrum is of considerable importance for therapeutic work since it appears to be possible to obtain the favorable depth–dose characteristics of neutrons of 10 MeV and above with low values of the oxygen enhancement ratio.

B. Thermal Neutrons (Neutron Capture Therapy)

The techniques for clinical applications of thermal neutrons are presently being reappraised and improved. In treatments using the neutron capture therapy technique, the "capture compound" used is most frequently administered by intravenous injection some 5–30 min prior to irradiation with thermal neutrons. Boron in the form of sodium perhydrodecaborate or pentaborate decahydrate has been used frequently. The concentrations have ranged in the order of 10–50 μg of ^{10}B per gram body weight [27–29]. Basic radiobiological data required to evaluate the effectiveness of this mode of therapy include: the effectiveness of the neutrons alone in the absence of an injected "capture compound," the effectiveness of the combined radiations from the neutrons and the products of neutron capture on normal and neoplastic tissue, the microdistribution of boron in order to evaluate further the microdistribution of dose from the capture process, and if possible, the OER for the radiations from the neutron capture process.

Comparative studies of the effects of thermal neutrons after injection of ^{10}B and ^{6}Li compounds, of thermal neutrons alone and of electromagnetic

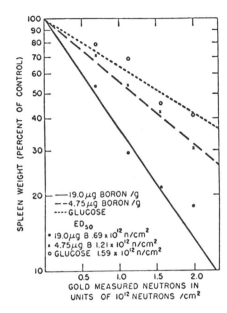

FIG. 12. RBE of thermal neutrons determined from weight loss of the spleen of mice. For the same neutron flux, the effective dose rises with concentration of injected boron. Injection of glucose (no boron) was done for reference [from V. P. Bond, O. D. Easterday, E. E. Stickley, and J. S. Robertson, *Radiology* **67**, 650 (1956)].

radiations, have been done in mice [22, 30]. Large amounts of capture material were injected intravenously, and a very large proportion of the total radiation dose was due to the products of the neutron capture process. RBE values were determined using the biological end point of weight loss in the spleen and thymus, as shown in Fig. 12. The values for RBE are shown in Table I. It was found that an integrated flux or fluence of 0.5×10^{10} neutrons/cm^2 corresponds in effectiveness to 1 R of ^{60}Co γ rays, and 0.63 R of 250 kVp x rays.

Mortality studies in mice were done with both ^{10}B and ^{6}Li capture compounds. Mice so irradiated tended to die early, around 3–5 days. This is in contrast to the time rate of mortality following exposure to x or γ radiation, in which deaths tend to center around the tenth or the twelfth day (Fig. 13). Similar results are seen following exposure of mice to fast neutrons, and since early deaths are usually associated with damage to the gastrointestinal

TABLE I

RBE Values in Studies Using Thermal Neutrons in Combination With Injected ^{10}B
(RBE values are in comparison to 250 kVp x rays)

	Spleen weight	Thymus weight
α particles and ^{7}Li nuclei from neutron capture		
(For 19 μg ^{10}B per gram body weight)	1.1	1.1
α particles only, from neutron capture	1.8	1.7

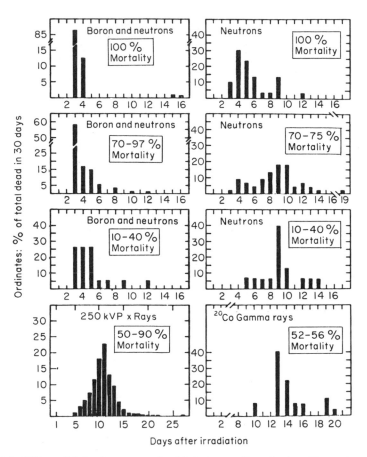

Fig. 13. Differential death patterns for 30-day mortality of mice. Thermal neutrons alone and, even more so, in combination with heavy particles from capture in ^{10}B cause a tendency to early deaths. In the case of electromagnetic radiation the highest death rate is observed about one week later [from V. P. Bond and O. D. Easterday, *Rad. Res.* **10,** 20 (1959)].

tract, the effect is presumed to be due to a selectively greater effectiveness of high LET radiation on the bowel as opposed to the bone marrow and other tissues.

Lithium-6 compounds were found to be more effective than boron compounds, per unit of injected capture element, in producing mortality. Values for the LD_{50} of mice using various treatments were as follows: lithium-injected animals, 0.84×10^{12} neutrons/cm^2, boron-injected animals, 1.30×10^{12} neutrons/cm^2, thermal neutrons alone, 4.77×10^{12} neutrons/cm^2, and ^{60}Co γ rays, 789 R. Although more effective per unit of injected capture element, inorganic lithium is less suitable than inorganic boron for neutron capture therapy because of its higher toxicity.

Experiments were done on the skin of pigs and the cortex of dogs [27], to determine the doses of thermal neutrons alone, and of thermal neutrons in combination with injected capture elements that would damage normal tissues. It was shown that, following the injection of 35 μg of ^{10}B per gram of body weight, an integrated flux of approximately 5×10^{12} particles/cm^2 caused nonhealing necrosis in the skin of pigs and in the brain cortex of dogs. The calculated absorbed dose was approximately 1500 rads. These values serve as limits, beyond which serious difficulty from damage to normal tissue can be expected.

The rate of falloff of thermal neutron flux in tissue depth, using a variety of geometries has been determined. The flux falls off approximately exponentially, with half-value layers of 1.8–2.4, depending on the tissue and the geometry [22]. This rapid rate of falloff constitutes a serious disadvantage, and for this reason "epithermal" neutron beams have been investigated ("epithermal" in this context frequently refers to a well-degraded fission spectrum, such as might be obtained in the resultant beam after a ^{10}B or cadmium filter is placed in the usual external thermal beam from a reactor). A number of studies [8, 31–34] have shown that the depth–flux pattern of thermal neutrons can be markedly improved by this technique, and an example is shown in Fig. 14. Neutrons in the energy range from about 0.3 keV to 300 keV appear most promising in this respect [35]. Both their ability to penetrate and the relative ease of collimating make them attractive for therapy.

A number of efforts are being made to produce a compound to which a capture element can be attached, that will produce a more favorable ratio of capture elements in the tumor to that in the normal tissue, that is now attainable. Experiments with $B_{12}H_{11}SH^{2-}$, paracarboxybenzenoboronic acid and $NaB_{10}H_9NH(CH_2CH_2OH)_2$ have been reported with concentrations of up to 50 μg/g body weight of boron.* Tumor-to-brain concentrations of such compounds of nearly 7/1, for an extended period of time, have been obtained in mouse gliomas. It is not only the ratio of capture element in tumor/capture element in normal tissue that is of importance. With inorganic compounds particularly, under most circumstances the absolute concentration in the blood remains higher than that in either tumor or normal tissues, during the period when the ratio of capture element in the tumor to that in the blood is most favorable. Thus, during irradiation with thermal neutrons, the walls of the small blood vessels of both normal and tumor tis-

* Information on boron compounds can be obtained from a paper by Muetterties and Knoth [36].

sue are irradiated. This fact may well account for blood vessel damage noted in animals and patients treated with the neutron capture therapy technique using inorganic boron compounds.

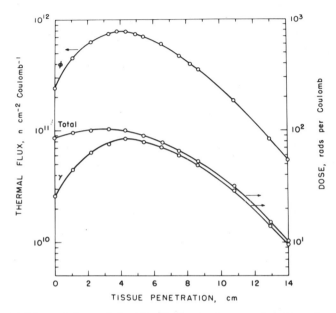

FIG. 14. Thermal flux and dose distribution in tissue produced by a beam of intermediate-energy neutrons (60 keV).

There is evidence to indicate that specific proteins or protein groups may be found to concentrate to a significant degree in tumors. For instance, serum albumin, from an unrelated individual or preferably from another species, is taken up to a significant degree by tumors. It has been shown that significant amounts of boron can be attached to proteins by means of the diazo linkage. Initial experiments with serum albumin have shown a promising uptake of arsenazoproteins by murine tumors, and tolerance for the injected material has been favorable [8]. Many problems have to be solved, however, before this approach can be used in the human being.

C. Protons and α Particles

High-energy charged particles are suitable for accurate irradiation of targets located deep in the body. Thus, patients at the Donner Laboratory in Berkeley have been treated with 910-MeV α particles from the 184-in. synchrocyclotron [11], and protons from the Harvard 160-MeV machine [37] have been used for treatment of patients at the Massachusetts General Hospital. Protons of 187 MeV have been applied at the Gustav Werner Institute at Uppsala, Sweden [38]. Radiobiological information required includes not only the RBE and OER for radiations in the plateau region of the Bragg curve, but values for the region of the peak of the curve as well.

A number of experiments have been conducted to evaluate the effects of the plateau region of the beams, many in connection with studies designed to yield information on the possible hazards of space flight. Whole-body irradiation of animals has been employed frequently, and the usual end point is mortality. Single and fractionated doses have been administered at average dose rates ranging from approximately 1–20 rads/sec. Basic elements of an animal irradiation facility are outlined in Fig. 15 [39].

A number of experiments involving animal mortality have been done at a number of laboratories [40–45], and some of the results are summarized in Table II. For particles of enough energy to be interesting from the clinical viewpoint (penetration in tissue from a few centimeters to perhaps 20 centimeters), RBE values of the order of 0.6–1.0, compared to x or γ radiation, have been found (Table II). Data obtained in the Soviet Union with 500- and 660-MeV protons are in good agreement [46–49]. Thus, the findings with respect to RBE are in general what one might expect from the physics involved. The beams in the plateau region are minimally ionizing and therefore the effect should be quantitatively similar to those obtained with x and γ radiation. Similar results have been obtained with very-high-energy α-particle beams [40, 41, 45] and RBE values of somewhat less than unity have been obtained.

Some evidence exists that a higher LET component of the high-energy minimally ionizing beams may play a role in determining the degree of biological effects under some circumstances. The findings involve mainly mortality, and center principally about the occasional findings of early deaths that have been associated with high LET radiations (see Section V,2), although a few studies [40, 43, 45] have been reported showing RBE values greater than unity. There is no consistent thread throughout all of the experiments [50]; however, there is some evidence to indicate that the phenomenon is not seen with small animals such as mice irradiated with a narrow beam. The phenomenon of early deaths, reported by American and Russian investigators, may be associated with total body irradiation of larger specimens such as the monkey, or small specimens such as the mouse irradiated in a larger phantom such that the degree of scatter, especially lateral, is appreciable. Additional work is required to be certain that the phenomenon is real, and the reasons for it if it does, in fact, exist.

Other experiments using other criteria of effect indicate that under most conditions of exposure the plateau portion of the Bragg curve from very-

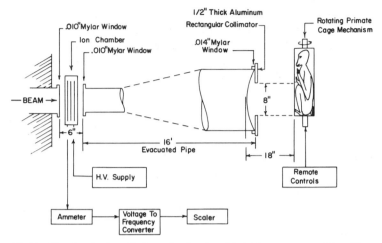

FIG. 15. Facility for the irradiation of animals with high-energy particles. The animal cage rotates for uniform exposure. Dose monitoring is done with a parallel-plate ionization chamber.

TABLE II
RBE FOR PROTON EXPOSURE[a]

| Energy (MeV) | Protons | | Standard radiation | | | |
	Dose rate (rads/min)	LET (keV/μ)	Quality	Dose rate (rads/min)	Criterion	RBE
138	86	0.64	^{60}Co	86	LD$_{50(30)}$ (mouse)	1.07 ± 0.04
	256	0.64	^{60}Co	256	LD$_{50(30)}$ (mouse)	0.97 ± 0.04
	550	0.64	^{60}Co	550	LD$_{50(30)}$ (mouse)	1.06 ± 0.04
157	250	0.55	250-kVp x rays	80	LD$_{50(8)}$ (mouse)	0.77 ± 0.1
315	—	0.4	200-kVp x rays	—	LD$_{50(30)}$ (mouse)	Approx. 1
440	40–80	0.29	250-kVp x rays (1.01 mm Cu HVL)	40	LD$_{50(30)}$ (mouse)	0.72
440	18	0.3	125-kVp x rays	18	LD$_{50(30)}$ (mouse)	0.7 ± 0.2
592	361	0.25	250-kVp x rays	80	LD$_{50(30)}$ (mouse)	09.8
660	300–450	0.3	180-kVp x rays	—	LD$_{50(30)}$ (mouse)	0.55
					LD$_{50(30)}$ (rat)	0.65
730	500–1000	0.3	200-kVp x rays (1 mm Cu HVL)	30	LD$_{50(30)}$ (mouse)	0.75
	100, 300, 1000	—	100-kVp x rays	20, 100	LD$_{50(30)}$ (mouse)	1.3
	100, 300, 1000	—	250-kVp x rays	20, 100	LD$_{50(30)}$ (mouse)	0.8
	1000	—	250-kVp x rays	100	LD$_{50(6)}$ (mouse)	1.4
	1000	—	250-kVp x rays	100	LD$_{50(6)}$ (mouse)	1.2
	100, 1000	—	250-kVp x rays	20, 100	LD$_{50(12)}$ (mouse)	0.89

[a] Data from G. V. Dalrymple et al., Rad. Res. **28**, 489 (1966).

high-energy proton or α beams act as minimally ionizing radiation. Thus, the rate of recovery from a single exposure has been examined in mice, using mortality as the criterion of effect, and appreciable recovery was observed when the dose fractions were separated by as little as 3 hr [41]. This rapid rate of recovery is obtained with x rays and would not be consistent with an appreciable contribution from high LET radiations in the beam. Similarly, experiments using life expectancy and weight loss of spleen, thymus, and testicles of mice [42, 44] have shown results compatible with the low LET nature of the beam. Similarly, investigations on the hematological effects of whole-body exposure to 660-MeV protons administered to mice have been compatible with this thesis [48, 49, 51]. Blood count changes, as well as changes in mitotic activity of the bone marrow, were studied. For equal doses of high-energy protons and γ radiation, the degree of effect was found to be less for proton irradiation. Thus, the RBE was less than unity, in agreement with results indicated above for early mortality and other end points.

Although not suitable for radiotherapy because of their great penetration, 2.2-GeV protons have been studied using a variety of biological end points. The results are summarized in Table III and are compatible (with the possible exception of the data on the rat) with that expected from low LET radiation. These experiments were done under conditions of minimal scatter.

TABLE III
EARLY AND LATE EFFECTS OF 2.2 GeV PROTONS

Criterion of effect	Dose at ED_{50}		Protons/cm^2 (units of 10^7) at ED_{50}, per rad of x rays at ED_{50}	RBE (normalized to 1.0 for mortality)
	Protons/cm^2	250-kVp x rays		
30-day mortality, mouse	1.81×10^{10}	557	3.24	1.0
Spleen weight loss, mouse	0.55×10^{10}	200	2.75	1.18
Thymus weight loss, mouse	0.49×10^{10}	180	2.72	1.19
Bursa weight loss, chick	1.57×10^{10}	500	3.14	1.03
Spleen weight loss, chick	1.31×10^{10}	450	2.91	1.11
Thymus weight loss, chick	1.18×10^{10}	400	2.95	1.09
Neoplasia induction, rat	0.26×10^{10}	150	1.75	1.85
Lens opacification, mouse	0.14×10^{10} to 0.21×10^{10}	50	2.86–4.06	0.80–1.13

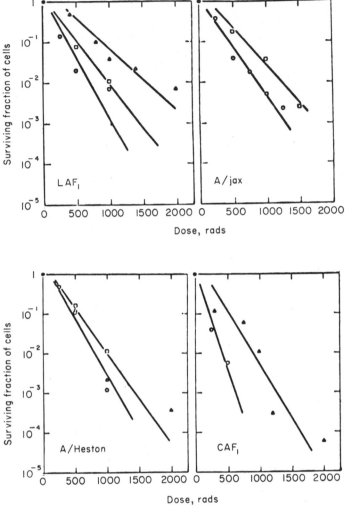

FIG. 16. Dose–response curves of lymphoma cells irradiated *in vivo* with 200-keV x rays and 910-MeV α particles. Irradiation in the ionization peak of the α beam shows an elevated RBE. Different strains of mice were used for carriers of the cancer cells.

Other experiments [50] done under conditions of appreciable scatter showed some results compatible with the contributions from possible high LET components in the beam (see above).

Studies on the RBE in the peak portion of the Bragg peak are difficult because the peak is narrow and because the LET of the radiation in this area changes rapidly as a function of depth. The RBE of the peak versus the plateau portion of an α particle beam from the Berkeley synchrocyclotron has been investigated using measurements of the proliferative capacity of neoplastic cells in mice [52]. Surviving factions of the lymphoma cells are shown in Fig. 16. The RBE values for the Bragg peak ranging from 1.5 to 1.9 were reported. These RBE's are higher, of course, than the less-than-unity figures reported for the plateau region of the same beam [41]. Additional studies of this nature were carried out by Larsson and Kihlman [53]. Thin layers of material were exposed at various depths in the beam. The RBE appeared to remain low until actually the peak of the Bragg curve was reached with apparently a rapid increase in RBE beyond this point. The OER was low even well beyond the Bragg peak. Thus, in a beam of protons or α particles, the RBE and OER change appreciably only at the very end of the beam track. At this point, one has "run out of dose," and the high values thus are of no practical use in radiotherapy.

The above results at the end of the range of a beam of α or proton particles can be contrasted with the effects of relatively low-energy α particles from radioactive sources. These effects have been studied using inhibition

TABLE IV[a]

D_{37} Values and OER's of Different Radiations for Impairment of the Reproductive Capacity of Human Cells

Radiation	LET[c] (keV/μ)	D_{37} in air		D_{37} in nitrogen		OER[f]
		Low dose region[d] (rads)	High dose region[e] (rads)	Low dose region[d] (rads)	High dose region[e] (rads)	
2.5-MeV α particles	166 ± 20	79	79	79	79	1.0 ± 0.1
3.4-MeV α particles	140 ± 20	64	64	70	70	1.1 ± 0.05
4.0-MeV α particles	110 ± 10	57	57	71	71	1.3 ± 0.1
5.1-MeV α particles	88 ± 6	70	70	118	118	1.7 ± 0.15
8.3-MeV α particles	61 ± 5	84	84	172	172	2.05 ± 0.25
25.0-MeV α particles	26 ± 2	165	120	455	270	2.4 ± 0.3
3.0-MeV deuterons	20 ± 2	165	145	440	350	2.4 ± 0.2
14.9-MeV deuterons	5.6 ± 0.3	400	160	1250	380	2.6 ± 0.3
250 kVp x rays	1.3 (av)	450	180	1300	410	2.7 ± 0.2
Fast neutrons[b]	20 (av)	105	80	170	130	1.6 ± 0.2

[a] G. W. Barendsen, D. K. Bewley, S. B. Field, C. J. Koot, C. J. Parnell, and G. R. van Kersen, *Int. J. Rad. Biol.* **10**, 317 (1966).
[b] Produced by bombarding Be with 16-MeV deuterons.
[c] For directly ionizing particles: LET$_\infty$ in keV/μ of unit density tissue.
[d] Values correspond to part of survival curve between surviving fractions 1.0 and 0.5.
[e] Values correspond to part of survival curve between surviving fractions 0.1 and 0.01.
[f] Determined as the ratio of two doses required for a given surviving fraction, mean value for all levels of survival.

of clone formation of human kidney cells by ^{210}Po α particles. Values for RBE as high as 8 were obtained for radiation in air [54]. The lowest OER measured was not significantly different from 1.0, for 2.5-MeV α particles with an LET$_x$ of 166 keV/μ to tissue (Table IV). Thus, while irradiation with low-energy α particles does yield a high RBE and a low OER, and while high-energy α particles of course do slow down at the end of the range, usable high values of RBE are not attainable with high energy, penetrating beams of protons, or α particles.

D. Heavy Ions

The low energies available to date make it impossible to use heavy ions in any experiments which require depth of penetration. Therefore, biological studies with heavy ion linear accelerators such as those at Berkeley and Yale have to be confined to thin layers of cell cultures. However, investigations of the RBE, the oxygen effect, and effects being caused by electron pickup can be done. The question of whether different heavy particles with the same dE/dx (or LET$_x$) but with presumably different patterns of δ rays produce the same degree of biological effect (have the same RBE) has not been decided as yet [1, 55–56].

Irradiations of human cancer cells with 40-MeV α particles, 69-MeV ^{7}Li, 105-MeV ^{12}C, and 130-MeV ^{16}O ions with the Yale accelerator gave multi-target-type survival curves (Fig. 2). The inactivation cross sections* for C and O allow one to conclude that the sensitive area of the nucleus is a fairly large fraction of its geometrical cross section. From the LET of the ions, information was obtained on how much energy must be dissipated locally to produce a single hit curve. No dose rate effects were observed in the range used, 150–500 rads/min. The peak in the LET–RBE relationship was found between lithium and carbon "where the extrapolation number drops to one."

Studies of the reversible and irreversible biological effects caused by heavy ions were done at the Berkeley HILAC with human kidney and Chinese hamster cells [55]. The proliferative ability of the cells *in vitro* was used as the biological end point. A wide spectrum of LET's was available by the use of ions up to ^{40}Ar. Irradiations took place in aerobic and unaerobic environments. Reversible-sigmoid and irreversible-exponential responses were inferred from survival curves. The two components of inactivation were found to occur independently from each other, with the LET governing the percentage distribution. Figure 17 is a plot of inactivation cross sections versus LET for various cell cultures. The differences between irradiation in air and

---

* Inverse of the incident particle fluence which corresponds to a $1/e$ reduction in cell survival. The unit is area per inactivating particle.

nitrogen disappear when the LET reaches values of about 3000 MeV·cm²/g of tissue (300 keV/μ of tissue). The component distribution becomes obvious from Fig. 18. The diminution of the reversible, sublethal component of radiation damage relative to the irreversible, lethal component becomes obvious by a gradual disappearance with increasing LET of the shoulder of the survival curves. Maximal sensitivity of the cells (highest RBE) was found for ¹²C ions of an LET of 220 keV/μ. It is significant that these ions also represent the borderline between sigmoid and strictly exponential dose re-

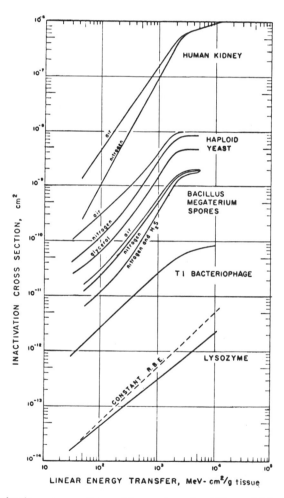

FIG. 17. Inactivation cross sections of heavy ions for different biological indicators. Irradiation in air and anoxic environment becomes equally effective when the LET reaches values of about 300 keV/μ of tissue (3000 MeV·cm²/g).

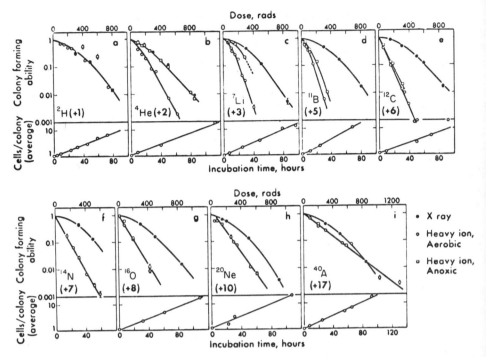

FIG. 18. Effects of heavy ions on human kidney cells. The steepest dose-effect curve was found for ^{12}C ions (LET, 220 keV/μ). A perfect overlap of aerobic and anaerobic response (disappearance of the oxygen effect) was found for ions heavier than ^{12}C.

sponse curves. The RBE of the carbon ions was measured and found to be about 2 for reversible lesions and about 8 for irreversible injury. The OER decreased systematically from about 2.8 for x rays to approximately 1 for ^{16}O ions.

Other experiments with the Berkeley HILAC dealt particularly with the oxygen effect [52]. Lithium-7, ^{11}B, and ^{20}Ne ions of linear energy transfers from 4.5 to 1.717 keV/μ were used to study the proliferative capacity of mouse lymphoma and mammary carcinoma cells after *in vitro* irradiation in oxygen and nitrogen atmosphere. Doses ranging from 200 to 2000 rads were administered at rates of 300–600 rads/min. Surviving fractions of the cells were determined by comparison of the LD_{50} of mice injected with irradiated tumor cells to the LD_{50} of mice inoculated with nonirradiated cells. An oxygen effect was still observable when lithium ions were used, but none was observed with ^{11}B and ^{20}Ne. From this the conclusion was drawn that the oxygen effect might not disappear before the LET values reach appreciably beyond 20 keV/μ.

E. Negative π Mesons

Low-beam intensities and the disturbing presence of μ mesons and electrons have made it impossible so far to do major biological experiments with π^- mesons. However, a series of investigations has shed more light on what might be expected from this kind of particle. Measurements of the absorbed dose allow estimates of the effectiveness of the peak region of the pion ionization depth curve, in comparison to the effects of the plateau region of the curve.

Measurements at the Cosmotron were done in order to compare the characteristics of low-energy pions with protons [4, 57]. Very low intensities and a heavy background of muons and electrons made it necessary to limit the experiments to comparative dose measurements, with electronic discrimination against undesired particles. Ionization curves were measured by analysis of the pulse-height spectra from scintillation counters. It was found that the relative height of the ionization peak amounts to about 3 for π^- mesons and does not exceed the peak height of protons of the same range. However, the meson peak appeared somewhat broader and slightly skewed by comparison (Fig. 8).

At CERN, measurements were made using pion beams from the 600-MeV synchrocyclotron. With a tissue-equivalent recombination chamber a value of 2.2 was determined for the differential dose deposition in peak and plateau of the ionization curve.* An RBE of 2.7 was estimated for the top, and 3.4 for the downslope portions of the peak, respectively [58]. A LET chamber was used for comparison of spectra from positive and negative pions [2, 59]. A well-pronounced maximum in the LET distribution was found in the peak of the negative beam at about 70 keV/μ of tissue (Fig. 19). This corresponds to an α energy of 6–7 MeV. From LET–RBE relationships, an RBE of 3 could be attributed to this part of the LET spectrum.

A series of dose measurements and biological experiments has been done with 90-MeV pions from the Berkeley synchrocyclotron [16–18, 60–62]. Calculation of the best obtainable differential dose peak/plateau gave a value of about 3.4, correcting for beam contamination [17].

Effects of the beam on ascites lymphoma cells were studied at Berkeley by exposing mice in the plateau and peak of the ionization curve. The highest dose rates in the beam used were of the order of 10 rads/hr, and total doses

* The response of this chamber depends on the ion collection efficiency and permits an estimate of RBE values.

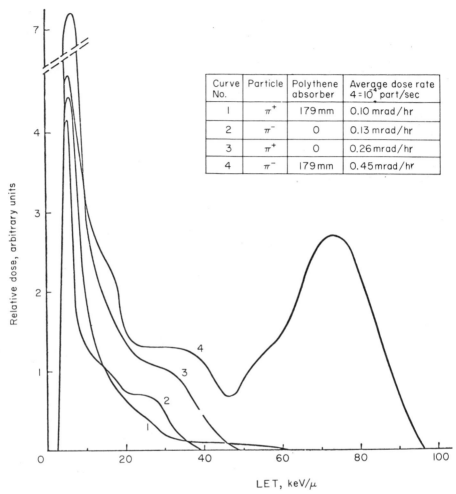

Curve No.	Particle	Polythene absorber	Average dose rate 4=10⁴ part/sec
1	π^+	179 mm	0.10 mrad/hr
2	π^-	0	0.13 mrad/hr
3	π^+	0	0.26 mrad/hr
4	π^-	179 mm	0.45 mrad/hr

FIG. 19. LET spectrum of a beam of negative π mesons measured with a tissue-equivalent proportional counter. Absorber depth, 18 cm of polyethylene. The high LET peak corresponds to an α particle energy of 6–7 MeV.

from about 50 to 350 rads were administered. Intraperitoneal fluid was withdrawn after irradiation and injected into female LaF₁ mice at various dilutions. The proliferative capacity of the tumor cells was checked at the end of eight weeks by scoring the percentage of animals that developed tumors. An RBE of 4.6 for the peak portion of the curve was found, at the low dose rates used [58]. Additional cytological experiments were done [18, 61]. An increase in aneuploidy was observed as well as chromatid breaks and metacentric chromosomes. In Fig. 20 is shown that 3 days after irradiation the

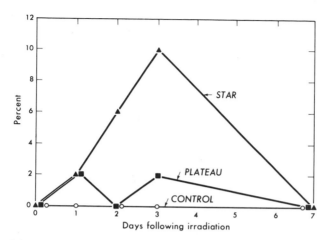

FIG. 20. Biological effectiveness of negative π mesons: anaphase bridges occurring in mouse lymphoma cells after irradiation in the plateau and the ionization peak of 90-MeV pions, compared to the occurrence in unirradiated cells.

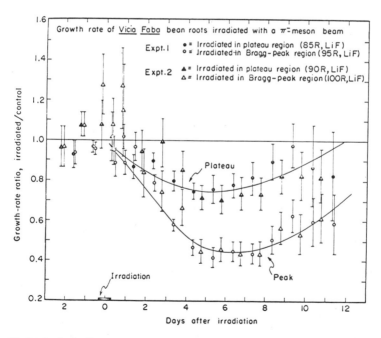

FIG. 21. Biological effectiveness of negative π mesons: reduction of the growth rate of *Vicia faba* room meristem, after exposure in the plateau and the ionization peak of 90-MeV pions [from S. P. Richman, C. Richman, M. R. Rajer & B. Schwartz. UCRL-16387 (1965)].

percentage of abnormal anaphase bridges produced in the region in which large numbers of slowed down mesons interacted with nuclei was roughly five times the effect caused in the plateau of the ionization curve. An estimate of 4.7 is given for the peak RBE of an essentially clean pion beam at the low dose rates used.

Bridges and fragments observed at anaphase also served as a biological end point in an investigation of irradiated *Vicia faba* (bean) root meristems [62]. A ratio of 2.6 to 1 was found for the "effective dose" in peak versus plateau regions of the pion beam. Results of studies on growth rate inhibition of bean roots are shown in Fig. 21. After 6 days post irradiation, the growth rate in the peak region had dropped to about 45% of the normal rate. Irradiation in the plateau region caused a drop to about 75%.

V. SUMMARY

It is believed that current radiotherapy could be improved if it were possible to (1) selectively deposit relatively greater amounts of radiation energy in the tumor and less in normal tissues that are unavoidably exposed in the process, and (2) overcome the significant protection afforded some cells within a tumor by virtue of the fact that they are hypoxic as a result of reduced blood supply to focal areas. With currently employed x, γ, or electron (low LET) radiations, severely hypoxic cells may be protected by a factor as great as 3. Significantly less protection is afforded such hypoxic cells by densely ionizing (high LET) radiations. Several approaches to cope with (1) and (2), utilizing high LET radiations, are summarized below (see Table V).

Fast neutrons (several million electron volts) represent the most readily available source of high LET radiations at present. They afford no advantage with respect to (1) and are, in fact, less desirable in this regard than some currently available low LET radiations. They are advantageous with respect to (2), and clinical trials with external beams are currently underway in England. The availability of the transplutonic isotope ^{252}Cf that fissions spontaneously provides a medium for the use of fast neutrons for implantation (source placed directly in the tumor) therapy, much in the manner that radium implants have been used.

Neutron capture therapy involves the attempt to selectively localize in a tumor, by physiological means, an isotope with a very high cross section for thermal neutrons (that is, ^{10}B), and then irradiating the tumor and adjacent normal structures with a thermal neutron beam. Because of the short range of particles from the ^{10}B (n, α) ^{7}Li reaction, the tumor would in principle

TABLE V

	Protons, deuterons, alphas	Thermal neutrons	Epithermal and intermediate neutrons	Fast neutrons	Negative pions	Heavy ions
Availability:	Fair	Good	Limited	Good	No clean high-intensity beams	Low energies only up to date
Energy (range):	Good	Poor	Fair	Good	Good	Much too low
Beam design and operation:	Easy	Easy	Difficult	Spectrum: easy Monoenergetic: very difficult	Very difficult	Moderately difficult
Focusing, collimation:	Good	Very poor	Poor	Fair	Good	Good
Scattering and/or diffusion:	Moderate	Heavy	Heavy	Moderate	Moderate	Very little
Beam contamination:	None	γ rays	γ rays,	γ rays	Muons Electrons	None
Density and dose rate:	Good	Good	Good	Good	Very low	Good
Bragg peak:	Good	No	No	No	Good	Fair (?)
Exit dose:	Not necessarily	Yes	Yes	Yes	Yes	Yes
Dose to surrounding tissue:	Low	Very high	High	Medium	Medium	Low
Dosimetry:	Easy	Moderately difficult	Very difficult	Moderately difficult	Very difficult	Easy
LET:	Low	High	High	High	Medium	High

be selectively irradiated. Several practical problems have precluded successful application of the approach to date. The approach has advantages with respect to (1), probable limited advantage with respect to (2), since it is unlikely that the isotope would gain access to regions of tumor rendered hypoxic because of poor blood supply.

High-energy proton and α particles have advantages with respect to (1); however, the increased ionization at the end of the Bragg (dE/dx vs. x) curve affords only limited advantage because the width of the "peak" is unacceptably narrow for radiotherapy. These beams are of no benefit with respect to (2), since at the high energies required for adequate penetration the LET is unacceptably low through essentially all of the particle range.

Beams of high-energy heavy ions have some advantages with respect to (1); more so with respect to (2). Beams with sufficient energy to achieve usable penetration are not currently available.

In principle, negative π mesons have distinct advantages with respect to both (1) and (2). The ionization–depth curves show a "peaking" at the end of the range that extends over several centimeters, adequate in width to accommodate a number of tumors. The π^- mesons act similarly to electrons as regards dE/dx losses, and thus the bulk of the beam track represents low LET radiation. Because of star formation at the end of the range, however, the peak of the ionization–depth curve has a large component of high LET radiations. Beams of π^- mesons of sufficient intensity and purity for radiotherapeutic trials are not available at present.

REFERENCES

1. P. H. Fowler, *Proc. Phys. Soc.* **85,** 1051–1066 (1965).
2. H. H. Rossi, *IAEA Tech. Rep. Ser.* **58,** 81–95 (1966).
3. J. F. Fowler, *Biol. Effects Neutron Proton Irradiation* **2,** 185–212, IAEA, Vienna (1964).
4. V. P. Bond, C. V. Robinson, R. Fairchild, and G. M. Tisljar-Lentulis, *Nat. Cancer Inst. Monogr.* **24,** 23–43 (1967).
5. M. L. Mendelsohn, *Nat. Cancer Inst. Monogr.* **24,** 157–165 (1967).
6. L. H. Gray, *Rad. Res. Suppl.* **1,** 73–101 (1959).
7. R. S. Stone, J. H. Lawrence, and P. C. Aebersold, *Radiology* **35,** 322 (1940); *Radiology* **39,** 608 (1942); R. S. Stone, *Amer. J. Roentgen* **59,** 771 (1948).
8. N. A. Frigerio and M. J. Shaw, ANL-7136, 182–184 (1965).
9. W. H. Sweet, TID 20856 (1964); NYO 3267-3 (1966); NYO 3267-4 (1967).
10. J. H. Lawrence and C. A. Tobias, *Progr. Atom. Med.* **1,** 127–146 (1965).
11. C. A. Tobias, J. H. Lawrence, J. Lyman, J. L. Born, A. Gottschalk, J. Linfoot, and J. McDonald, *Int. Symp. Response Nervous System to Ionizing Radiation, 2nd, Boston* 19–35 (1964).

12. H. H. Heckman, B. L. Perkins, W. G. Simon, F. M. Smith, and W. H. Barkas, *Phys. Rev.* **117**, 544–556 (1960).
13. A. B. Wittkower and H. B. Gilbody, *Proc. Phys. Soc.* **90**, 353–359 (1967).
14. P. H. Fowler and D. H. Perkins, *Nature* **189**, 524–528 (1961).
15. J. P. Blaser and H. A. Willax, *IEEE Trans. Nucl. Sci. NS-13*, No. 4, 194–214 (1966).
16. C. Richmond, H. Aceto, M. R. Raju, and B. Schwartz, *Am. J. Roentgen* **96**, 777–790 (1966).
17. S. B. Curtis and M. R. Raju, UCRL-17606 (1967).
18. W. D. Loughman, H. S. Winchell, M. R. Raju, and J. H. Lawrence, UCRL-16898, 11–13 (1966).
19. G. E. Stapleton, ORNL-TM-924, ORNL-P-758 (1964).
20. J. L. Bateman and V. P. Bond, *Ann. N.Y. Acad. Sci.* **114**, 32–47 (1964).
21. J. L. Bateman, V. P. Bond, and H. H. Rossi, *Biol. Effects Neutron Proton Irradiation* **2**, 321–336, IAEA, Vienna (1964).
22. V. P. Bond, O. D. Easterday, E. E. Stickley, and J. S. Robertson, *Radiology* **67**, 650–663 (1956).
23. V. P. Bond, Biophys. Aspects Radiation Quality, 2nd Panel Rep., 149–160, IAEA, Vienna (1968).
24. D. K. Bewley and S. Hornsey, *Biol. Effects Neutron Proton Irradiation* **2**, 173–181, IAEA, Vienna (1964).
25. S. B. Field, T. Jones, and R. H. Thomlinson, *Brit. J. Radiol.* **41**, 597–607 (1968).
26. J. F. Fowler in "Biological Interpretation of Dose from Accelerator-Produced Radiation," CONF 670305, 128–142 (1967).
27. Archambeau, O. J., V. Alcober, W. Calvo, and H. Brenneis, *IAEA Symp. Biol. Effects Neutron Proton Irradiation* **2**, 55–72, Vienna (1964).
28. W. Entzian, A. H. Soloway, M. R. Raju, W. H. Sweet, and G. L. Brownell, *Acta Radiol. Theoret. Phys. Biol.* **5**, 95–100 (1966).
29. L. E. Farr and T. Konikowski, *Nature* **215**, 550–552 (1967).
30. V. P. Bond and O. D. Easterday, *Rad. Res.* **10**, 20–29 (1959).
31. R. G. Fairchild, *Phys. Med. Biol.* **10**, 492–504 (1965).
32. R. G. Fairchild, *Radiology* **85**, 555–564 (1965).
33. R. G. Fairchild and L. J. Goodman, *Phys. Med. Biol.* **11**, 15–30 (1966).
34. N. A. Frigerio, *Phys. Med. Biol.* **6**, 541–549 (1962).
35. N. A. Frigerio, ANL-7136, 175–182 (1965).
36. E. L. Muetterties and W. H. Knoth, *Chem. Eng. News* 88–98 (May 9, 1966).
37. R. N. Kjellberg, A. M. Koehler, W. M. Preston, and W. H. Sweet, *Int. Symp. Response Nervous System to Ionizing Radiation, 2nd, Boston,* 36–53 (1964).
38. B. Larsson, AD 632 702 (1966); *Rad. Res. Suppl.* **7**, 304–311 (1967).
39. G. V. Dalrymple, J. J. Ghidoni, J. D. Hall, R. Jacobs, H. L. Kundel, I. R. Lindsay, J. C. Mitchell, I. L. Morgan, E. T. Still, G. H. Williams, and T. L. Wolfle, *Rad. Res.* **28**, 365–566 (1966).
40. J. K. Ashikawa, C. A. Sondhaus, C. A. Tobias, A. G. Greenfield, and V. Paschkes, *Symp. IAEA Biol. Effects Neutron Proton Irradiation* **1**, 249–260, Vienna (1964).
41. J. K. Ashikawa, C. A. Sondhaus, C. A. Tobias, L. L. Kayfetz, S. O. Stephens, and M. Donovan, *Rad. Res. Suppl.* **7**, 312–324 (1956).
42. J. Baarli and P. Bonet-Maury, *Nature* **205**, 361–364 (1965).
43. V. P. Bond, *Conf. AEC Div. Operational Safety Res.* No. 651109, 617–623 (1966).
44. P. Bonet-Maury, J. Baarli, T. Kahn, G. Dardenne, M. Frilley, and A. Deysine, *Symp. IAEA Biol. Effects Neutron Proton Irradiation* **1**, 261–277 (1964).

45. C. A. Sondhaus, NASA-SP-71, 97–103 (1965).
46. E. S. Gaidova, V. N. Ivanov, and S. P. Yarmonenko, *Radiobiol. USSR (Engl. Transl.)* **5**, 89–100 (1965).
47. Yu. G. Nefedov, Problems of radiation safety in space flights: Physical and biological studies with high-energy protons, NASA TTF-353 (1965).
48. V. L. Ponomareva, *Radiobiol. USSR (Engl. Transl.)* **5**, 38–45 (1965).
49. N. L. Shmakova, *Radiobiol. USSR (Engl. Transl.)* **3**, 215–219 (1963); **5**, 170–175 (1965).
50. J. E. Jesseph, W. H. Moore, R. F. Straub, G. M. Tisljar-Lentulis, and V. P. Bond, *Rad. Res.* **36**, 242–253 (1968).
51. Yu. I. Moskalev and I. K. Petrovich, *Bull. Exp. Biol. Med. USSR (Engl. Transl.)* **55**, 542–545 (1964); BNL-TR-143 (1967).
52. K. Sillesen, J. H. Lawrence, and J. T. Lyman, *Acta Isotopica* **3**, 107–126 (1963).
53. B. Larsson and B. A. Kihlman, *Int. J. Rad. Biol.* **2**, 8–19 (1960).
54. G. W. Barendsen, *Int. J. Rad. Biol.* **8**, 453–466 (1964); G. W. Barendsen, D. K. Bewley, S. B. Field, C. J. Koot, C. J. Parnell, and G. R. van Kersen, *Int. J. Rad. Biol.* **10**, 317–327 (1966).
55. P. Todd, UCRL-11614 (1964); NASA-SP-71, 105–114 (1965).
56. R. A. Deering and R. Rice, *Rad. Res.* **17**, 774–786 (1962).
57. G. M. Tisljar-Lentulis, V. P. Bond, J. S. Robertson, and W. H. Moore, BNL 12035 (1968).
58. J. Baarli and K. Goebel, CERN DI/HP/77 (1965); J. Baarli, *Rad. Res. Suppl.* **7**, 10–19 (1967).
59. T. R. Overton, CERN 66–33 (Oct. 1966).
60. J. M. Feola, C. Richman, M. R. Raju, and J. H. Lawrence, UCRL-16613, 23–26 (1965).
61. W. D. Loughman, H. S. Winchell, H. Aceto, C. Richman, M. R. Raju, and J. H. Lawrence, UCRL 16246, 100–102 (1965).
62. S. P. Richman, C. Richman, M. R. Raju, and B. Schwarz, *Rad. Res.* **7**, 182–189 (1967).

7

Interactions between Elementary Particle Research and Engineering

J. P. Blewett   *Brookhaven National Laboratory*
Upton, New York

I. INTRODUCTION

The particle physicist of today is almost completely dependent on the successful operation of multibillion-electron-volt particle accelerators. For example, the 33-GeV alternating-gradient synchrotron at the Brookhaven National Laboratory—a marvel of electrical and mechanical engineering—is the result of thirty-five years of engineering advances in accelerator design. The numerous discoveries and inventions necessary to make such accelerators possible evolved in the high-energy physics laboratories, and later became important parts of industrial technology.

In the early days of the accelerator art the transition from the laboratory to general use was slow. During the early 1930's physicists may be said to have lived in ivory towers; they had very little financial support and built their own apparatus. However, leading industries became aware of the possible impact of the new techniques and increasingly participated in accelerator development. World War II brought the physicists out of their laboratories to apply the technologies they had created to the development of radar and atomic weapons. The further sophistication of the engineering arts developed during the war allowed for the design of accelerators of ever increasing capabilities and, as such, elementary particle physics has become a multimillion dollar operation. New engineering developments from high-energy physics laboratories have proliferated, and have been rapidly accepted by industry for its own purposes. Indeed, accelerator construction is now an important industry in itself.

The increasing capabilities of accelerators opened the way for more complex and revealing experiments and have made the task of studying particle interactions more difficult. Experimenters using the new accelerators have been forced to develop advanced methods of particle detection and identification. In many cases these methods proved to have applications far beyond the dreams of their designers. This catalysis of industrial technology continues at an accelerated rate, and within two years from the date of this writing the reader will certainly be aware of new industrial developments, now unsuspected, which are based on developments in elementary particle physics.

In 1936, a visit to the Cavendish Laboratory in Cambridge, England, offered an intriguing preview of things to come. Under Lord Rutherford's inspired direction, Cockcroft and Walton had constructed the first successful accelerator and, in 1932, had succeeded in the artificial disintegration of several atomic nuclei. The homemade apparatus is shown in Fig. 1. A voltage-multiplying circuit, which was not invented at the Cavendish Laboratory but is now generally known as a Cockcroft–Walton system, was used in the accelerator. Cockcroft and Walton solved the following problems: sustaining voltages of over 700,000 V, generation of protons (hydrogen nuclei) in a gas discharge maintained at the high-voltage terminal, and means for identifying the nuclear reactions. To cope with the stream of events requiring analysis, they, with their associates C. E. Wynn-Williams and W. B.

FIG. 1. Cockcroft and Walton's original 700-kV set at the Cavendish Laboratory.

Lewis (now a director of the Canadian atomic energy laboratory at Chalk River), had developed circuits using thyratron tubes, which counted in the scale of two. Here, an incoming event triggered one thyratron of a pair and extinguished the other; a second event extinguished the fired thyratron and triggered the second, at the same time triggering one of the next pair—for example, with five pairs, the final thyratron yielded a signal for every 2^5 or 32 events. This was the first of the bistable counting circuits, which now, in transistorized versions, form the heart of every digital computer. Wynn-Williams piped the outputs of the counting circuits across the hall to another room where they were connected to a typewriter which he had electrified in a system of marvelous complexity to have it record the results. This very system was the precursor of the modern on-line computer. (In 1936 Wynn-

Williams left the laboratory, and with his departure the electric typewriter ceased to work—no one else was able to diagnose its peculiarities.)

By 1936 the original Cockcroft–Walton set had become obsolete and was about to be replaced by the first commercially built accelerator—another Cockcroft–Walton set built by the Philips Company of Eindhoven, Holland, and designed to hold over 1,000,000 V. This accelerator was the first of a series of high-voltage sets built at Philips. One of the latest, shown in Fig. 2, is a 750,000-V set that serves as the preinjector for the 33-GeV machine at Brookhaven.

FIG. 2. 750-kv Cockcroft–Walton generator built by Philips for use as the preinjector of the Brookhaven 33-GeV alternating-gradient synchrotron.

After a capsule history of accelerator development, several of the techniques and devices which have resulted from the technologies of nuclear and particle physics will be discussed. This is not a complete list by any means. No single author could hope, from his experience, to comprehend all of the inventions and discoveries during the past decades. The reader should regard the sections that follow as a reasonably typical sampling of the technological by-products of the efforts of the physicist to understand the inner workings of the atom and its nucleus.

II. REVIEW OF ACCELERATOR HISTORY

Accelerator development has proceeded through four major stages. The first accelerators were direct-voltage machines, some of which used voltage-multiplying circuits like that of Cockcroft and Walton. Another system developed almost simultaneously was the Van de Graaff electrostatic generator, in which charge was sprayed by a corona discharge onto a moving belt, which transported it to the inside of a high-voltage terminal. In air at atmospheric pressure the direct-voltage machines were limited by voltage breakdown to about 2 MV. Higher voltages were obtained by enclosing the accelerator in a pressure tank; even with this method about 10 MV for many years seemed to be the upper energy limit.

The second phase of accelerator development began almost immediately, the method of repetitive acceleration by radiofrequency fields in the cyclotron. Here ions cross a gap where they are accelerated by a radio-frequency field, enter a shielded region, and are bent back by a magnetic field to recross another part of the same gap. The magnetic field and frequency are chosen so that the return to the gap is timed simultaneously with reversal of the field, and as a result, acceleration again takes place across the gap. This process is repeatd many times. As the energy of the particle increases, its radius of curvature in the magnetic field increases, and one sees that the path of the particle results in a spiral of ever increasing radius.

The first cyclotrons were built by Lawrence and Livingston and their associates of the University of California, and by 1936 particle energies of 8 MeV had been attained. The cyclotron was enormously popular. The energy output of successive models was increased continually, toward what appeared to be a theoretical upper limit of a few tens of mega electron volts, a limit set by the fact that relativistic effects eventually caused the ions to fall out of synchronism with the accelerating field.

This limit was not superseded until 1945 when, almost simultaneously, Veksler in the Soviet Union and McMillan of the University of California discovered a principle now known as "phase stability." According to this concept, if the frequency of the accelerating field in the cyclotron is varied according to an appropriate program, ions can be kept in synchronism to any desired energy. The "new" cyclotron, now the "synchrocyclotron," was rapidly developed to attain energies of several hundred mega electron volts. However, the size of such cyclotrons was enormous, and the magnets of later models weighed several thousands of tons. Hence, accelerator designers turned to an invention proposed by Veksler and McMillan at the time of their discovery, who had suggested that a machine be designed to give a circular orbit rather than the spiral orbit of the cyclotron. This could be accomplished by making the magnetic field increase with time from a low value, consistent with the initial ion energy, to a final value appropriate for the final energy after acceleration. Acceleration was accomplished by radio-frequency fields whose frequency was varied to maintain synchronism. This machine was called a "synchrotron" by McMillan; Veksler chose the name of "synchrophasotron." Both names are still in use in the respective countries of the two inventors.

Synchrotrons are used both for electrons and protons. The "Cosmotron" at Brookhaven, a proton synchrotron, was the first accelerator to produce energies past 1 GeV (the abbreviation BeV is often used instead of GeV in the United States where 1000 million = 1 billion). The largest synchrotron of that type was built in the Soviet Union, with an energy output of 10 GeV; its magnet weighs 36,000 tons.

Again it appeared that an economic limit was being approached as the magnet weight increased as about the cube of the particle energy produced. Yet, again the day was saved by a new invention. The difficulty with existing synchrotrons lay in the fact that, during their many revolutions around the machine, ions tend to wander from the orbit which they are supposed to travel. It was necessary that some restoring force be built into the accelerator design, and this had been done by shaping the magnetic field. At best, the restoring force produced in such a manner was rather weak and enough space had to be designed into the machine to allow a rather large area for the particle excursions. In 1952, Courant, Livingston (of cyclotron fame), and Snyder at Brookhaven discovered a new way of shaping magnetic fields to give much greater restoring forces—termed "alternating-gradient focusing." Later it was found that the same method had been invented and patented a couple of years earlier in Greece by Christofilos, who, however, had failed to publish his ideas.

The Brookhaven group immediately began the design and construction of an AGS (alternating-gradient synchrotron) with an energy output of about 30 GeV, while in Europe a multi-nation organization began work on a similar machine, both of which are now in full operation. The economic importance of this improvement can be shown from the fact that the AGS magnet weighs only about 4000 tons. Moreover, magnet weight in AGS-type accelerators scales only *linearly* with particle energy, and higher energies can be contemplated with confidence. For example, a 76-GeV machine has been completed in the Soviet Union, and construction is in progress of a 200-GeV accelerator to be built near Chicago. In Europe plans are well advanced for a 300-GeV machine. For construction in the more distant future, speculative studies are in progress both in the United States and the Soviet Union for a synchrotron producing energies of the order of 1000 GeV.

In this brief summary, two important machine types have been neglected. One is the betatron, which has a guiding magnetic field system similar to that of the synchrotron, which depends, for acceleration, however, not on radio-frequency fields but rather on the electric field associated with a changing magnetic flux through a steel core that links the orbit. Betatrons are used for electron acceleration to energies of up to 300 MeV.

The second machine neglected thus far is the linear accelerator, in which particles are accelerated down the axis of a pipe by radiofrequency fields generated by one of several possible esoteric methods. Proton linear accelerators have been built only for energies of up to 200 MeV; 200-MeV machines are in operation at Brookhaven and Chicago. Electron linear accelerators, however, now operate in the gigavolt range; the largest to date is the 2-mile machine at Stanford, which presently operates at an energy of about 20 GeV.

III. HIGH-VOLTAGE TECHNIQUE

Solutions to the problems involved in maintenance of increasingly higher direct voltages have led to (1) a much better understanding of the mechanisms of high-voltage breakdown, (2) the development of methods for inhibition of breakdown, and (3) invention of methods for measurements of high voltages. This knowledge can be applied to many industrial developments, perhaps the most spectacular of which is the high-voltage transmission line. To transmit power over increasingly greater distances, it is necessary to use increasingly higher voltages. At present, seven-hundred-thousand volt lines are in use and transmission lines are under design to operate at 1,000,000 V or more. Ordinary insulation techniques used at 200,000 V or so are no longer adequate; instead, insulation for high-voltage transmission lines consists of voltage-dividing corona rings which are copies of the voltage-dividing rings on accelerating columns of Cockcroft–Walton sets or electrostatic machines, and the Cockcroft–Walton sets are being commercially built for sale to power companies for high-voltage testing purposes.

Voltages of over 1 MV are now employed in commercially available electron microscopes, in which all of the procedures developed to make high-voltage accelerators operable and reliable are used.

Early in the development of high-voltage accelerators it was noticed that the high-voltage terminal of the accelerator seemed to collect inordinately large quantities of dust. Soon it was realized that the ionization of dust particles in corona discharges around the terminal made this inevitable. This phenomenon was found to be applicable in the development of electrostatic dust precipitators, which are now widely used for air purification and conditioning systems, in the control of smoke emission from chimneys, and in many other areas where fine dusts or smoke particles must be controlled.

Still another important application of high-voltage technology is found in the spraying of paint. Here, the application of high electric fields between the spray nozzle and the object being painted serve to guide the spray precisely to the desired areas.

IV. SOURCES OF X RAYS AND OTHER TYPES OF RADIATION

Coolidge's original x-ray tubes were no more than low-energy electron accelerators. In standard practice they are rarely run at voltages above 100 kV. However, with the advent of the betatron, new vistas were opened for applications in industrial radiography. Although the betatron was developed primarily as a tool for nuclear physics, its industrial importance was immediately appreciated by the electrical industry. It seems quite fitting that a good deal of betatron evolution took place in Coolidge's laboratory at the General Electric Company. After he had built the first successful betatron at the University of Illinois, Kerst collaborated with the General Electric Research Laboratory in the design of a 100-MeV betatron. When the accelerated electron beam was brought to a target, a sharply collimated beam of very high-energy x rays emerged, which had a half-width of only about 2°. When the machine was operated at 60 accelerations/sec, the intensity of the x-ray beam was measured to be 2600 R/min. Needless to say, the machine was operated by remote control from a station protected by a concrete wall 3 ft thick. In the first publication describing this machine, Westendorp and Charlton show a radiograph of a clock taken through 4 in. of steel.

Soon several companies in the United States and abroad were manufacturing betatrons for use in radiography. Techniques were rapidly developed for taking stereographic x-ray pictures, and it is impressive to view these pictures through a stereoviewer and to see all of the internal details of a heavy piece of steel machinery.

Other particle accelerators have been found to be of major importance as sources of hard x rays. In 1947 several of the pioneers in the development of electrostatic accelerators joined to form the High Voltage Engineering Corporation in Cambridge, Massachusetts, with the objective of producing electrostatic machines commercially; this group included Van de Graaff, who had built the first machine of this type. Although many of the more academic accelerator physicists regarded this enterprise with some skepticism, the corporation was an immediate success, which has been expanding continually and now is a major supplier of electrostatic x ray machines. Figure 3 shows one of the High Voltage Engineering machines used in radiography.

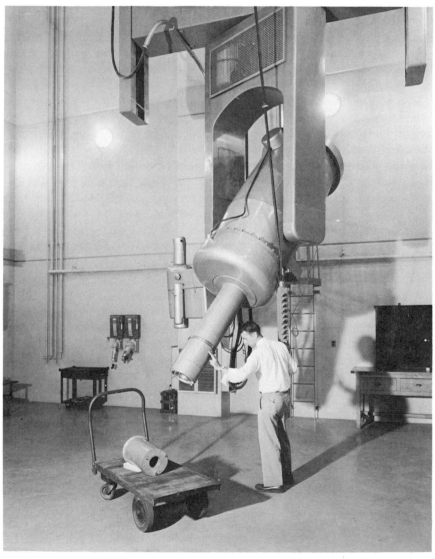

FIG. 3. Two-million-volt Van de Graaff generator built by the High Voltage Engineering Corporation for use in radiography.

More recently the electron linear accelerator also has become important as an x-ray source for energies somewhat higher than those easily produced by electrostatic machines. Electron linac x-ray sources, built commercially by several companies, are widely used in industrial radiography and in medical applications. It is possible to extract the electron beam either from a betatron or from a linear accelerator. The extracted beam can be sharply focused and used for selective irradiation of parts of the brain in the treatment of tumors, and x-ray and electron beams are now used in the sterilization of drugs and sutures, of foods to permit long shelf lives, and in the pasteurization of milk.

Nuclear particles have sharply defined ranges and are suitable for local irradiation. Protons and α particles (helium nuclei) deposit most of their energy at the end of their range and thus can pass through normal tissue to an internal tumor. The effect of deposition of energy at the end of the particle range is even more striking for negative π mesons which are produced by the higher-energy accelerators; their use in cancer therapy is just now under study. Deuteron beams incident on beryllium targets provide useful beams of neutrons which have been used for many purposes including reactor design, chemical activation analysis, and solid state research.

V. VACUUM TUBES FOR HIGH POWER LEVELS

During the early days of cyclotron operation, the excitation of radio-frequency accelerating systems operating at 10–20 MHz required more power than that obtainable from commercially available vacuum tubes. For this reason, cyclotrons were powered during the early 1930's by homemade, continually pumped triodes. Figure 4 is a photograph of a demountable triode built in 1934 by Sloan, Thornton, and Jenkins to power a cyclotron at the University of California; within a few years suitable, sealed-off subes were made available through industry. Since that date, designers of transmitting and power tubes have been in close touch with accelerator builders, and in several cases triodes and tetrodes originally designed to meet accelerator designs have found new major applications in industry, in communications, and in radar.

At the end of World War II, Hansen and his associates at Stanford were anxious to build a large electron linear accelerator. The power required was several megawatts, to be supplied in short pulses at radar frequencies. The magnetrons used in radar did not present sufficient stability and reliability for this application and Hansen's group began to think of the klystron which had been developed some ten years previously for use at energy levels from

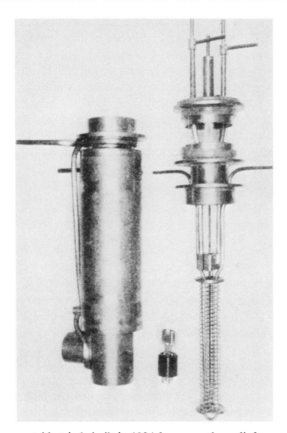

FIG. 4. Demountable triode built in 1934 for use as the radiofrequency power source for a cyclotron.

milliwatts up to a few watts. Other physicists and engineers were skeptical but the Stanford group proceeded with their own development. After some years of work they were successful. The results were klystrons designed for the megawatt range and finally, after many difficulties were overcome, for operation and use to power a 1-GeV electron LINAC.

These developments came to the immediate attention of industry. Two of Hansen's associates, the Varian brothers, set up their own company for manufacturing klystrons, known as Varian Associates, which is now a major supplier of electronic equipment, including klystrons and electron linacs.

When researchers in the United States Army initiated the observation of radar echoes from the planet Venus, they chose the klystron tube for the power source. An enormous 400-MHz klystron was built by the Eimac Company just for this purpose and the experiment was performed successfully.

A new proton linear accelerator for use in production of intense beams of mesons and high fluxes of thermal neutrons is under design at the Atomic Energy Commission's Los Alamos laboratory. It requires very high levels of radio frequency power. To supply such power the Raytheon company is now developing a new power tube known as the "Amplitron." The development of this tube has met with some success, and it is possible that this tube will be an important addition to the technology of the electronic and communications industries. High-frequency power tubes also have been used in successful applications to dielectric heating and its by-product, electronic cooking.

VI. VACUUM TECHNIQUE

In 1930 there were very few industrial applications of vacuum techniques. Mercury diffusion pumps, home-built of glass, were found almost entirely in research laboratories, and such laboratory vacuum systems, for the most part, were made from glass by amateur glass blowers. But small mercury pumps did not produce vacuums high enough for the cyclotron, with its demountable metal systems of relatively large volume. Therefore, cyclotron physicists turned to oil as a diffusion pump fluid, and increasingly larger oil pumps were built with increasingly higher speeds to cope with the leaks and outgassing of the "dirty" vacuum chambers of the larger cyclotrons. As more rugged and manufacturable pumps were developed, their industrial utility became more diversified. Particularly important were metallurgical applications where the preparation of important alloys became possible *in vacuo*.

Many by-products were the result of this effort. With the development of "O-ring" elastomer gaskets it was possible to design demountable systems that could be reassembled with the assurance that they would be vacuum tight. The further refinement of such systems continues today, mainly in industrial and space laboratories, with a parallel development of metal gaskets usable in high radiation fields where elastomers would rapidly be destroyed by radiation damage.

Furthermore leak detection became important mainly as a result of developments in accelerator laboratories. Cyclotron physicists built leak detectors, based on ionization gauges or mass spectrometers designed to emit a whistle when a leak was flooded with helium from a jet in the hand of the leak hunter. Such leak detectors are now widely used in the laboratory and industry.

During the past two decades a new type of vacuum pump based on the gas absorbing properties of titanium vapor has been developed. This pump was first used in the electrostatic accelerator laboratory of Herb at the University of Wisconsin. Industrial variations and improvements on Herb's pumps have resulted in a multimillion-dollar industry, which continues to expand today.

More recently a need has developed for "supervacuums" with pressures lower by a factor of 10,000 than those achievable thirty years ago. Such vacuums are imperative in the chambers of "colliding beam" systems with which particle physicists hope to make available more energy than that possible with conventional accelerators. Researchers in accelerator laboratories have produced such vacuums with a variety of technological improvements. Since simulation of outer space also requires the production of supervacuums, vacuum engineers in accelerator laboratories and their colleagues in the space, missile, and rocket industries maintain close contact on these developments.

The new vacuum technology has had a considerable impact on the gas transport industry, where leak-tight seals either under vacuum or pressure are a requirement. O-ring seals have proved to be reliable in this connection. Even liquid oxygen and hydrogen are now transportable over great distances, thus making possible the operation of such space centers as Cape Kennedy at points remote from gas liquefaction plants.

Vacuum technique has now been used in innumerable industrial applications: pure metals prepared by smelting under vacuum; electron beam welding, a powerful new method, necessarily carried on *in vacuo,* and the now common vacuum processing and freezing of foods and juices.

VII. MAGNET DESIGN

Magnetic fields with precisely specified characteristics are essential to the operations of cyclotrons, betatrons, and synchrotrons. Hence, accelerator builders have devoted a great deal of study to the design of electromagnets. The first cyclotron magnets were based on very rough design procedures and then were "shimmed" with bits of iron until the field presented the desired configuration. Some improvement in this procedure was made by the use of scaled models whose fields were measured carefully. The magnet of the 3-GeV Cosmotron (the first proton synchrotron) built at Brookhaven during 1948–1952 was based on measurements of a series of four models: ranging from the first, one-twelfth scale, to the last, one-quarter scale. Some attempts were made to compute the field patterns in the cosmotron by using relaxation techniques, but this method was not used extensively until the design period of the 33-GeV AGS, where the shape of the magnet poles was computed to a precision of 0.002 in. Measurements on the models agreed precisely with the predictions: these calculations took into account the effects of the nearby exciting coils; but they did not include the effects of the decreasing permeability of iron at high fields. The latter effect was finally covered in a program worked out at the MURA accelerator laboratory in Madison, Wisconsin. The AGS calculations originally were done tediously by hand, but now the whole procedure of magnet design can be performed rapidly on a computer in several accelerator laboratories.

The full impact of these rather recent developments has not yet been felt. It is inevitable that they will be enormously valuable to industry since the methods apply equally well to the design of transformers, motors, or generators. In transformers the power is transferred from winding to winding through the medium of magnetic fields in the core, while motors and generators work by the interaction of electromagnets, of which the stator and rotor of these machines consist.

In high-energy physics laboratories these methods are widely applied at present to the design of the many bending magnets and quadrupole lenses that make up the beam transport systems that are part of every experiment in particle physics.

VIII. DESIGN OF ROTATING MACHINES

Although rotating machines for continuous operation as motors or generators are conventional and reliable, there are some intermittent applications where loads are suddenly applied or removed and place severe strains on conventional machines. Mine hoist motors and generators for steel mills and heavy welding perform the most strenuous of these tasks. A similar problem is encountered in the use of large generators which supply power at levels of tens of megawatts to the magnets of multi-giga electron volt accelerators. At the peak of the accelerating cycle, voltage on the magnet is reversed and stored energies of many megajoules are dumped back into the generator, reversing the torque on the generator's stator in a fraction of a second. A number of generators have failed in this service necessitating lengthy and expensive repairs. The problem has been analyzed by several accelerator engineers, the reason for the failure has been pinpointed, and methods have been found for correcting it. The results of this study have been made available to the electrical industry and will be helpful in designing the large rotating machines of the future.

IX. FERRITES

As part of the radiofrequency accelerating system of the Brookhaven Cosmotron, it was necessary to supply a ferromagnetic material which retained its properties at high frequencies (up to 5 MHz). Powdered iron and very thinly laminated ferromagnetic alloys were studied and proved to be either unsatisfactory or prohibitively expensive. The difficulty was associated with the fact that all ferromagnetic materials seemed to be conducting and thus supported high eddy currents with associated disastrously high losses at high frequencies. At this point the Cosmotron's designers became aware of a new development at the Philips Company in Holland. A scientist at Philips had developed a ferromagnetic material that was not conducting—a black ceramic made by the combination of iron oxide with other oxides. Brookhaven became one of Philips' first customers and, as a result, played an important part in the introduction of ferrites to the United States. Philips worked out new manufacturing procedures to provide the large ferrite blocks needed at Brookhaven. At the same time, to acquire an American source of ferrites, Brookhaven encouraged an American ceramics company to begin the manufacture of such material, and the final accelerating system included ferrites from both Philips and the General Ceramics and Steatite Corporation of New Jersey.

Ferrites were accepted immediately by industry for applications in all sorts of electronic apparatus, where they are used in large quantities in radios, television sets, and in computers. Ferrites with a wide variety of properties are now available, some of which are very useful as permanent magnets, and many American companies now manufacture them in large quantities. Once again the builders of accelerators played an active part in the founding of a multimillion-dollar industry.

X. SCALERS AND COMPUTERS

As mentioned in the introduction, the basic element of the modern digital computer was developed in nuclear physics laboratories during the early 1930's. Computers depend on counting in the scale of two and use bistable circuits; a one-stage scaler gives one count for every two input pulses, a two-stage scaler gives one count for every four input pulses, etc. The first of these, mentioned previously, was developed by Wynn-Williams and was quite satisfactory for the first experiments on nuclear disintegration. However, such a system was limited by the time of recovery of the first stage, which depended on the deionization time of the thyratrons used. This set a limit of a fraction of a millisecond to the resolving power of the system. Since the events being counted had a random distribution and were not evenly spaced in time, counting rates were limited to somewhat less than a thousand per second.

These rates were no longer adequate as accelerators were being built to reach higher energies and intensities than previously. Almost simultaneously in 1937 Stevenson and Getting in the United States and Lewis in England announced the development of scalers using vacuum tubes. The basic circuit used for each stage was the bistable "multivibrator" circuit, with which the resolution time was cut by more than a factor of 10.

During the years since 1937 increasingly faster circuits have been designed. It was obvious that digital computers, in which counting is the basic operation, could use these circuits. Digital computers were built using hundreds and thousands of vacuum tubes; a computer thus designed in order to include adequate complexity to perform all of the necessary operations and provide adequate memory was enormous in size and cost. Moreover, the relatively short life of vacuum tubes made it very difficult to keep all of several thousand vacuum tubes continually operative. These drawbacks disappeared with the advent of transistors and miniature circuits. Operation at high speeds with a high degree of reliability became feasible, resulting in the major computer industry of today.

Some computers have returned, so to speak, to their birthplace. Many experiments in high-energy physics include "on-line" computers that are part of the experimental apparatus. Formerly the significance of the experimental results could not be evaluated until exhaustive calculations were completed. The results might indicate desirable modifications of the experiment some months after its finish when the apparatus had been removed from the experimental area. Now the computations are performed virtually instantaneously and serve as an immediate guide for the next experimental steps.

The computer is also an integral part of the control system of the accelerator itself where a rapid evaluation of accelerator operation and quick modification of malfunctioning components is thus made possible.

The computer industry today is estimated at several billion dollars annually, in which myriad possibilities for future development are being discovered. For example, the control of satellite orbits would be impossible without fast computers, applications to business procedure are universal; and many scientific programs could not be completed in a man's lifetime without computer assistance.

XI. RADIATION

It has been shown in satellite studies of outer space that a rather high level of radiaiton of various sorts exists of sufficient intensity to be hazardous both to astronauts and their apparatus. Fortunately the characteristics of these radiations and their biological effects have been studied in great detail by high-energy physicists using the new multi-giga electron volt accelerators. As a result, much is known about the penetrating power of such radiation and methods for shielding against it are now being developed.

The importance of the tools of high-energy physics to the space effort is indicated by the fact that NASA has built an accelerator laboratory in Virginia which includes both a large synchrocyclotron and an electron linear accelerator. Devices which will be used in outer space can now be exposed to massive doses of radiation in this laboratory, and one can observe the undesirable effects before they cause failure in an actual space mission.

These radiations were first produced in the nuclear physics laboratory where they were originally studied. Now they are applied in the most unlikely places, for example, in oil well logging, studies of engine wear, and tracing of the paths of ingested chemicals through the human body.

XII. PATTERN RECOGNITION

We now turn to three developments in high-energy physics laboratories which have not yet realized practical importance but which give promise of yielding important technological advances during the next decades. The first of these results from the laboratories of those physicists using bubble chambers. Laboratories such as those centered on the Brookhaven AGS operate at high efficiency and produce bubble chamber photographs at the rate of several million a year. Each of these photographs shows a number of tracks of high-energy particles, which often result in complex secondary tracks corresponding to scattering by or production of new particles from collisions with the nuclei of the liquid hydrogen in the bubble chamber. In the past these pictures were projected on measuring screens and carefully measured by a team of scanners, from which calculation revealed the nature of the event. Although these pictures from Brookhaven are distributed among a number of laboratories throughout the United States and other countries, the increasing flood of photographs cannot be analyzed rapid!y enough. Hence, in a number of laboratories intensive efforts have been made to scan the pictures using electronic detection equipment that feeds data directly into a computer. After some years of tedious development, these techniques are beginning to become effective.

The above problem is closely allied to that of pattern recognition faced by engineers who have been trying for many years to develop a machine that can read printing or handwriting. It would seem probable that the further development of bubble chamber photograph scanners will lead to a solution of the more general problem of mechanical pattern recognition. Devices capable of "recognizing" patterns will be enormously valuable in many commercial operations—to mention only one, the keeping of records in banks and post offices.

XIII. REMOTE HANDLING

Researchers handling modern accelerators of high energy and intensity, experience considerable difficulty with residual radioactivity and radiation damage in those parts of the machine which have experienced bombardment by high-energy particle beams. Parts which fail from radiation damage must be removed and replaced in areas where radiation levels are so intense that a man can safely remain for only a few minutes. Hence, it becomes desirable to use remote-control handling equipment which can perform replacement operations while manipulated from a safe distance. Some remote-handling gear has been developed for use in "hot laboratories" but this equipment operates only at one location and is only capable of very simple operations. Special-purpose gear also has been developed for single repetitious operations such as scooping up earth on the surface of the moon. A general-purpose, mobile, remote-controlled manipulator of sufficient capability is not yet available, however, and for this reason such manipulators are under study in several accelerator laboratories. It seems probable that at least one of these projects will result in a useful device within a few years.

A versatile remote-controlled manipulator could be applied in many areas: for example, the performance of mechanical or mining operations in any hazardous area where noxious gases are present, there is danger of explosion, high radiation levels are present, or underwater or in a vacuum, the manipulators under development would "see" through television systems and would transmit "feel" to an operator through electrical links.

XIV. SUPERCONDUCTING DEVICES

Recent discoveries in the field of superconductivity are of great interest to high-energy physicists. It has been shown that a few alloys become superconducting (lose their resistivity and hence operate without losses) at temperatures of a few degrees Kelvin and are able to maintain this condition at very high magnetic fields—the so-called "hard superconductors," in contrast to the "soft" ones known for many years in which superconductivity is destroyed by relatively low magnetic fields. Most of the presently available hard superconductors are alloys or compounds which include niobium. The highest fields yet obtained are obtained in coils of Nb_3Sn. Fields already have been produced of 140 kg with promise of still higher fields. This is in contrast to fields of only slightly more than 20 kg achievable with conventional iron and copper electromagnets.

The new superconducting devices are of great interest for beam transport systems to be used in high-energy physics because (1) they will make major reductions in the million-dollar annual power bills incurred at major high-energy installations and (2) they also will make available higher fields for the bending and focusing of high-energy particle beams. Such superconducting bending magnets and focusing quadrupoles have already been built and tested in the laboratory, and within a couple of years it is probable that they will be in general use in beam transport systems. It seems also possible that superconducting magnets can be used at high fields in accelerators, thus making important reductions in the enormous size and cost of super-energy accelerators.

Laboratory work on superconducting devices is being closely watched by industry and new, improved superconductors are under development in industrial laboratories. In particular, the electrical industry has already published many speculations about the possibility of virtually lossless superconducting transformers and transmission lines which would make possible transport of power over great distances from remote power sources to centers of population. This development should have an impact on many phases of industrial operation during the next twenty years.

XV. CONCLUSION

The slow pace of the last century, during which Faraday died at an advanced age without suspecting that an enormous industry would be founded on his electrical discoveries, has been enormously accelerated. Technical advances and discoveries in the high-energy physics laboratory of today are almost immediately available to an enthusiastic industry.

The accelerators now being built by physicists and engineers require a stream of technological advances before they can be made to work; the necessary progress is continually being made. If evidence of industrial interest is needed, it is necessary to inspect the roster of attendees at the 1967 1969 and 1971 National Accelerator Conferences in Washington. From a total of about 800 engineers and scientists in attendance, almost 150 were from industry and 50 more came from military installations. It would appear that contact between high-energy physics and the builders of the nation's industry is in a favorable state and many further industrial advances based on high-energy physics can be expected in the future.

Not the least important development has been the education of physicists and engineers to appreciate each other and borrow heavily from each other's disciplines. The pool of versatile, trained manpower is one of the nation's most valuable assets.

Subject Index

A

Ablation, 122–124
Acceleration particles
  cancer induction vs. cancer therapy in, 213–214
  future research with, 232–235
  with high rate of energy loss, 226–227
  nerve cultures and, 224–226
  physical penetration by, 199–201
Accelerator(s)
  computer and, 303–304
  high-voltage technique in, 295–296
  history of, 293–295
  increased capabilities of, 290–291
  nuclide production and, 191
  rotating machine design and, 302
  vacuum technique in, 300–301
Accelerator materials, induced activities in, 187–188
Achondrites, 82
Acromegaly, 204
ACTH, 204
Alpha particles, 180
  radiobiological experiments with, 272–278
  in radiotherapy, 255–257
Alternating-gradient synchrotron, 23–24
Antimatter, 66–67
Antineutrinos, 67
Apollo lunar missions, 136, 223
Astrophysics
  elementary particles and, 191–192
  temperatures and densities in, 64
Atomic bomb, 34, 36, 39
Atomic energy, 36
Atomic Energy Commission, 233, 300

B

Beam monitoring in elementary particles, 186–187
Betatron, 23, 295–296
Biological research, accelerated particles in, 196–237
Blood–brain barrier, 217

Bragg ionization curves, 199–200
Bragg ionization peak area, 257–258, 261
  laminar lesions and, 219
Brain cells, radiation damage to, 217
Brain injuries, 201
Brain lesions, scar tissue in, 218
Brain organization as related to brain function, 201–202
Brain tissue, anoxia of, 218
Brain tumor, radiation therapy of, 227–232

C

Cancer
  hypophysectomy and, 210
  mammary, see Mammary cancer
Cancer cells, effects of heavy ions on, 246
Cancer therapy
  vs. cancer induction, 213–214
  heavy particles in, 227–232
Cascade calculations, intranuclear, 183–185
Cell types, neural lesions and ecology of, 217–222
Charge dispersion curves, 171
Chemist vs. physicist, 160–161
Chemistry, elementary particles and, 159–162
Chondrites, 82
  bronzite, 136
Cockcroft–Walton system, 290–292
Coherent light, 37
Colliding beam systems, 301
Computers, scalers and, 303–304
Coolidge tube, 296
Cosmic dust, 81–83
  radionuclides and, 138–140
Cosmic radiation, 79
  composition of, 121
Cosmic ray(s), 61–63
  energy spectrum and effective energy of, 120–121
  light sensation from, 223
  research in, 23